# Current Perspective on the Study of Liquid-Fluid Interfaces: From Fundamentals to Innovative Applications

# Current Perspective on the Study of Liquid-Fluid Interfaces: From Fundamentals to Innovative Applications

Editor

**Eduardo Guzmán**

MDPI • Basel • Beijing • Wuhan • Barcelona • Belgrade • Manchester • Tokyo • Cluj • Tianjin

*Editor*
Eduardo Guzmán
Complutense University of Madrid
Spain

*Editorial Office*
MDPI
St. Alban-Anlage 66
4052 Basel, Switzerland

This is a reprint of articles from the Special Issue published online in the open access journal *Coatings* (ISSN 2079-6412) (available at: https://www.mdpi.com/journal/coatings/special_issues/Fluid_II).

For citation purposes, cite each article independently as indicated on the article page online and as indicated below:

LastName, A.A.; LastName, B.B.; LastName, C.C. Article Title. *Journal Name* **Year**, *Volume Number*, Page Range.

ISBN 978-3-0365-4723-7 (Hbk)
ISBN 978-3-0365-4724-4 (PDF)

# Contents

Editorial

# Current Perspective on the Study of Liquid–Fluid Interfaces: From Fundamentals to Innovative Applications

**Eduardo Guzmán** [1,2]

1  Departamento de Química Física, Facultad de Ciencias Químicas, Universidad Complutense de Madrid, Ciudad Universitaria s/n, 28040 Madrid, Spain; eduardogs@quim.ucm.es; Tel.: +34-9-1394-4107
2  Instituto Pluridisciplinar, Universidad Complutense de Madrid, Paseo Juan XXIII 1, 28040 Madrid, Spain

**Citation:** Guzmán, E. Current Perspective on the Study of Liquid–Fluid Interfaces: From Fundamentals to Innovative Applications. *Coatings* 2022, *12*, 841. https://doi.org/10.3390/coatings12060841

Received: 2 June 2022
Accepted: 13 June 2022
Published: 16 June 2022

**Publisher's Note:** MDPI stays neutral with regard to jurisdictional claims in published maps and institutional affiliations.

Liquid–fluid interfaces are ubiquitous systems, having a paramount importance for daily life as well as for academia, providing the basis for the study of different aspects of interest for medicine, biology, and physics. Moreover, liquid–fluid interfaces emerge as very promising platforms enabling the fabrication of a broad range of functional materials as a result of the novel properties resulting from quasi-2D confinement [1,2]. This drives extensive research activity trying to shed light on the most fundamental physico-chemical aspects underlying the formation of liquid–fluid interfaces, as well as the properties emerging as a consequence of the quasi-2D confinement forced by the presence of a liquid–fluid interface [3,4]. For instance, the flows emerging from the presence of an interface play a central role in a broad range of industrial and biological aspects in which liquid–fluid interfaces are involved. Thus, the reorganization of materials within a liquid–fluid interface during the compression–expansion of the alveoli as well as the exchange of material from such interfaces and the adjacent liquid layer are essential for breathing, and any dysfunction on the interfacial flows occurring during expiration–inspiration cycles can result in an acute respiratory distress syndrome [5,6]. On the other hand, interfacial flows also play a very important role in the foaming and detergency abilities of most detergents and shampoos [7,8] and in many other processes of industrial and technological relevance, including icing-deicing processes, fouling, tertiary oil recovery, drug delivery, diffusion in porous matrices, ink-jet printing, and tissue engineering [9,10].

According to the above discussion, the study of liquid–fluid interfaces is a broad field with multiple implications. Therefore, the study of this type of system deserves importance, and this Special Issue tries to bring together different studies, providing a general overview of the current perspectives offered in the study of liquid–fluid interfaces. This is only possible in the context of combining a series of research papers and reviews dealing with different experimental and theoretical studies involving liquid–fluid interfaces, expanding on the exploitation of different phenomena occurring in liquid–fluid interfaces to understand specific phenomena of biophysical relevance, e.g., the impact of inhaled pollutants on normal respiratory function [11], and the use of advances in characterization techniques for the evaluation of phenomena and processes occurring within the interface [12,13].

Moreover, liquid–fluid interfaces play a very important role in the control of the flows occurring under different boundary conditions and their implications in different processes with technological and industrial interest. For instance, an accurate modelling of the flows occurring within porous systems can help in the optimization of different processes, including tertiary oil recovery and froth flotation [14–17]. Furthermore, the interfacial flows also contribute to spreading and evaporation phenomena, influencing the performance of different industrial processes, lubrication, and heat dissipation [18–21].

In summary, the study of the phenomena and applications involving liquid–fluid interfaces requires a broad perspective, which in current years is structured as a multidisciplinary challenge involving theoretical and experimental aspects to solve very complex problems. This Special Issue, together with the previous one entitled "Fluid Interfaces" [22] and the topic entitled "Insight into Liquid-Fluid Interfaces" [23], are focused to provide a comprehensive perspective on the current understanding of the study of liquid/fluid interfaces, contributing to open new avenues that close the gap between the most fundamental aspects of liquid–fluid interfaces and their potential applications.

**Funding:** This research received no external funding.

**Conflicts of Interest:** The authors declare no conflict of interest.

## References


1. Forth, J.; Kim, P.Y.; Xie, G.; Liu, X.; Helms, B.A.; Russell, T.P. Building Reconfigurable Devices Using Complex Liquid–Fluid Interfaces. *Adv. Mat.* **2019**, *31*, 1806370. [CrossRef] [PubMed]
2. Bayles, A.V.; Vermant, J. Divide, Conquer, and Stabilize: Engineering Strong Fluid–Fluid Interfaces. *Lanmguir* **2022**, *21*, 6499–6505. [CrossRef] [PubMed]
3. Llamas, S.; Guzmán, E.; Akanno, A.; Fernández-Peña, L.; Ortega, F.; Campbell, R.A.; Miller, R.; Rubio, R.G. Study of the Liquid/Vapor Interfacial Properties of Concentrated Polyelectrolyte–Surfactant Mixtures Using Surface Tensiometry and Neutron Reflectometry: Equilibrium, Adsorption Kinetics, and Dilational Rheology. *J. Phys. Chem. C* **2018**, *122*, 4419–4427. [CrossRef]
4. Llamas, S.; Fernández-Peña, L.; Akanno, A.; Guzmán, E.; Orteg, V.; Ortega, F.; Csaky, A.G.; Campbell, R.A.; Rubio, R.G. Towards understanding the behavior of polyelectrolyte–surfactant mixtures at the water/vapor interface closer to technologically-relevant conditions. *Phys. Chem. Chem. Phys.* **2018**, *20*, 1395–1407. [CrossRef] [PubMed]
5. Guzmán, E.; Santini, E. Lung surfactant-particles at fluid interfaces for toxicity assessments. *Curr. Opin. Colloid Interface Sci.* **2019**, *39*, 24–39. [CrossRef]
6. Castillo-Sánchez, J.C.; Cruz, A.; Pérez-Gil, J. Structural hallmarks of lung surfactant: Lipid-protein interactions, membrane structure and future challenges. *Arch. Biochem. Biophys.* **2021**, *703*, 108850. [CrossRef]
7. Fernández-Peña, L.; Guzmán, E.; Leonforte, F.; Serrano-Pueyo, A.; Regulski, K.; Tournier-Couturier, L.; Ortega, F.; Rubio, R.G.; Luengo, G.S. Effect of molecular structure of eco-friendly glycolipid biosurfactants on the adsorption of hair-care conditioning polymers. *Colloids Surf. B* **2020**, *185*, 110578. [CrossRef]
8. Fernández-Peña, L.; Guzmán, E.; Fernández-Pérez, C.; Barba-Nieto, I.; Ortega, F.; Leonforte, F.; Rubio, R.G.; Luengo, G.S. Study of the Dilution-Induced Deposition of Concentrated Mixtures of Polyelectrolytes and Surfactants. *Polymers* **2022**, *14*, 1335. [CrossRef]
9. Ferrari, M.; Cirisano, F. High transmittance and highly amphiphobic coatings for environmental protection of solar panels. *Adv. Colloid Interface Sci.* **2020**, *286*, 102309. [CrossRef]
10. Perrin, L.; Pajor-Swierzy, A.; Magdassi, S.; Kamyshny, A.; Ortega, F.; Rubio, R.G. Evaporation of Nanosuspensions on Substrates with Different Hydrophobicity. *ACS Appl. Mater. Interfaces* **2018**, *10*, 3082–3093. [CrossRef]
11. Guzmán, E. Fluid Films as Models for Understanding the Impact of Inhaled Particles in Lung Surfactant Layers. *Coatings* **2022**, *12*, 277. [CrossRef]
12. Yano, A.; Hamada, K.; Amagai, K. Evaluation of Coating Film Formation Process Using the Fluorescence Method. *Coatings* **2021**, *11*, 1076. [CrossRef]
13. Salum, P.; Güven, O.; Aydemir, L.Y.; Erbay, Z. Microscopy-Assisted Digital Image Analysis with Trainable Weka Segmentation (TWS) for Emulsion Droplet Size Determination. *Coatings* **2022**, *12*, 364. [CrossRef]
14. Khan, S.; Selim, M.M.; Khan, A.; Ullah, A.; Abdeljawad, T.; Ikramullah; Ayaz, M.; Mashwani, W.K. On the Analysis of the Non-Newtonian Fluid Flow Past a Stretching/Shrinking Permeable Surface with Heat and Mass Transfer. *Coatings* **2021**, *11*, 566. [CrossRef]
15. Usman, A.H.; Shah, Z.; Kumam, P.; Khan, W.; Humphries, U.W. Nanomechanical Concepts in Magnetically Guided Systems to Investigate the Magnetic Dipole Effect on Ferromagnetic Flow Past a Vertical Cone Surface. *Coatings* **2021**, *11*, 1129. [CrossRef]
16. Shoaib, M.; Zubair, G.; Raja, M.A.Z.; Nisar, K.S.; Abdel-Aty, A.-H.; Yahia, I.S. Study of 3-D Prandtl Nanofluid Flow over a Convectively Heated Sheet: A Stochastic Intelligent Technique. *Coatings* **2022**, *12*, 24. [CrossRef]
17. Shah, Z.; Vrinceanu, N.; Rooman, M.; Deebani, W.; Shutaywi, M. Mathematical Modelling of Ree-Eyring Nanofluid Using Koo-Kleinstreuer and Cattaneo-Christov Models on Chemically Reactive AA7072–AA7075 Alloys over a Magnetic Dipole Stretching Surface. *Coatings* **2022**, *12*, 391. [CrossRef]
18. Li, D.; Zhao, N.; Feng, Y.; Xie, Z. Numerical Investigation on the Evaporation Performance of Desulfurization Wastewater in a Spray Drying Tower without Deflectors. *Coatings* **2021**, *11*, 1022. [CrossRef]
19. Veronesi, F.; Guarini, G.; Corozzi, A.; Raimondo, M. Evaluation of the Durability of Slippery, Liquid-Infused Porous Surfaces in Different Aggressive Environments: Influence of the Chemical-Physical Properties of Lubricants. *Coatings* **2021**, *11*, 1170. [CrossRef]
20. Jiang, B.; Shen, Y.; Tao, J.; Xu, Y.; Chen, H.; Liu, S.; Liu, W.; Xie, X. Patterning Configuration of Surface Hydrophilicity by Graphene Nanosheet towards the Inhibition of Ice Nucleation and Growth. *Coatings* **2022**, *12*, 52. [CrossRef]
21. Liu, W.-J.; Chang, Y.-H.; Fern, C.-L.; Chen, Y.-T.; Jhou, T.-Y.; Chiu, P.-C.; Lin, S.-H.; Lin, K.-W.; Wu, T.-H. Annealing Effect on the Contact Angle, Surface Energy, Electric Property, and Nanomechanical Characteristics of $Co_{40}Fe_{40}W_{20}$ Thin Films. *Coatings* **2021**, *11*, 1268. [CrossRef]

22. Guzmán, E. Eduardo Guzmán. *Coatings* **2020**, *10*, 1000. [CrossRef]
23. Insight into Liquid/Fluid Interfaces. Available online: https://www.mdpi.com/topics/liquid_fluid_interfaces (accessed on 30 May 2022).

**Short Biography of Author**

**Eduardo Guzmán**, Associate Professor at the Physico-Chemistry Department and researcher at the Multi-disciplinary Institute in the Complutense University of Madrid (Spain), received his MSc in Chemistry and in Science and Technology of Colloids and Interfaces, and his PhD in Science at the Complutense University of Madrid (Spain). After his PhD, he worked for a period of four years at the Istituto per l'Energetica e le Interface in Genoa (Italy), after which he returned to his alma mater. He has published over 100 papers in JCR journals and 12 chapters in books (https://orcid.org/0000-0002-4682-2734), H-index 32, and has co-authored more than 100 contributions to different national and international conferences. His main research interests are LbL assembly, interfacial rheology, drug delivery, biophysics, cosmetics, and pest control. He has been the supervisor of 3 PhD students, 10 Master students and 25 undergraduate students. He has been involved in 2 EU and 6 Spanish-funded founded I+D grants and has been scientifically responsible for 4 cooperation projects between academia and industry. He is a member of the editorial board of different scientific journals, including Advances in Colloid and Interface Science, Colloids and Interfaces, Coatings (Editor in Chief of the Section "Liquid-Fluid Interfaces"), Polymers and Current Cosmetic Science, and has edited special issues in Coatings, Processes, Polymers, and Advances in Colloid and Interface Science.

*Article*

# Entropy Optimization on Axisymmetric Darcy–Forchheimer Powell–Eyring Nanofluid over a Horizontally Stretching Cylinder with Viscous Dissipation Effect

Muhammad Rooman [1,2], Muhammad Asif Jan [2], Zahir Shah [1,*], Narcisa Vrinceanu [3,*], Santiago Ferrándiz Bou [4], Shahid Iqbal [5] and Wejdan Deebani [6]

[1]  Department of Mathematical Sciences, University of Lakki Marwat, Lakki Marwat 28420, Pakistan; roomankhan31@gmail.com
[2]  Department of Mathematics, Kohat University of Science and Technology KUST, Kohat 26000, Pakistan; majan.math@gmail.com
[3]  Department of Industrial Machines and Equipments, Faculty of Engineering, "Lucian Blaga" University of Sibiu, 10 Victoriei Boulevard, 550024 Sibiu, Romania
[4]  Department of Mechanical and Materials Engineering, Higher Polytechnic School of Alcoy, Universitat Politècnica de València, 46022 Valencia, Spain; sferrand@mcm.upv.es
[5]  Department of Physics, Albion College, Albion, MI 49224, USA; siqbal@albion.edu
[6]  Department of Mathematics, College of Science & Arts, King Abdulaziz University, Rabigh 21911, Saudi Arabia; wdeebani@kau.edu.sa
*  Correspondence: zahir@ulm.edu.pk (Z.S.); vrinceanu.narcisai@ulbsibiu.ro (N.V.)

**Abstract:** The effect of entropy optimization on an axisymmetric Darcy–Forchheimer Powell–Eyring nanofluid flow caused by a horizontally permeable stretching cylinder, as well as non-linear thermal radiation, was investigated in this research work. The leading equations of the current problem were changed into ODEs by exhausting appropriate transformations. To deduce the reduced system, the numerical method bvp4c was used. The outcome of non-dimensional relevant factors on velocity, entropy, concentration, temperature, Bejan number, drag force, and Nusselt number is discussed and demonstrated using graphs and tables. It is perceived that, with a higher value of volume fraction parameter, the skin friction falls down. Likewise, it is found that the Nusselt number drops with enhancing the value of the volume fraction. Moreover, the result reveals that the entropy generation increases as the volume fraction, curvature parameter, and Brinkman number increase.

**Keywords:** heat transfer; stretching cylinder; nonlinear radiation; Powell–Eyring; nanofluid; porous medium; Darcy–Forchheimer

**Citation:** Rooman, M.; Jan, M.A.; Shah, Z.; Vrinceanu, N.; Ferrándiz Bou, S.; Iqbal, S.; Deebani, W. Entropy Optimization on Axisymmetric Darcy–Forchheimer Powell–Eyring Nanofluid over a Horizontally Stretching Cylinder with Viscous Dissipation Effect. *Coatings* **2022**, *12*, 749. https://doi.org/10.3390/coatings12060749

Academic Editor: Eduardo Guzmán

Received: 12 April 2022
Accepted: 26 May 2022
Published: 30 May 2022

**Publisher's Note:** MDPI stays neutral with regard to jurisdictional claims in published maps and institutional affiliations.

## 1. Introduction

It is a well-known fact that stretching flows have acquired a lot of attention because of their numerous applications. In industrial and mechanical engineering progressions (rubber and plastic sheets, cooling of electronic chips, glass blowing, metal spinning, production of glass fiber, liquid film crystallization during condensation, etc.), stretching surfaces are used extensively. In addition, various research of boundary layer flow in conjunction with a plane extending surface has already been carried out. However, research journals offer just a few experiments on horizontally stretching sheets with the axisymmetric flow. Crane [1] first studied the flow of viscous materials caused by a stretched sheet. Nadeem and Haq [2] investigated the convective flow of viscous nanoparticles using radiation beyond a stretched sheet. Ahmad et al. [3] explored power-law fluid in the occurrence of axisymmetric flow and heat transfer. Ariel [4] looked at the classic problem of an axisymmetric flow caused by a stretched sheet and gave perturbed, asymptotic, exact, and numerical solutions. Hsiao [5] investigated MHD heat transfer across a stretching surface utilizing Maxwell fluid flow by means of radiative and viscous dissipation properties.

Researchers are still interested in studying non-Newtonian fluids because they are more suitable for industrial areas such as power engineering, polymer solution industries, food engineering, and petroleum production. A linear association flanked by stress and rate of strain cannot be used to represent non-Newtonian fluids. Because of their various features, non-Newtonian fluids are much more complex than Newtonian fluids. Powell–Eyring fluids [6] are a type of non-Newtonian fluid with several compensations upon the power-law model, for example, they are built on liquid kinetic theory and behave Newtonian at both low and high shear rates. Patel et al. [7] used asymptotic boundary conditions to study a numeric solution of MHD Powell–Eyring fluid flow. Hayat et al. [8] discovered the radiative effects in electrically conducting Eyring–Powell fluid in three dimensions. Ara et al. [9] inspected the effect of radiation on Eyring–Powell fluid boundary layer flow across an exponentially shrinking sheet. Hayat et al. [10] showed boundary layer stagnation-point flow of Powell–Eyring fluid including dissolving heat transfer. In the existence of a double-stratified medium, Rehman et al. [11] numerically measured a flow study with heat generation/absorption influences of Powell–Eyring fluid mixed-convection flow around a stretching cylinder. According to Hayat et al. [12], these fluids had a variety of complicated features that provided them an edge and various applications over Newtonian fluids. Improvements in mud house renovation and the production of clay pots, gels, medical syrups, and fruit juices, such as Delmonte, yoghurt, Afya, and energy drinks are just a few of the benefits. Additionally, they are used in the production of pseudo-plastic fluids, paints, and medications in the pharmaceutical industry.

In thermodynamics, entropy is a key term. The concept of irreversibility is inextricably linked to the concept of entropy. Irreversibility is something that everyone instinctively understands. We may easily comprehend the irreversibility phenomenon by watching a movie in both forward and reverse sequences. Many progressive processes in ordinary life cannot be reversed, such as plastic deformation, pouring water into a glass, unrestrained fluid expansion, gas rising from a chimney, egg unscrambling, and so on. Originally, the term entropy was manipulated to define the loss of energy in numerous mechanical systems and heat engines that could not efficiently transform the energy into work. Many engineers and scientists are working hard in this modern period to find novel ways to control or limit the waste of valuable energy. This energy loss in thermodynamic systems can cause a lot of chaos. Using Bejan number and entropy creation, any system's efficiency can be boosted. Bejan [13] studied if heat transfer and flow mechanism abnormalities might be scrutinized in expressions of entropy formation. Many investigators have inspected entropy production results in heat flow and transmission to back up his claim. In a dissipative Blasius flow, But et al. [14] looked at entropy creation as well as radiative flux. Their findings show that as the heat radiation variable rises, entropy decreases. Entropy formation for mass and heat transmission over an isothermal medium was proposed by San and Laban [15]. Tamayol et al. [16] looked at how entropy affects heat transmission and fluid flow past a leaky material on a stretchy surface. Rashidi et al. [17] used the homotopy approach to entropy production in hydromagnetic flow across a spinning disk. Shit et al. [18] studied the irreversibility of hydromagnetic nanoparticle flow and heat transit on an exponentially speeded sheet. In the existence of radiative heat flux, convective boundary conditions, and MHD, the flow was explored. But and Ali [19] used a radially stretched surface to scrutinize the impact of a magnetic force on entropy formation in heat transfer and flow processes. Munawar et al. [20] deliberate the formation of entropy in viscid flow via an oscillated stretching cylinder. Khan et al. [21] used a radially stretched disk to evaluate the influence of entropy formation on Carreau nanofluid due to nonlinear thermal radiation.

Furthermore, nanotechnology is regarded as one of the most important conduits for the advancement in key manufacturing rebellion in our sector. Nanofluids are mostly employed due to their enhanced thermal properties. They are manipulated as coolants in heat transfer devices such as electronic cooling systems (such as flat plates), radiators, and heat exchangers. Nanofluid is made up of nanoparticles ranging in size from 1 to 100 nanometers. Choi and Eastman [22] were the first to propose the term "nanfluid".

Iqbal et al. [23] used the Newtonian Carreau model to do a computational investigation of thin-film flow through a moving surface. Khan et al. [24] investigated the inspiration of Cattaneo–Christov heat flux on Maxwell nanofluid boundary layer hydromagnetic flow using the two-phase Buogiorno model. Ali et al. [25] created the mathematical model of the unsteady and laminar couple stress nanofluid flow using engine oil and molybdenum disulphide nanomaterial as the base fluid and nanoparticles, respectively. They discovered that adding molybdenum disulphide nanoparticles to the base fluid improves the heat transfer rate of engine oil by up to 12.38 percent. Acharya et al. [26], investigated the effect of entropic production of a time-independent radioactive combination nanoliquid flowing through a slip spinning disk. Acharya et al. [27], used an entropy approach to evaluate mixed convection and radiation impacts in non-Newtonian-flowing fluid by a flexible cylinder. Verra Krishna and Chamkha [28] examined the impact of Hall and ion slip on the MHD convective flow of elastico-viscous fluid via a permeable channel between two rigidly rotating parallel plates. Takhar et al. [29], characterized the free stream of a vertically moving cylinder, as well as mass and heat transfer. A significant amount of noteworthy work has recently been accomplished [30–41].

Several scholars have looked into entropy propagation effects in the context of heat and mass transport on stretching surfaces. Although, there are just a few papers on the subject of entropy generation's impacts on inflow on a stretching disk. The consequences of entropy formation in Powell–Eyring nanofluid caused by mass and heat transport on a horizontally stretched disk are investigated in this article. The heat equation was modeled using several factors such as viscous dissipation, heat radiation, thermophoresis, and Brownian diffusion. The equations are numerically solved by the bvp4c method. The velocity, Bejan number, concentration, temperature, and entropy are all graphically explained.

## 2. Mathematical Formulation

The flow of Powell–Eyring nanofluid in a two-dimensional axisymmetric flow across a horizontally stretched sheet is assumed. We are using a system of cylindrical coordinates in which the z-axis, is chosen parallel to the cylinder's axis and the r-axis is chosen perpendicular to the cylindrical surface, as revealed in Figure 1. The cylinder is porous and continually stretches horizontally at $u = u_\omega = \frac{U_0 z}{L}$, where $L$ is a characteristic length and $U_0 > 0$. Despite the fact that the moving fluid temperature is set to $T_\infty$, the cylindrical surface is kept at $T_\omega$, with the assumption that $T_\omega > T_\infty$. The Buongiorno model and nanofluid contain important slip mechanisms such as thermophoresis diffusion and Brownian motion. The velocity profile for the assumed flow is $V = [w(r,z), 0, u(r,z)]$.

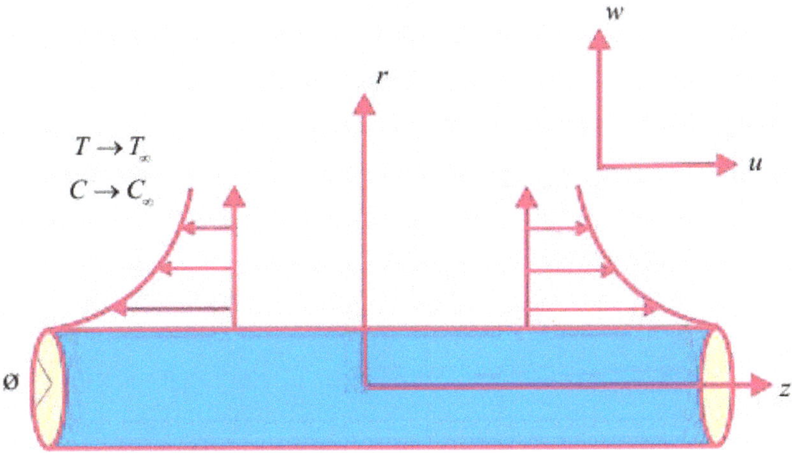


**Figure 1.** Sketch for the flow field.

The equations that govern the flow are as follows [10,11]

$$\frac{\partial(rw)}{\partial r} + \frac{\partial(ru)}{\partial z} = 0 \tag{1}$$

$$w\frac{\partial u}{\partial r} + u\frac{\partial u}{\partial z} = \frac{\mu_{nf}}{\rho_{nf}}\left(\frac{\partial^2 u}{\partial r^2} + \frac{1}{r}\frac{\partial u}{\partial r}\right) + \frac{1}{\rho_{nf}\beta c}\left(\frac{\partial^2 u}{\partial r^2} + \frac{1}{r}\frac{\partial u}{\partial r}\right)$$
$$-\frac{1}{6\rho_{nf}\beta c^3}\left(\frac{1}{r}\left(\frac{\partial u}{\partial r}\right)^3 + 3\left(\frac{\partial u}{\partial r}\right)^2\left(\frac{\partial^2 u}{\partial r^2}\right)\right) - \frac{\mu_{nf}}{\rho_{nf}K}u - \frac{c_b}{\rho_{nf}\sqrt{K}}u^2 \tag{2}$$

$$w\frac{\partial T}{\partial r} + u\frac{\partial T}{\partial z} = \frac{k_{nf}}{\left(\rho c_p\right)_{nf}}\left(\frac{\partial^2 T}{\partial r^2} + \frac{1}{r}\frac{\partial T}{\partial r}\right) + \tau\left(D_B\frac{\partial C}{\partial r}\frac{\partial T}{\partial r} + \frac{D_T}{T_\infty}\left(\frac{\partial T}{\partial r}\right)^2\right)$$
$$+\frac{1}{\left(\rho c_p\right)_{nf}}\left(\mu_{nf}\left(\frac{\partial u}{\partial r}\right)^2 + \frac{1}{\beta c}\left(\frac{\partial u}{\partial r}\right)^2 - \frac{1}{6\beta c^3}\left(\frac{\partial u}{\partial r}\right)^4\right) - \frac{1}{\left(\rho c_p\right)_{nf}}\frac{\partial}{\partial r}\left(-\frac{16\sigma^*}{3k^* k_{nf}}T^3\frac{\partial T}{\partial r}\right) \tag{3}$$

$$w\frac{\partial C}{\partial r} + u\frac{\partial C}{\partial z} = D_B\left(\frac{\partial^2 C}{\partial r^2} + \frac{1}{r}\frac{\partial C}{\partial r}\right) + \frac{D_T}{T_\infty}\left(\frac{\partial^2 T}{\partial r^2} + \frac{1}{r}\frac{\partial T}{\partial r}\right) \tag{4}$$

With boundary conditions

$$u = u_\omega = \frac{U_0 z}{L}, w = 0, T = T_\omega, C = C_\omega \; at \quad r = R$$
$$u \to 0, T \to T_\infty, C \to C_\infty \; at \quad r \to \infty \tag{5}$$

Illustrations of the description of the various symbols are shown in Table 1. The fundamental equations can be transformed using the following transforms [21]:

$$\psi(r,z) = \sqrt{u_\omega \mu_f z} R f(\eta), \eta = \frac{r^2 - R^2}{2R}\sqrt{\frac{u_\omega}{\mu_f z}}, \theta(\eta) = \frac{T - T_\infty}{T_\omega - T_\infty}, \phi(\eta) = \frac{C - C_\infty}{C_\omega - C_\infty} \tag{6}$$

**Table 1.** A description of the multiple symbols that appear in the governing equations is illustrated.

| Symbols | Description | Symbols | Description |
|---|---|---|---|
| $u, v, w$ | Velocity in $r, \theta, z$ direction | $\beta, c$ | Fluid parameter |
| $\mu_f$ | Base fluid dynamic viscosity | $\mu_{nf}$ | Nanofluid dynamic viscosity |
| $v_f$ | Base fluid kinematic viscosity | $c_p$ | Specific heat |
| $c_b$ | Drag factor | $K$ | Permeability of porous medium |
| $k_f$ | Base fluid thermal conductivity | $k_{nf}$ | Nanofluid thermal conductivity |
| $k_s$ | Nanoparticles' thermal conductivity | $D_B$ | Brownian motion |
| $\rho_f$ | Base fluid density | $\rho_{nf}$ | Nanofluid density |
| $L$ | Characteristic length | $\left(\rho c_p\right)_f$ | Heating capacity of base fluid |
| $\left(\rho c_p\right)_{nf}$ | Heating capacity of nanofluid | $\left(\rho c_p\right)_s$ | Heating capacity of nanoparticles |
| $T$ | Temperature | $T_\infty$ | Temperature at free stream |
| $k^*$ | Absorption coefficient | $D_T$ | Thermo-coefficient |
| $C$ | Concentration | $C_\infty$ | Ambient concentration |

Equation (7) identifies the components of velocity

$$u = \frac{1}{r}\frac{\partial \psi(r,z)}{\partial r}, w = -\frac{1}{r}\frac{\partial \psi(r,z)}{\partial z} \tag{7}$$

The nanofluid expressions are given by [25]:

$$\left.\begin{aligned}
\mu_{nf} &= \mu_f(1 - \phi)^{-2.5} \\
\rho_{nf} &= \rho_f(1 - \phi) + \phi\rho_s \\
\left(\rho c_p\right)_{nf} &= \left(\rho c_p\right)_f(1 - \phi) + \phi\left(\rho c_p\right)_s \\
k_{nf} &= k_f\left[\frac{k_s + 2k_f - 2\phi\left(k_f - k_s\right)}{k_s + 2k_f + 2\phi\left(k_f - k_s\right)}\right]
\end{aligned}\right\} \tag{8}$$

The base fluid and nanofluid dynamic viscosity are denoted by $\mu_f$ and $\mu_{nf}$, respectively, where $\phi$ signifies volume fraction.

The equation of incompressibility is fulfilled identically, whereas Equations (2)–(5) are reduced to

$$(1 + 2\eta\gamma)\left(\frac{1}{(1-\phi_1)^{2.5}} + \alpha\right)f''' - \lambda\alpha(1 + 2\eta\gamma)^2 f''^2 f''' + 2\gamma\left(\frac{1}{(1-\phi_1)^{2.5}} + \alpha\right)f''$$
$$-\tfrac{4}{3}\alpha\lambda\gamma(1 + 2\eta\gamma)f''^3 - \frac{\beta_0}{(1-\phi_1)^{2.5}}f' - F_r f'^2 + \left((1 - \phi_1) + \phi_1\frac{\rho_s}{\rho_f}\right)(ff'' - f'^2) = 0 \tag{9}$$

$$\frac{k_{nf}/k_f}{Pr}\left((1 + 2\eta\gamma)\theta'' + \gamma\theta'\right) + Ec(1 + 2\eta\gamma)\left(\left(\frac{1}{(1-\phi_1)^{2.5}} + \alpha\right)f''^2 - \tfrac{1}{3}\lambda\alpha(1 + 2\eta\gamma)f''^4\right)$$
$$+ \left((1 - \phi_1) + \phi_1\frac{(\rho c_p)_s}{(\rho c_p)_f}\right)\left(Nt(1 + 2\eta\gamma)\theta'^2 + Nb(1 + 2\eta\gamma)\theta'\phi' + f\theta'\right) \tag{10}$$
$$+ \frac{Rd}{Prk_{nf}/k_f}(1 + 2\eta\gamma)\left((\theta(\theta_w - 1) + 1)^3\theta'' + 3(\theta(\theta_w - 1) + 1)^2(\theta_w - 1)\theta'^2\right) = 0$$

$$\frac{Nt}{Nb}\frac{1}{Sc}\left((1 + 2\eta\gamma)\theta'' + \gamma\theta'\right) + \frac{1}{Sc}\left((1 + 2\eta\gamma)\phi'' + \gamma\phi'\right) + f\phi' = 0 \tag{11}$$

$$f(0) = 0, f'(0) = 1, \theta(0) = 1, \phi(0) = 1$$
$$f'(\infty) \to 0, \theta(\infty) \to 0, \phi(\infty) \to 0 \tag{12}$$

where $\gamma = \sqrt{\frac{v_f L}{U_0 R^2}}$ represents the curvature parameter, $\alpha = \frac{1}{\mu_f \beta c}$ and $\lambda = \frac{U_0^3 z^2}{2L^3 c^2 v_f}$ are fluid parameters, $\beta_0 = \frac{Lv}{\rho_f U_0 K}$ represents the porosity parameter, $F_r = \frac{c_b z}{\rho_f \sqrt{K}}$ is the inertia coefficient, $Ec = \frac{u_w^2}{c_p(T_w - T_\infty)}$ is the Eckert number, $Rd = \frac{16\sigma^* T_\infty^3}{3k^* k_f}$ denotes the radiation parameter, $Pr = \frac{\mu c_p}{k_f}$ denotes the Prandtl number, $\theta_w = \frac{T_w}{T_\infty}$ is the temperature ratio, $Sc = \frac{V_f}{D_B}$ is the Schmidt number, $Nt = \frac{\tau D_B(T_w - T_\infty)}{v_f}$ and $Nb = \frac{\tau D_B(C_w - C_\infty)}{v_f}$ represent the thermophoresis and Brownian diffusion parameters, respectively.

*Skin Friction and Nusselt Number*

The following are the definitions of the skin friction and the local Nusselt number:

$$c_f = \frac{\tau_w}{\rho_f u_w^2}, N_u = \frac{zq_w}{k_f\left(T_f - T_\infty\right)} \tag{13}$$

where $\tau_w$ and $q_w$ are the surface shear stress and heat flux, respectively. These are defined as:

$$\tau_w = \left[\left(\mu_{nf} + \frac{1}{\beta c}\right)\left(\frac{\partial u}{\partial r}\right) - \frac{1}{6\beta c^3}\left(\frac{\partial u}{\partial r}\right)^3\right]_{r=R}, q_w = -\left(k_{nf} + \frac{16\sigma^* T_\infty^3}{3k^*}\right)\left(\frac{\partial T}{\partial r}\right)_{r=R} \tag{14}$$

In dimensionless form, the skin friction and Nusselt number are:

$$C_f Re_z^{1/2} = \left(\frac{1}{(1-\phi_1)^{2.5}} + \alpha\right)f''(0) - \frac{\lambda}{3}\alpha(f''(0))^3, N_u Re_z^{-1/2} = -\left(\frac{k_{nf}}{k_f} + Rd\right)\theta'(0) \tag{15}$$

## 3. Entropy Optimization and Bejan Number

There are three causes of entropy Optimization in the current problem. Heat transport, viscous dissipation, and mass diffusion all generate entropy. The entropy equation is written as follows [18,21]:

$$S_g = \frac{k_{nf}}{T_\infty^2}\left(1 + \frac{16\sigma^* T^3}{3k^* k_{nf}}\right)\left(\frac{\partial T}{\partial r}\right)^2 + \frac{1}{T_\infty}\left(\mu_{nf}\left(\frac{\partial u}{\partial r}\right)^2 + \frac{1}{\beta c}\left(\frac{\partial u}{\partial r}\right)^2 - \frac{1}{6\beta c^3}\left(\frac{\partial u}{\partial r}\right)^4\right)$$
$$+ \frac{RD}{T_\infty}\left(\frac{\partial T}{\partial r}\frac{\partial C}{\partial r}\right) + \frac{RD}{C_\infty}\left(\frac{\partial C}{\partial r}\right)^2 \tag{16}$$

By using the typical entropy generation rate to convert Equation (16) to dimensionless form, we obtain:

$$N_G = \left(\frac{k_{nf}}{k_f} + Rd(\theta(\theta_\omega - 1) + 1)^3\right)(1 + 2\eta\gamma)\alpha_1\theta'^2 + L_1(1 + 2\eta\gamma)\theta'\phi'$$
$$+ L_1(1 + 2\eta\gamma)\frac{\alpha_2}{\alpha_1}\phi'^2 + Br(1 + 2\eta\gamma)\left(\left(\frac{1}{(1-\phi_1)^{2.5}} + \alpha\right)f''^2 - \frac{1}{3}\alpha\lambda f''^4\right) \tag{17}$$

where $N_G = \frac{S_g}{S_0} = \frac{S_g}{k_f U_0(T_\omega - T_\infty)/T_\infty \mu_f L}$ denotes the total entropy production, $\alpha_1 = \frac{T_\omega - T_\infty}{T_\infty}$ and $\alpha_2 = \frac{C_\omega - C_\infty}{C_\infty}$ are the temperature ratio and concentration ratio variables, respectively, and $Br = \frac{\mu_f U_0^2 z^2}{k_f L^2(T_\omega - T_\infty)}$ and $L_1 = \frac{RD(C_\omega - C_\infty)}{k_f}$ are Brinkman number and diffusion parameter, respectively.

Equation (17) can be written using the pattern shown below:

$$N_G = N_H + N_f + N_m \tag{18}$$

Here, $N_H$ represents entropy production because of heat transmission, $N_f$ represents entropy production because of fluid friction, and $N_m$ represents entropy production. The Bejan number is defined as:

$$Be = \frac{N_H + N_m}{N_G} \tag{19}$$

$$Be = \frac{\left(\frac{k_{nf}}{k_f} + R(\theta(\theta_\omega - 1) + 1)^3\right)(1 + 2\gamma\eta)\theta'^2\alpha_1 + L(1 + 2\gamma\eta)\theta'\phi' + L(1 + 2\gamma\eta)\frac{\alpha_2}{\alpha_1}\phi'^2}{\left(\begin{array}{c}\left(\frac{k_{nf}}{k_f} + Rd(\theta(\theta_\omega - 1) + 1)^3\right)(1 + 2\eta\gamma)\alpha_1\theta'^2 + Br(1 + 2\eta\gamma)\left(\left(\frac{1}{(1-\phi_1)^{2.5}} + \alpha\right)f''^2 - \frac{1}{3}\alpha\lambda f''^4\right) \\ + L_1(1 + 2\eta\gamma)\theta'\phi' + L_1(1 + 2\eta\gamma)\frac{\alpha_2}{\alpha_1}\phi'^2\end{array}\right)} \tag{20}$$

## 4. Numerical Method

The solution mechanism for the currently constructed model is computed in this part manipulating the bvp4c technique (shooting scheme). The bvp4c technique (shooting scheme) in the MATLAB tool is used to solve the ODEs (7)–(11) via (12) numerically. First, we transform a higher-order system to a first-order system for this technique. To accomplish this, we follow the steps below:

$$f = y_1, f' = y_2, f'' = y_3, f''' = y_3'$$
$$\theta = y_4, \theta' = y_5, \theta'' = y_5' \tag{21}$$
$$\phi = y_6, \phi' = y_7, \phi'' = y_7'$$

$$y_3' = \frac{\left(\begin{array}{c}\frac{4}{3}\alpha\lambda\gamma(1 + 2\eta\gamma)y_3^3 - 2\gamma\left(\frac{1}{(1-\phi_1)^{2.5}} + \alpha\right)y_3 \\ + \frac{\beta_0}{(1-\phi_1)^{2.5}}y_2 + Fry_2^2 - \left((1 - \phi_1) + \phi_1\frac{\rho_s}{\rho_f}\right)(y_1y_3 - y_2^2)\end{array}\right)}{\left((1 + 2\eta\gamma)\left(\frac{1}{(1-\phi_1)^{2.5}} + \alpha\right) - \lambda\alpha(1 + 2\eta\gamma)^2y_3^2\right)} \tag{22}$$

$$y_5' = \frac{\left(\begin{array}{c}Ec(1 + 2\eta\gamma)\left(\frac{1}{3}\lambda\alpha(1 + 2\eta\gamma)y_3^4 - \left(\frac{1}{(1-\phi_1)^{2.5}} + \alpha\right)y_3^2\right) - \frac{k_{nf}/k_f}{Pr}\gamma y_5 \\ -\left((1 - \phi_1) + \phi_1\frac{(\rho c_p)_s}{(\rho c_p)_f}\right)(Nt(1 + 2\eta\gamma)y_5^2 + Nb(1 + 2\eta\gamma)y_5y_7 + y_1y_5) \\ -3\frac{Rd}{Prk_{nf}/k_f}(1 + 2\eta\gamma)(y_4(\theta_\omega - 1) + 1)^2(\theta_\omega - 1)y_5^2\end{array}\right)}{\frac{k_{nf}/k_f}{Pr}(1 + 2\eta\gamma) + \frac{Rd}{Prk_{nf}/k_f}(1 + 2\eta\gamma)(y_4(\theta_\omega - 1) + 1)^3} \tag{23}$$

$$y_7' = \frac{-\frac{Nt}{Nb}((1 + 2\eta\gamma)y_5' + \gamma y_5) - (\gamma y_7) - Scy_1y_7}{(1 + 2\eta\gamma)} \tag{24}$$

with

$$y_1(0) = 0, y_2(0) = 1, y_4(0) = 1, y_6(0) = 1$$
$$y_2(\infty) \to 0, y_4(\infty) \to 0, y_6(\infty) \to 0 \tag{25}$$

## 5. Results and Discussions

This section's focus is on examining the effects of velocity, temperature, concentration profile, entropy generation, and Bejan number. The influence of fluid parameter $\alpha$, porosity parameter $\beta_0$, volume fraction $\phi_1$, curvature parameter $\gamma$, and inertia coefficient $Fr$ on fluid velocity is examined in Figure 2a–e. The effect of fluid parameter $\alpha$ on the velocity profile is presented in Figure 2a. With the higher value of $\alpha$, the fluid velocity and boundary layer thickness increase. In reality, as $\alpha$ increases, the viscosity of the fluid declines, resulting in an improvement in the velocity profile. The impression of the porosity factor $\beta_0$ on the velocity profile of Powell–Eyring nanofluid is shown in Figure 2b. The velocity of the nanofluid declines when permeability of the fluid rises, which is in line with authenticity. Moreover, the permeability of the border has no consequence on the fluid velocity as we move away from it. The impact of volume fraction $\phi_1$ on velocity profile is plotted in Figure 2c. By raising the volume fraction, the Powell–Eyring nanofluid velocity decreases. The science here seems to be that when the volume fraction increases, the flow becomes more vicious, resulting in friction forces that slow the nanofluid velocity. The feature of the curvature parameter $\gamma$ on the velocity profile is presented in Figure 2d. As $\gamma$ goes up, the fluid velocity shrinks at the surface and escalates further from the cylinder, according to the results. In reality, as the curvature parameter is increased, the cylinder's radius decreases. As a result, the cylinder's contact surface with the fluid lowers, providing less confrontation to fluid motion. Consequently, the velocity profile rises. The impression of inertia coefficient $Fr$ on the velocity profile is shown in Figure 2e. The velocity profile drops as the inertia coefficient rises.

The impact of temperature profile $\theta(\eta)$ against relevant flow parameters, such as fluid parameter $\alpha$, porosity parameter $\beta_0$, curvature parameter $\gamma$, Prandtl number Pr, temperature ratio $\theta_\omega$, thermal radiation $Rd$, Eckert number $Ec$, thermophoresis constraint $Nt$, and Brownian motion constraint $Nb$, are depicted in Figure 3a–j. Figure 3a signifies the inspiration of the fluid constraint $\alpha$ on the temperature profile. Higher $\alpha$ values result in a lower temperature profile, according to the findings. The viscosity of the thermal boundary layer continues to shrink, as the higher value of fluid parameter $\alpha$ viscosity of the fluid decreases.

Accordingly, the temperature profile declines. The variation in temperature profile with $\eta$. over a range of the porous parameter $\beta_0$ is shown in Figure 3b. This illustration clearly demonstrates that the heat distribution is a weak function of $\beta_0$ and that it changes little when it passes through the thermal boundary layer. Therefore, increasing $\beta_0$ causes a modest thickening of the thermal boundary layer. The outcome of volume fraction parameter $\phi_1$ on the temperature profile is depicted in Figure 3c. It is observed that the increasing volume fraction raises the temperature profile. The science behind this mounting temperature pattern is because the temperature rises as the smash flanked by the molecules of the Powell–Eyring nanofluid rises. Figure 3d exhibits the impressions of the curvature constraint on the temperature profile, with temperature showing an increasing trend via $\gamma$. As $\gamma$ increases, the surface contact area exposed to fluid particles decreases, resulting in less resistance for particles and an increase in their average velocity. The temperature rises because the Kelvin temperature is expressed by an average kinetic energy. The characteristics of the Prandtl number Pr on the temperature distribution are exposed in Figure 3e. The temperature profile and thickness of the thermal boundary layer are found to diminish as the Prandtl number rises. It connects thermal diffusivity to momentum diffusivity. Accordingly, a higher Prandtl number correlates to a reduced thermal diffusivity; consequently, temperature distribution rises up before dropping down. Figure 3f depicts the effects of temperature ratio $\theta_\omega$ on the temperature distribution. It has been realized that advanced temperature ratio enhances the temperature distribution. Figure 3g depicts the behavior of the radiation parameter $Rd$ on a temperature profile. For bulky levels of the radiation parameter, temperature and the accompanying boundary layer thickness rise. Higher values of the radiation parameter reduce the mean absorption coefficient, resulting in an upturn in the temperature distribution. Figure 3h depicts the difference in fluid temperature caused by the change in Eckert number $Ec$. The figure depicts how the fluid temperature rises as the value of $Ec$ rises. This happens since frictional heating produces heat in the fluid as the value of $Ec$ rises. Physically, the Eckert number is explained as the ratio of kinetic energy to the difference in specific enthalpy flanked by the wall and the fluid. As a result of the effort exerted against the viscous fluid pressures, an upsurge in Eckert number converts kinetic energy into internal energy. As a result, as $Ec$ rises, the fluid's temperature rises. As shown in Figure 3i, the thermophoresis parameter $Nt$ has an outcome on the temperature. The graph shows that as the number of thermophoresis parameters $Nt$ increases, so does the temperature. The temperature of the fluid rises as the temperature variance amid the surface and ambient heat grows. The stimulus of the Brownian motion $Nb$ on the temperature is revealed in Figure 3j. This graph shows that augmented $Nb$ raises the temperature.

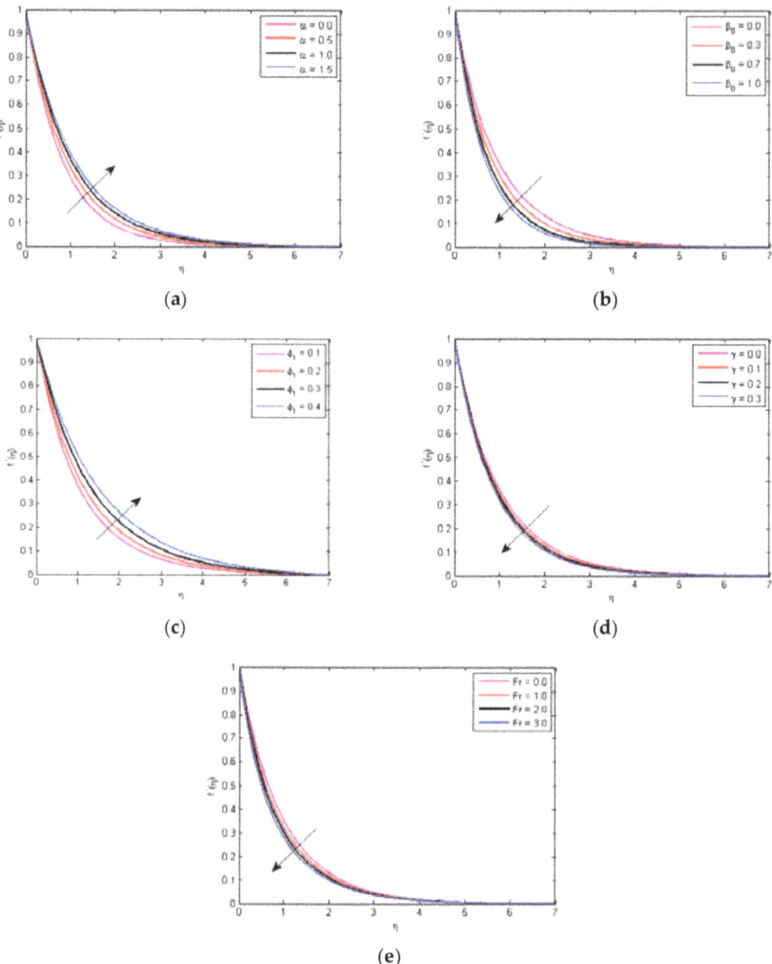

**Figure 2.** Velocity profile variation for distinct values of (**a**) is $\alpha$, (**b**) is $\beta_0$, (**c**) is $\phi_1$, (**d**) is $\gamma$, (**e**) is $Fr$.

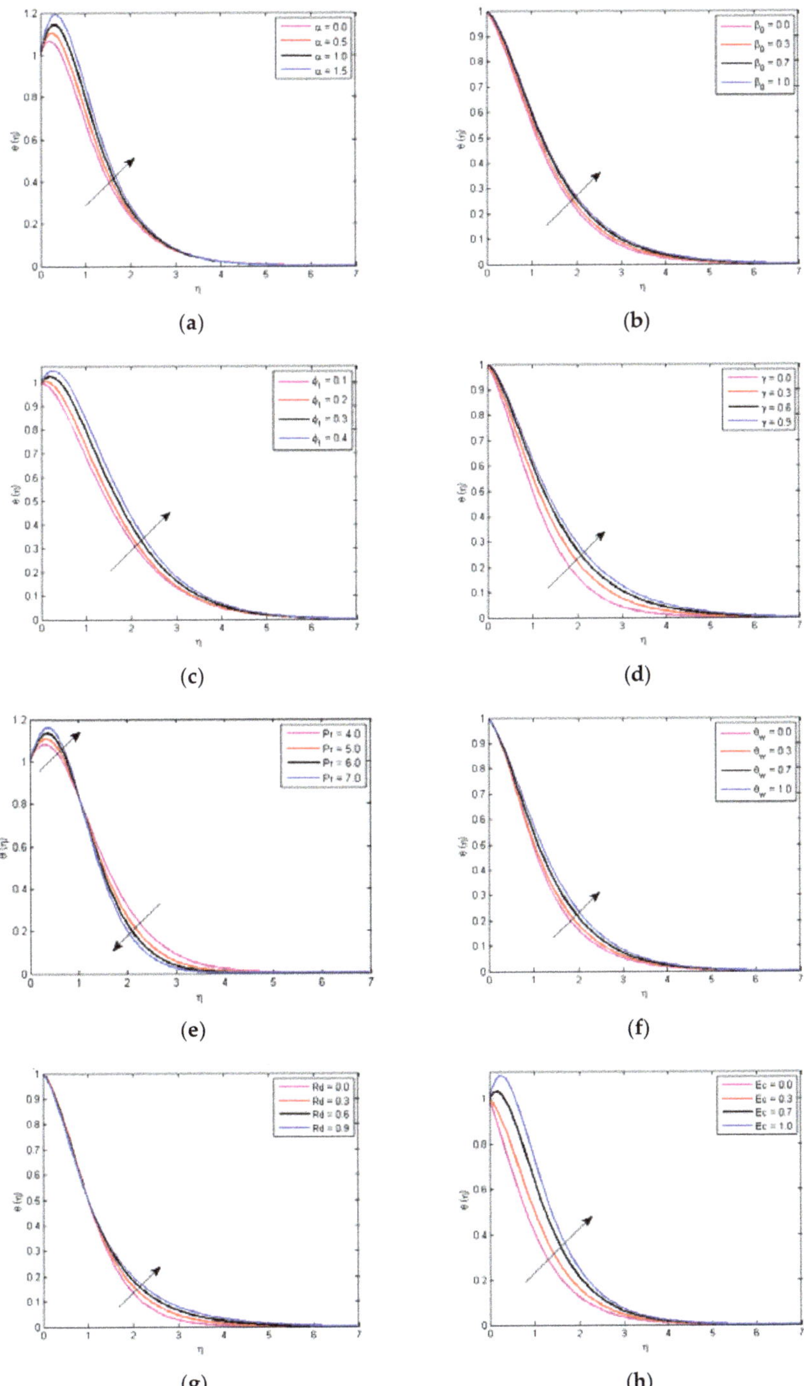

**Figure 3.** *Cont.*

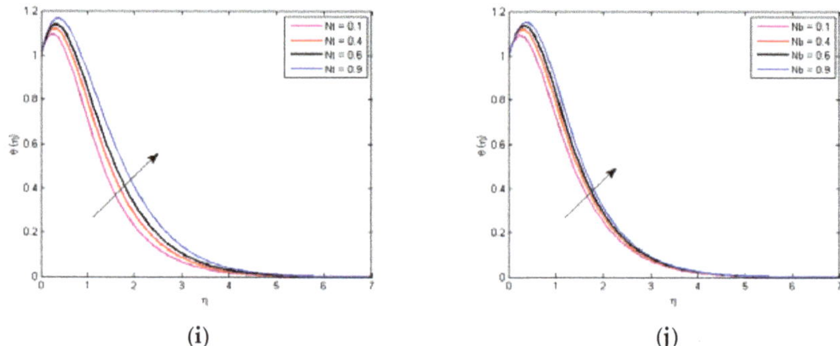

**Figure 3.** Temperature profile variation for distinct values of (**a**) is $\alpha$, (**b**) is $\beta_0$, (**c**) is $\phi_1$, (**d**) is $\gamma$, (**e**) is Pr, (**f**) is $\theta_\omega$, (**g**) is $Rd$, (**h**) is $Ec$, (**i**) is $Nt$, (**j**) is $Nb$.

The consequence of significant flow parameters such as the fluid parameter $\alpha$, curvature parameter $\gamma$, the Schmidt number $Sc$, and volume fraction $\phi_1$ on the concentration profile $\alpha$ are exposed in Figure 4a–d. Figure 4a demonstrates the influence of fluid parameter $\alpha$ on concentration distribution $\phi(\eta)$. It is detected that the concentration profile declines by rising the value of $\alpha$. The inspiration of curvature constraint $\gamma$ on the concentration profile $\phi(\eta)$ is shown in Figure 4b. Proof is provided through $\gamma$ rising along with the fluid concentration and the thickness of the resulting boundary layer. The impacts of Schmidt number $Sc$ and volume fraction $\phi_1$ on the distribution of concentrations are depicted in Figure 4c,d, respectively. As can be observed in both of these diagrams, the concentration distribution diminishes for substantial values of $Sc$ and $\phi_1$. The reason for this phenomenon is that viscous forces increase as the concentration slows down. $Sc$ is the ratio of mass diffusion to viscous forces at each end of the scale. As $Sc$ increases, viscous forces grow and mass diffusion declines, causing the concentration distribution to decrease.

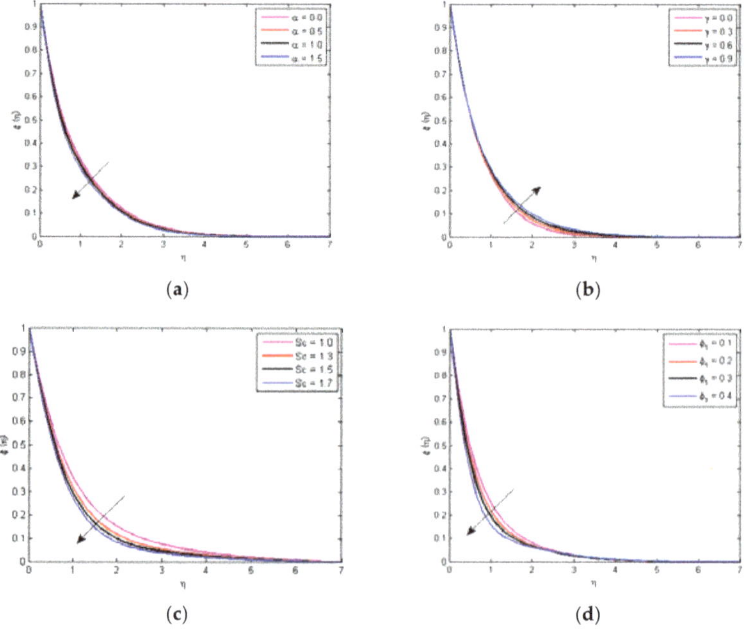

**Figure 4.** Concentration profile variation for distinct values of (**a**) is $\alpha$, (**b**) is $\gamma$, (**c**) is $Sc$, (**d**) is $\phi_1$.

The behavior of entropy optimization and Bejan number is shown in Figures 5 and 6, respectively, for various parameters such as porosity parameter $\beta_0$, volume fraction $\phi_1$, curvature parameter $\gamma$, and Brinkman number $Br$. Figures 5a and 6a show the nature of entropy formation and Bejan number for the rising porosity parameter $\beta_0$. As the values of porosity parameter $\beta_0$ upsurge, the values of entropy generation escalations near the wall slightly decrease; the Bejan number, on the other hand, exhibits the opposite pattern. As the porosity parameter tends to oppose the fluid flow, as a result, it increases the rate of total entropy formation. The influence of nanoparticles' volume fraction parameter $\phi_1$ on entropy production and Bejan number is exposed in Figures 5b and 6b. These figures show that entropy production increases, while Bejan number drops with an escalation in volume fraction parameter $\phi_1$. The increase in thermal conductivity and temperature of nanofluid caused by nanoparticles is directly related to this phenomenon. Figures 5c and 6c show the stimuli of the curvature parameter on entropy formation and Bejan number, respectively. As $\gamma$ upturns, the value of Bejan number and entropy generation increases, because less resisting force is offered when the contact surface of a cylinder with particles is reduced. This allows for more nanoparticle movement, increasing the rate of entropy formation. As a result, more curved bodies produce more entropy. Figures 5d and 6d show the nature of entropy formation and Bejan number for the rising Brinkman number $Br$. The outcomes of entropy formation and Bejan number are utterly opposite when Brinkman number $Br$ is changed. Entropy production increases as the Brinkman number increases, as shown in Figure 5d. The ratio of heat creation via viscous heating transition for conduction is known as the Brinkman number. More heat is created in the system to reimburse for increasing Brinkman number. As a result, the overall system's level of disturbance grows. For a high Brinkman number $Br$, Figure 6d shows the exact opposite characteristics.

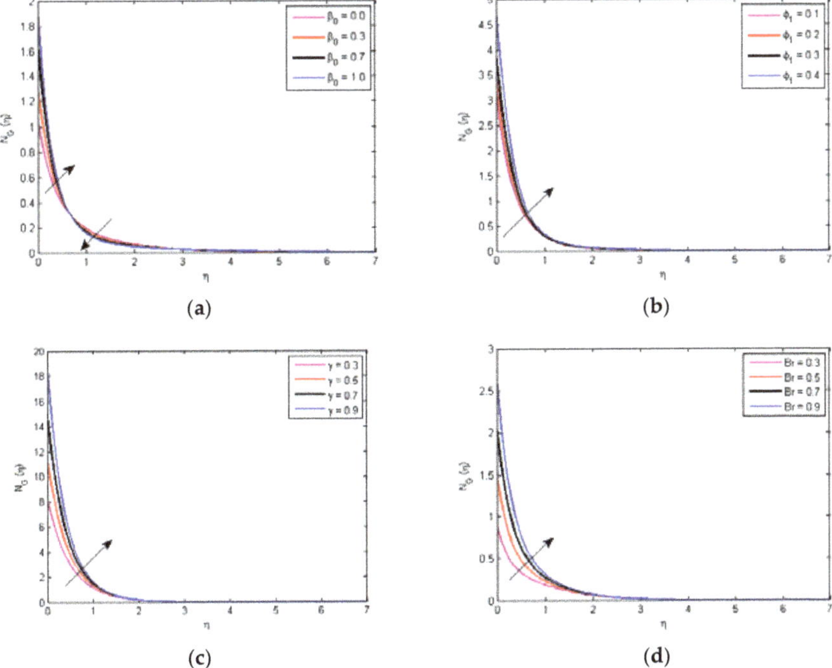

**Figure 5.** Entropy optimization variation for distinct values of (**a**) is $\beta_0$, (**b**) is $\phi_1$, (**c**) is $\gamma$, (**d**) is $Br$.

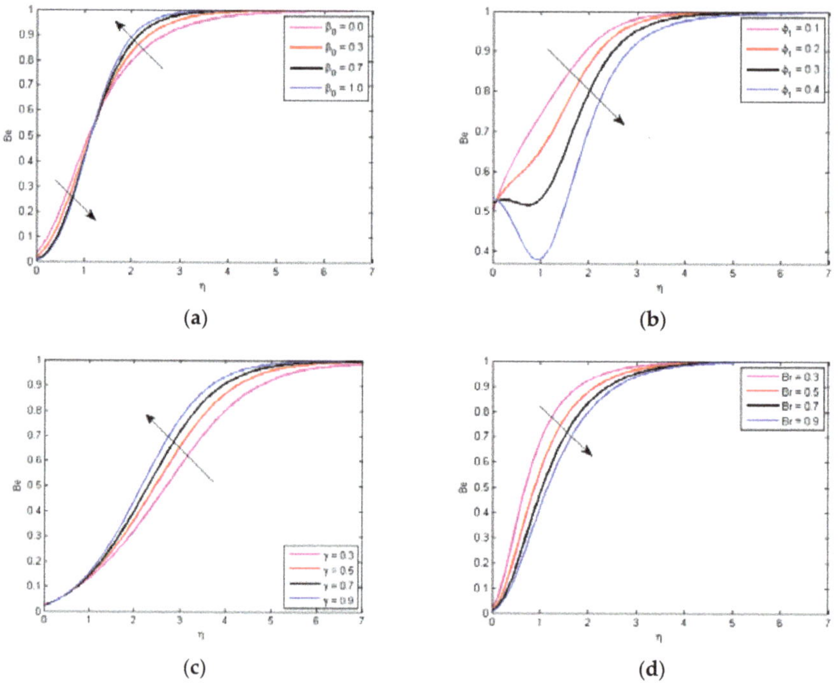

**Figure 6.** Bejan number variation for distinct values of (**a**) is $\beta_0$, (**b**) is $\phi_1$, (**c**) is $\gamma$, (**d**) is $Br$.

Table 2 indicates how the numerous constraints affect the skin friction. It is determined that skin friction coefficient decreases for large values of the volume fraction, fluid parameter, porosity parameter, curvature parameter, and inertia coefficient. Table 3 indicates the numeric data of the Nusselt number for a variety of constraints. It is determined that Nusselt number decreases for large values of the volume fraction, fluid parameter, porosity parameter, and temperature ratio, whereas it upsurges for superior values of the curvature parameter. Table 4 compares the numeric values of skin fraction with the pervious result. Both outcomes are noticed to be highly congruent.

**Table 2.** The skin friction coefficient variation for numerous values of $\phi_1$, $\alpha$, $\beta_0$, $\gamma$, and $Fr$.

| $\phi_1$ | $\alpha$ | $\beta_0$ | $\gamma$ | $Fr$ | $C_f \mathrm{Re}_z{}^{1/2}$ |
|---|---|---|---|---|---|
| 0.1 | - | - | - | - | $-2.463269$ |
| 0.2 | - | - | - | - | $-2.919301$ |
| 0.3 | - | - | - | - | $-3.6675$ |
| 0.1 | 0.1 | - | - | - | $-2.002416$ |
| - | 0.3 | - | - | - | $-2.223522$ |
| - | 0.5 | - | - | - | $-2.463269$ |
| - | - | 0.1 | - | - | $-2.272982$ |
| - | - | 0.3 | - | - | $-2.368182$ |
| - | - | 0.5 | - | - | $-2.463269$ |
| - | - | - | 0.1 | - | $-1.893775$ |
| - | - | - | 0.3 | - | $-2.045572$ |
| - | - | - | 0.5 | - | $-2.186751$ |
| - | - | - | - | 0.1 | $-1.948242$ |
| - | - | - | - | 0.5 | $-2.045572$ |
| - | - | - | - | 0.9 | $-2.138431$ |

**Table 3.** The variation in Nusselt number for numerous values of $\phi_1$, $\alpha$, $\beta_0$, $\gamma$, and $\theta_\omega$.

| $\phi_1$ | $\alpha$ | $\beta_0$ | $\gamma$ | $\theta_\omega$ | $N_u Re_z^{-1/2}$ |
|---|---|---|---|---|---|
| 0.1 | - | - | - | - | 0.416374 |
| 0.2 | - | - | - | - | 0.413638 |
| 0.3 | - | - | - | - | 0.3887 |
| - | 0.1 | - | - | - | 0.449055 |
| - | 0.3 | - | - | - | 0.43289 |
| - | 0.5 | - | - | - | 0.416374 |
| - | - | 0.1 | - | - | 0.441787 |
| - | - | - | - | - | 0.428625 |
| - | - | - | - | - | 0.416374 |
| - | - | - | 0.1 | - | 0.330731 |
| - | - | - | 0.3 | - | 0.357808 |
| - | - | - | 0.5 | - | 0.380167 |
| - | - | - | - | 0.1 | 0.419923 |
| - | - | - | - | 0.3 | 0.418526 |
| - | - | - | - | 0.5 | 0.409504 |
| - | - | - | - | - | - |
| - | - | - | - | - | - |

**Table 4.** Comparison of $f''(0)$ skin friction values for several fluid parameter $\alpha$ values.

| $\alpha$ | Present Result | Hayat et al. [12] |
|---|---|---|
| 0.2 | $-0.7749406$ | $-0.91287$ |
| 0.4 | $-0.7282394$ | $-0.84516$ |
| 0.6 | $-0.6891753$ | $-0.79057$ |
| 0.8 | $-0.6558923$ | $-0.74536$ |
| 1.0 | $-0.6270961$ | $-0.70711$ |

## 6. Conclusions

In this research, an entropy generation interpretation for axisymmetric flow of Powell–Eyring nanofluid via a horizontal porous stretching cylinder was performed. By using a similarity variable transformation system, the nonlinear equation describing the flow problem is changed to nonlinear ODEs, which are then solved using bvp4c. The impacts of several factors in the model problem on velocity, temperature, concentration, entropy optimization, Bejan number, drag force, and Nusselt number are analyzed. The following conclusions were derived from the study's findings:

- It is indicated that with growing value of fluid parameter $\alpha$, velocity increased, while with rising values of porosity parameter $\beta_0$, the volume fraction $\phi_1$, curvature parameter $\gamma$, and inertia coefficient $Fr$ velocity profile declined.
- The temperature declined with the mounting value of fluid parameter $\alpha$, whereas it increased with increasing values of the porosity parameter $\beta_0$, volume fraction $\phi_1$, curvature parameter $\gamma$, temperature ratio $\theta_\omega$, thermal radiation $Rd$, Eckert number $Ec$, thermophoresis parameter $Nt$, and Brownian motion $Nb$. It is noted that the temperature of the fluid rose up and then dropped down when we increased the Prandtl number.
- The concentration profile declined with rising values of fluid parameter $\alpha$, curvature parameter $\gamma$, the Schmidt number $Sc$, and volume fraction $\phi_1$.
- Entropy optimization rose up for the values of volume fraction $\phi_1$, curvature parameter $\gamma$, and Brinkman number $Br$, whereas for the rising value of porosity parameter $\beta_0$, entropy optimization first increased and then decreased.
- Bejan number decayed down for greater $\phi_1$ and Brinkman number, while for a higher value of curvature parameter, Bejan number rose up. It is indicated that if we increased the porosity parameter $\beta_0$, Bejan number decayed down first and then rose up.

- It is concluded that skin friction decreased for a large number of volume fraction, fluid parameter, porosity parameter, curvature parameter, and inertia coefficient, whereas Nusselt number decreased for a cumulative number of volume fraction, fluid parameter, porosity parameter, and temperature ratio, whereas it rose for a cumulative number of curvature parameter.

**Author Contributions:** Conceptualization, M.R. and N.V.; methodology, Z.S. and M.A.J.; software, M.R.; validation, N.V. and W.D.; formal analysis, S.F.B. and S.I.; investigation, Z.S. and M.R.; resources, N.V.; data curation, W.D.; writing—original draft preparation, M.R.; writing—review and editing, Z.S., N.V. and S.I.; visualization, W.D.; supervision, M.A.J. and Z.S.; project administration, N.V. and S.F.B.; funding acquisition, N.V. and S.F.B. All authors have read and agreed to the published version of the manuscript.

**Funding:** The project was financed by Lucian Blaga University of Sibiu and Hasso Plattner Foundation research Grants LBUS-IRG-2021-07.

**Institutional Review Board Statement:** Not applicable.

**Informed Consent Statement:** Not applicable.

**Data Availability Statement:** The data that support the findings of this study are available from the corresponding author upon reasonable request.

**Conflicts of Interest:** The authors declare no conflict of interest.

## References

1. Crane, L.J. Flow past a stretching plate. *Z. Angew. Math. Phys. ZAMP* **1970**, *21*, 645–647. [CrossRef]
2. Nadeem, S.; Ul Haq, R. Effect of thermal radiation for megnetohydrodynamic boundary layer flow of a nanofluid past a stretching sheet with convective boundary conditions. *J. Comput. Theor. Nanosci.* **2014**, *11*, 32–40. [CrossRef]
3. Ahmed, J.; Mahmood, T.; Iqbal, Z.; Shahzad, A.; Ali, R. Axisymmetric flow and heat transfer over an unsteady stretching sheet in power law fluid. *J. Mol. Liq.* **2016**, *221*, 386–393. [CrossRef]
4. Ariel, P.D. Axisymmetric flow of a second grade fluid past a stretching sheet. *Int. J. Eng. Sci.* **2001**, *39*, 529–553. [CrossRef]
5. Hsiao, K.L. Combined electrical MHD heat transfer thermal extrusion system using Maxwell fluid with radiative and viscous dissipation effects. *Appl. Therm. Eng.* **2017**, *112*, 1281–1288. [CrossRef]
6. Powell, R.E.; Eyring, H. Mechanisms for the Relaxation Theory of Viscosity. *Nature* **1944**, *154*, 427–428. [CrossRef]
7. Patel, M.; Timol, M.G. Numerical treatment of Powell-Eyring fluid flow using Method of Satisfaction of Asymptotic Boundary Conditions (MSABC). *Appl. Numer. Math.* **2009**, *59*, 2584–2592. [CrossRef]
8. Hayat, T.; Awais, M.; Asghar, S. Radiative effects in a three-dimensional flow of MHD Eyring-Powell fluid. *J. Egypt. Math. Soc.* **2013**, *21*, 379–384. [CrossRef]
9. Ara, A.; Khan, N.A.; Khan, H.; Sultan, F. Radiation effect on boundary layer flow of an Eyring–Powell fluid over an exponentially shrinking sheet. *Ain Shams Eng. J.* **2014**, *5*, 1337–1342. [CrossRef]
10. Hayat, T.; Farooq, M.; Alsaedi, A.; Iqbal, Z. Melting heat transfer in the stagnation point flow of powell-eyring fluid. *J. Thermophys. Heat Transf.* **2013**, *27*, 761–766. [CrossRef]
11. Rehman, K.U.; Malik, M.Y.; Salahuddin, T.; Naseer, M. Dual stratified mixed convection flow of Eyring-Powell fluid over an inclined stretching cylinder with heat generation/absorption effect. *AIP Adv.* **2016**, *6*, 075112. [CrossRef]
12. Hayat, T.; Hussain, Z.; Farooq, M.; Alsaedi, A. Magnetohydrodynamic flow of powell-eyring fluid by a stretching cylinder with Newtonian heating. *Therm. Sci.* **2018**, *22*, 371–382. [CrossRef]
13. Bejan, A. A Study of Entropy Generation in Fundamental Convective Heat Transfer. *J. Heat Transfer* **1979**, *101*, 718–725. [CrossRef]
14. Butt, A.S.; Munawar, S.; Ali, A.; Mehmood, A. Entropy generation in the Blasius flow under thermal radiation. *Phys. Scr.* **2012**, *85*, 035008. [CrossRef]
15. San, J.Y.; Worek, W.M.; Lavan, Z. Entropy Generation in Convective Heat Transfer and Isothermal Convective Mass Transfer. *J. Heat Transfer* **1987**, *109*, 647–652. [CrossRef]
16. Tamayol, A.; Hooman, K.; Bahrami, M. Thermal Analysis of Flow in a Porous Medium Over a Permeable Stretching Wall. *Transp. Porous Med.* **2010**, *85*, 661–676. [CrossRef]
17. Rashidi, M.M.; Ali, M.; Freidoonimehr, N.; Nazari, F. Parametric analysis and optimization of entropy generation in unsteady MHD flow over a stretching rotating disk using artificial neural network and particle swarm optimization algorithm. *Energy* **2013**, *55*, 497–510. [CrossRef]
18. Shit, G.C.; Haldar, R.; Mandal, S. Entropy generation on MHD flow and convective heat transfer in a porous medium of exponentially stretching surface saturated by nanofluids. *Adv. Powder Technol.* **2017**, *28*, 1519–1530. [CrossRef]
19. Butt, A.S.; Ali, A. Effects of Magnetic Field on Entropy Generation in Flow and Heat Transfer due to a Radially Stretching Surface. *Chin. Phys. Lett.* **2013**, *30*, 024701. [CrossRef]

20. Munawar, S.; Ali, A.; Mehmood, A. Thermal analysis of the flow over an oscillatory stretching cylinder. *Phys. Scr.* **2012**, *86*, 065401. [CrossRef]
21. Khan, M.; Ahmed, J.; Rasheed, Z. Entropy generation analysis for axisymmetric flow of Carreau nanofluid over a radially stretching disk. *Appl. Nanosci.* **2020**, *10*, 5291–5303. [CrossRef]
22. Choi, S.U.S.; Eastman, J.A. Enhancing thermal conductivity of fluids with nanoparticles. In Proceedings of the International Mechanical Engineering Congress and Exhibition, San Francisco, CA, USA, 12–17 November 1995.
23. Iqbal, K.; Ahmed, J.; Khan, M.; Ahmad, L.; Alghamdi, M. Magnetohydrodynamic thin film deposition of Carreau nanofluid over an unsteady stretching surface. *Appl. Phys. A Mater. Sci. Process.* **2020**, *126*, 105. [CrossRef]
24. Khan, W.A.; Anjum, N.; Waqas, M.; Abbas, S.Z.; Irfan, M.; Muhammad, T. Impact of stratification phenomena on a nonlinear radiative flow of sutterby nanofluid. *J. Mater. Res. Technol.* **2021**, *15*, 306–314. [CrossRef]
25. Ali, F.; Ahmad, Z.; Arif, M.; Khan, I.; Nisar, K.S. A Time Fractional Model of Generalized Couette Flow of Couple Stress Nanofluid with Heat and Mass Transfer: Applications in Engine Oil. *IEEE Access* **2020**, *8*, 146944–146966. [CrossRef]
26. Acharya, N.; Maity, S.; Kundu, P.K. Entropy generation optimization of unsteady radiative hybrid nanofluid flow over a slippery spinning disk. *Proc. Inst. Mech. Eng. Part C J. Mech. Eng. Sci.* **2022**, 09544062211065384. [CrossRef]
27. Acharya, N.; Mondal, H.; Kundu, P.K. Spectral approach to study the entropy generation of radiative mixed convective couple stress fluid flow over a permeable stretching cylinder. *Proc. Inst. Mech. Eng. Part C J. Mech. Eng. Sci.* **2021**, *235*, 2692–2704. [CrossRef]
28. Krishna, M.V.; Chamkha, A.J. Hall and ion slip effects on MHD rotating flow of elastico-viscous fluid through porous medium. *Int. Commun. Heat Mass Transf.* **2020**, *113*, 104494. [CrossRef]
29. Venkata Ramana, K.; Gangadhar, K.; Kannan, T.; Chamkha, A.J. Cattaneo–Christov heat flux theory on transverse MHD Oldroyd-B liquid over nonlinear stretched flow. *J. Therm. Anal. Calorim.* **2021**, *147*, 2749–2759. [CrossRef]
30. Al-Mubaddel, F.S.; Farooq, U.; Al-Khaled, K.; Hussain, S.; Khan, S.U.; Aijaz, M.O.; Rahimi-Gorji, M.; Waqas, H. Double stratified analysis for bioconvection radiative flow of Sisko nanofluid with generalized heat/mass fluxes. *Phys. Scr.* **2021**, *96*, 055004. [CrossRef]
31. Naveen Kumar, R.; Jyothi, A.M.; Alhumade, H.; Gowda, R.J.P.; Alam, M.M.; Ahmad, I.; Gorji, M.R.; Prasannakumara, B.C. Impact of magnetic dipole on thermophoretic particle deposition in the flow of Maxwell fluid over a stretching sheet. *J. Mol. Liq.* **2021**, *334*, 116494. [CrossRef]
32. Lund, L.A.; Wakif, A.; Omar, Z.; Khan, I.; Animasaun, I.L. Dynamics of water conveying copper and alumina nanomaterials when viscous dissipation and thermal radiation are significant: Single-phase model with multiple solutions. *Math. Methods Appl. Sci.* **2022**. [CrossRef]
33. Song, Y.-Q.; Waqas, H.; Al-Khaled, K.; Farooq, U.; Gouadria, S.; Imran, M.; Khan, S.U.; Khan, M.I.; Qayyum, S.; Shi, Q.-H. Aspects of thermal diffusivity and melting phenomenon in Carreau nanofluid flow confined by nonlinear stretching cylinder with convective Marangoni boundary constraints. *Math. Comput. Simul.* **2022**, *195*, 138–150. [CrossRef]
34. Acharya, N.; Mabood, F.; Badruddin, I.A. Thermal performance of unsteady mixed convective Ag/MgO nanohybrid flow near the stagnation point domain of a spinning sphere. *Int. Commun. Heat Mass Transf.* **2022**, *134*, 106019. [CrossRef]
35. Chamkha, A.J.; Rashad, A.M. Unsteady heat and mass transfer by MHD mixed convection flow from a rotating vertical cone with chemical reaction and Soret and Dufour effects. *Can. J. Chem. Eng.* **2014**, *92*, 758–767. [CrossRef]
36. Seyedi, S.H.; Saray, B.N.; Chamkha, A.J. Heat and mass transfer investigation of MHD Eyring–Powell flow in a stretching channel with chemical reactions. *Phys. A Stat. Mech. Its Appl.* **2020**, *544*, 124109. [CrossRef]
37. Veera Krishna, M.; Ameer Ahamad, N.; Chamkha, A.J. Hall and ion slip effects on unsteady MHD free convective rotating flow through a saturated porous medium over an exponential accelerated plate. *Alexandria Eng. J.* **2020**, *59*, 565–577. [CrossRef]
38. Chakraborty, S.; Ray, S. Performance optimisation of laminar fully developed flow through square ducts with rounded corners. *Int. J. Therm. Sci.* **2011**, *50*, 2522–2535. [CrossRef]
39. Lorenzini, M.; Suzzi, N. The Influence of Geometry on the Thermal Performance of Microchannels in Laminar Flow with Viscous Dissipation. *Heat Transf. Eng.* **2016**, *37*, 1096–1104. [CrossRef]
40. Lorenzini, M.; Daprà, I.; Scarpi, G. Heat transfer for a Giesekus fluid in a rotating concentric annulus. *Appl. Therm. Eng.* **2017**, *122*, 118–125. [CrossRef]
41. Coelho, P.M.; Faria, J.C. On the generalized Brinkman number definition and its importance for Bingham fluids. *J. Heat Transfer* **2011**, *133*, 054505. [CrossRef]

*Article*

# Mathematical Modelling of Ree-Eyring Nanofluid Using Koo-Kleinstreuer and Cattaneo-Christov Models on Chemically Reactive *AA7072-AA7075* Alloys over a Magnetic Dipole Stretching Surface


**Zahir Shah [1], Narcisa Vrinceanu [2,*], Muhammad Rooman [1], Wejdan Deebani [3] and Meshal Shutaywi [3]**

[1] Department of Mathematical Sciences, University of Lakki Marwat, Lakki Marwat 28420, Pakistan; zahir@ulm.edu.pk (Z.S.); roomankhan31@gmail.com (M.R.)

[2] Department of Industrial Machines and Equipments, Faculty of Engineering, "Lucian Blaga" University of Sibiu, 10 Victoriei Boulevard, 550024 Sibiu, Romania

[3] Department of Mathematics, College of Science & Arts, King Abdulaziz University, P.O. Box 344, Rabigh 21911, Saudi Arabia; wdeebani@kau.edu.sa (W.D.); mshutaywi@kau.edu.sa (M.S.)

* Correspondence: vrinceanu.narcisai@ulbsibiu.ro



**Abstract:** In the current study, since nanofluids have a high thermal resistance, and because non-Newtonian (Ree-Eyring) fluid movement on a stretching sheet by means of suspended nanoparticles *AA7072-AA7075* is used, the proposed mathematical model takes into account the influence of magnetic dipoles and the Koo-Kleinstreuer model. The Cattaneo-Christov model is used to calculate heat transfer in a two-dimensional flow of Ree-Eyring nanofluid across a stretching sheet, and viscous dissipation is taken into account. The base liquid water with suspended nanoparticles *AA7072-AA7075* is considered in this study. The PDEs are converted into ODEs by exhausting similarity transformations. The numerical solution of the altered equations is then performed utilising the HAM. To examine the performance of velocity, temperature profiles, concentration profiles, skin friction, the Nusselt number, and the Sherwood number, a graphical analysis is carried out for various parameters. The new model's key conclusions are that the *AA7075* alloy outperforms the *AA7072* alloy in terms of thermal performance as the volume fraction and ferro-magnetic interaction constraint rise. Additionally, the rate of heat transmission and the skin friction coefficient improve as the volume fraction rises.

**Keywords:** Ree-Eyring nanofluid; magnetic dipole; viscous dissipation; Cattaneo-Christov model; Koo-Kleinstreuer model; chemical reaction



**Citation:** Shah, Z.; Vrinceanu, N.; Rooman, M.; Deebani, W.; Shutaywi, M. Mathematical Modelling of Ree-Eyring Nanofluid Using Koo-Kleinstreuer and Cattaneo-Christov Models on Chemically Reactive *AA7072-AA7075* Alloys over a Magnetic Dipole Stretching Surface. *Coatings* **2022**, *12*, 391. https://doi.org/10.3390/coatings12030391

Academic Editor: Eduardo Guzmán

Received: 15 February 2022
Accepted: 6 March 2022
Published: 15 March 2022

**Publisher's Note:** MDPI stays neutral with regard to jurisdictional claims in published maps and institutional affiliations.

## 1. Introduction

The progressive thermal patterns of nanoparticles have an extensive range of exploitations in the engineering, industrial, technical, and biomedical fields. Many thermal engineering and industrial processes employ nanofluids to increase their thermal efficiency. In recent decades, dynamic scientists have shown interest in nanoparticles with a small size (1–100 nm). Nanofluids are nanoparticle suspensions in base fluids. It is noted that these particles do not change the reaction process, but they do improve the fundamental thermal processes of base liquids at the peak level. Nanoparticles are used in sophisticated thermal extrusion systems, engineering heating devices, biomedical applications, cancer treatments, the chemotherapy process, energy resources, heat exchangers, manufacturing processes, thermal management equipment, and many other applications. Usually, these nanoparticles undergo aggregation so that a fluid can flow through a porous medium as a completely interconnected network (ideal porous pipe), formed by the constricted channel between each pore. Choi [1] proposed a ground-breaking study on the thermal characteristics of nanofluids, prompting other researchers to pay attention to the subject.

Here, we briefly highlight certain contributions due to innovative research on the subject. Kishan and Deepa [2] studied the immersion of nanoparticles in micropolar liquid and the stagnation point flow in a porous region. In a nanofluid material flow constrained by a vertical surface, Alim et al. [3] used the Joule heating process. Sheikholeslami et al. [4] conducted research on CuO nanoparticles contained in a heated chamber with a sensodial wall. The heat process in nanomaterial within a micro-channel with a sinusoidal double layer was studied by He et al. [5]. Mahdavi et al. [6] used nanoparticles to identify cooling applications in a hot jet surface. Abdelsalam et al. [7] addressed the thermal repercussions of hybrid nanofluids in blood flow when electro-osmotic forces are present. Nadeem et al. [8] used dual solution simulations to visualise the slip characteristics of nanofluid flow. Khan et al. [9] concentrated on the thermal characteristics of hybrid nanofluids in unsteady flow. Abbas et al. [10] investigated the influence of time-dependent viscosity on nanofluid flow over the Riga surface. Awais et al. [11] used the KKL model to investigate heat transfer in a suspension of nanomaterials containing (CuO and $Al_2O_3$).

Nanomaterials are important because of their high thermal and mechanical properties. The characteristics of the nanoliquids formed by each nanomaterial are considerably altered by these materials. Among nanomaterials, there is a substance known as aluminium alloy, in which aluminium plays a major role. Heat treatable and non-heat treatable alloys are the two main types of aluminium alloys. Aluminium alloys are widely utilised in the construction, testing, and production of spacecraft, aircraft parts, and other structures. Researchers have investigated numerous flow models consisting of aluminium alloys and discovered remarkable thermal transport behaviour due to the improved heat transport features of *AA7072* and *AA7075* aluminium alloys. Sandeep and Animasaun [12] reported an examination of heat transfer in nanoliquids consisting of *AA7072* and *AA7075* aluminium alloys while considering the impact of varying Lorentz forces. They discovered that nanoliquid made of *AA7075* alloy is superior in terms of heat transmission to nanoliquid made of *AA7072*. Kandasamy at al. [13] considered the electric field strength for the analysis of heat transport in magnetised *AA7075* alloys. Tlili et al. [14] investigated three-dimensional heat transfer characteristics in the hybrid colloidal model *AA7072-AA7075/Methanol* under various velocity conditions. They used a numerical approach to the model and described the results in terms of flow regimes.

Since MHD is commonly used in numerous fields, such as the polymer and petroleum industries, a significant amount of thought has been given to the approach of magnetic fields in liquid flow in recent decades. As we know that the pace of cooling is even more essential than in the standard processes, numerous fabrication processes have been used to regulate the rate of cooling for magneto-hydrodynamic liquids. Unifying metals in electric heaters, metal casting, and gem creating are some of the other functions of magneto-hydrodynamics. It also assists in the cooling of the atomic reactor's internal dividers. Magneto-hydrodynamic flows were first sculpted and highly valued in biodesign because they are used in a variety of symptomatic kinds of sickness. In this approach, studying magneto-hydrodynamic flow has a significant impact on several scientific fields. The convective circumstances for MHD Jeffrey flow on an elaborated sheet were examined by Ahmad et al. [15]. Khan et al. [16] studied MHD Falknar-Skan flow through a permeable material with a convective boundary condition. Malik et al. [17] explored MHD hyperbolic flow through an expanded cylinder via numerical methodology, the Kellor-Box method. By assuming magnetic field-dependent viscosity effects, Sheikholeslami et al. [18] described MHD nanofluid flow. The finite element method was used to address this problem.

Flow due to a stretchy surface has risen in prominence among researchers in recent years, owing to its widespread application in industry. Hot rolling, paper production, glass blowing, polymer extrusion, metal extrusion, and crystal growth are only a few of these uses. Crane [19] started flow research with an enlarged sheet. The fluid stream in an enlarged channel was examined by Brady and Acrivos [20]. Researchers discovered that there is a solution for a two-dimensional flow for any given Reynolds number. The movement of fluid past a stretchable cylinder was studied by Wang [21]. By changing the heat flux, Elbashbeshy [22] investigated heat transfer across a stretchy surface. By using a viscous fluid created by a stretchable surface, Nadeem et al. [23] examined a stagnation-point stream. Different fluid flow over stretched surfaces was inspected by Awan et al. [24]. The production of entropy for MHD Maxwell fluid across a stretchable and penetrable surface was investigated by Jawad et al. [25].

Because of its vast industrial use in nuclear reactor cooling, chemical engineering, geothermal reservoirs, and thermal oil recovery, the chemical reaction effect has gained a tremendous response. Generally, the relationship between mass transfer and chemical reaction is very important, and it can be studied in terms of reactant species deployment and creation at various speeds during nanofluid mass transfer. Bestman et al. [26] conducted ground-breaking work in defining these influences. Mustafa et al. [27] examined a hydromagnetic flow past a radial surface caused by chemical reaction and Arrhenius activation energy. They found that the concentration of a species rises as the activation energy of a chemical process rises. Mohyud-Din et al. [28] looked at how chemical reactions affected convergent/divergent channels. Aleem et al. [29] instigated alternative forms of water based nanofluids, such as titanium-oxide, aluminium-oxide, and copper-oxide, that arose in a porous media after a chemical reaction and Newtonian heating.

The majority of industrial applications necessitate non-Newtonian nanofluids with non-linearly related shear rates and shear stresses. The shear rate is heavily influenced by the timeframe of the shear stress. Thus, coefficients such as viscosity do not fully describe shear stress in such nanofluids. As a result, numerous mathematicians have debated the class of non-Newtonian models, one of which is the Ree-Eyring nanofluid model. Inks, molten polymers, adhesives, paints, organic materials, and other non-Newtonian fluids are some examples. These are used in food industries, drilling rigs, cooling systems, adhesive industries, and so on. Hayat et al. [30] conducted an entropy examination in the flow of Ree-Eyring nanofluid in this respect. Tanveer and Malik [31] investigated the thermal effectiveness of Ree-Eyring nanofluid peristaltic flow. Khan et al. [32] investigated the effect of Lorentz force on the velocity of a Ree-Eyring nanofluid flow past a paraboloid surface. Al-Mdallal et al. [33] investigated the thermal properties of *Cu-Water* nanofluid under the sway of radiation. Purna et al. [34] used the Darcy-Frochheimer law to examine the flow of Ree-Eyring nanofluid on a porous plate inclined at an angle, as well as the impact of the chemical reaction. Some recent studies about nanofluids and heat transfer properties are mentioned in Refs. [35–38].

The main goal of this article is to study the existence of a magnetic dipole and the Koo-Kleinstreuer model using different alloys over a stretching sheet. The Cattaneo-Christov model is used to calculate heat transfer in a two-dimensional flow of Ree-Eyring nanofluid across a stretching sheet. The mathematical formulation is created in the following section, utilising fluid flow assumptions. By applying appropriate similarity transformations, the physical flow phenomenon is represented and then translated into a non-dimensional form. The HAM is used to arrive at a solution. Through graphical demonstrations, the influences of a few key parameters on the temperature, velocity fields, and concentration profile are highlighted.

## 2. Mathematical Model and Formulation

We consider a Ree-Eyring nanofluid flow in a two-dimensional laminar boundary layer with the influence of magnetic dipoles. Furthermore, the Cattaneo-Christov heat flow model is used to analyse heat transmission. The magnetic dipole is located under the sheet, whereas the electrically non-conductive and incompressible Ree-Eyring nanofluid is located above the sheet in the half-space $y > 0$. Figure 1 depicts the flow geometry. By assuming two conflicting and comparable forces along the *x-axis*, the sheet is stretched at a proportional rate to the distance between it and the fixed origin $x = 0$. The dipole centre is located on the *y-axis* below the *x-axis*. It has a powerful magnetic field directed in the positive *x-direction*, which increases the magnetic field's intensity enough to feed the

Ree-Eyring nanofluid. The stretched sheet is kept at a temperature $T_w$ underneath Curie's temperature $T_c$, although the far-flung liquid elements are thought to be at $T = T_c$.

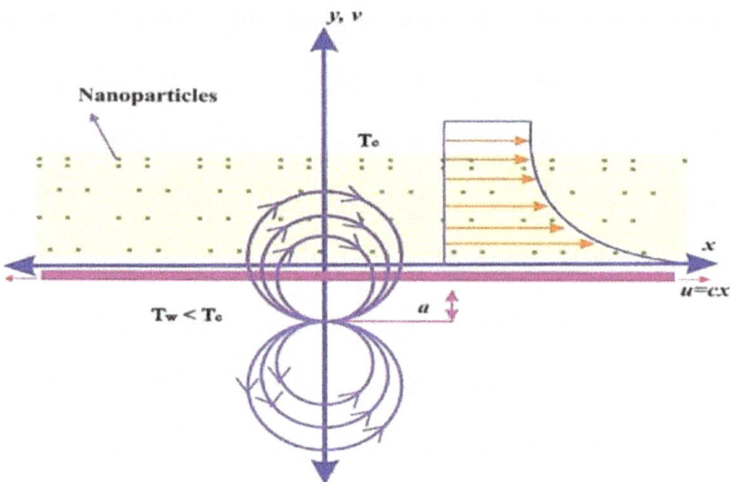


**Figure 1.** Physical description flow geometry.

Using the Koo-Kleinstreuer model, the efficacious $k_{nf}$ of nanofluids can be displayed as [11]

$$k_{nf} = k_{static} + k_{Brownian}$$

where

$$k_{static} = \frac{k_f\left(k_s + 2k_f\right) + 2\phi\left(k_s - k_f\right)}{k_f\left(k_s + 2k_f\right) - \phi\left(k_s - k_f\right)} \cdot k_{Brownian} = 5 \times 10^4 \gamma \varnothing (\rho C_p)_f \sqrt{\frac{k_\beta T}{2\rho_p r_p}} \Gamma(T, \phi),$$

where $k_\beta = 1.38 \times 10^{-23} \mathrm{m^2 kgs^{-2}k^{-1}}$ is the Boltzmann physical constant and $r_p$ is the nanoparticle radius.

Particularly,

$\gamma = 0.0137(100\phi)^{-0.8229}$, where $\phi < 1\%$;

$\gamma = 0.0011(100\phi)^{-0.7272}$, where $\phi > 1\%$;

$0.01 < \phi < 0.04\ 300K < T < 325K$.

Taking into account $\mu_{nf}$ reliance on particle volume fraction,

$$\mu_{nf} = \mu_{static} + \mu_{Brownian}$$

where

$$\text{and } \mu_{static} = \mu_f(1 - \phi)^{-2.5} \text{ and } \mu_{Brownian} = \frac{k_{Brownian}}{k_f} \frac{\mu_f}{Pr_f}$$

The governed equation is formulated as follows [32,36,37]:

$$\frac{\partial u}{\partial x} + \frac{\partial v}{\partial y} = 0 \tag{1}$$

$$u\frac{\partial u}{\partial x} + v\frac{\partial u}{\partial y} = \frac{1}{\rho_{nf}}\left(\frac{1}{\beta_1 \epsilon} + \mu_{nf}\right)\left(2\frac{\partial^2 u}{\partial x^2} + \frac{\partial^2 v}{\partial x \partial y} + \frac{\partial^2 u}{\partial y^2}\right) + \frac{\mu_0 M}{\rho_{nf}}\frac{\partial H}{\partial x} \tag{2}$$

$$u\frac{\partial T}{\partial x} + v\frac{\partial T}{\partial y} + \lambda_2\Omega_E = \frac{k_{nf}}{(\rho C_p)_{nf}}\left(\frac{\partial^2 T}{\partial y^2} + \frac{\partial^2 T}{\partial x^2}\right) + \frac{1}{(\rho C_p)_{nf}}\left(\frac{1}{\beta_1\epsilon} + \mu_{nf}\right)\left(\frac{\partial u}{\partial y}\right)^2$$
$$-\frac{\mu_0 T}{(\rho C_p)_{nf}}\frac{\partial M}{\partial T}\left(u\frac{\partial H}{\partial x} + v\frac{\partial H}{\partial y}\right) \tag{3}$$

$$u\frac{\partial C}{\partial x} + v\frac{\partial C}{\partial y} = D_{nf}\left(\frac{\partial^2 T}{\partial y^2} + \frac{\partial^2 T}{\partial x^2}\right) - k_r(C - C_c) \tag{4}$$

In the above equation, the term $\Omega_E$ is defined as

$$\Omega_E = u\frac{\partial u}{\partial x}\frac{\partial T}{\partial x} + v\frac{\partial v}{\partial y}\frac{\partial T}{\partial y} + u^2\frac{\partial^2 T}{\partial x^2} + v^2\frac{\partial^2 T}{\partial y^2} + 2uv\frac{\partial^2 T}{\partial x\partial y} + u\frac{\partial v}{\partial x}\frac{\partial T}{\partial y} + v\frac{\partial u}{\partial y}\frac{\partial T}{\partial y} \tag{5}$$

The associated boundary limitations are as follows:

$$\left.\begin{array}{l} u = cx,\ v = 0,\ T = T_w,\ C = C_w\ at\ y = 0 \\ u \to 0,\ T \to T_c,\ C \to C_c\ at\ y \to \infty \end{array}\right\} \tag{6}$$

The magnetic field affects the presumed liquid flow due to the magnetic dipole, and its magnetic scalar potential is given by

$$\phi_1 = \frac{x}{(y + a)^2 + x^2}\frac{\gamma}{2\pi} \tag{7}$$

$$H_y = -\frac{\partial\phi_1}{\partial y} = \frac{2(y + a)x}{\left((y + a)^2 + x^2\right)^2}\frac{\gamma}{2\pi},\ H_y = -\frac{\partial\phi_1}{\partial x} = -\frac{(y + a)^2 - x^2}{\left((y + a)^2 + x^2\right)^2}\frac{\gamma}{2\pi} \tag{8}$$

where

$$H = \left[\left(\frac{\partial\phi_1}{\partial y}\right)^2 + \left(\frac{\partial\phi_1}{\partial x}\right)^2\right]^{1/2} \tag{9}$$

We attain that

$$\frac{\partial H}{\partial y} = \left[\frac{4x^2}{(y + a)^5} - \frac{2}{(y + a)^3}\right]\frac{\gamma}{2\pi},\ \frac{\partial H}{\partial x} = \left[-\frac{2x}{(y + a)^4}\right]\frac{\gamma}{2\pi} \tag{10}$$

Supposing that the applied field $H$ is strong enough to saturate the supposed fluid and that the linear equation approximates the variance of magnetisation $M$ with temperature $T$,

$$M = K(T_c - T) \tag{11}$$

The following are some of the similarities:

$$(\eta, \xi) = \sqrt{\frac{c}{v_f}}(y, x),\ \psi(\eta, \xi) = \left(\frac{\mu_f}{\rho_f}\right)\xi f(\eta)$$
$$T = T_c - (T_c - T_w)\theta(\eta, \xi) = T_c - (T_c - T_w)\left[\theta_1(\eta) + \xi^2\theta_2(\eta)\right] \tag{12}$$

The stream function $\psi$ is given below:

$$u = \frac{\partial\psi}{\partial y} = cxf'(\eta),\ v = -\frac{\partial\psi}{\partial x} = -\sqrt{cv_f}f(\eta) \tag{13}$$

The continuity equation is easily satisfied, while the momentum, thermal equations, and mass transfer are transferred to the relating set of ODEs:

$$\left(\varepsilon_2 We + \varepsilon_1 + \frac{k_{Brownian}}{k_f Pr_f\varepsilon_2}\right)f''' - \varepsilon_2\frac{2\beta\theta_1}{(\eta + \alpha)^4} + ff'' - f'^2 = 0 \tag{14}$$

$$\varepsilon_3\frac{k_{nf}}{k_f}\frac{1}{Pr}(\theta_1'' + 2\theta_2) + f\theta_1' + \varepsilon_3\frac{1}{Pr}\frac{2\lambda\beta}{(\eta + \alpha)^4}f(\theta_1 - \varepsilon) - \delta_e\left(f^2\theta_1'' + ff'\theta_1'\right) = 0 \tag{15}$$

$$\varepsilon_3 \frac{k_{nf}}{k_f} \frac{1}{Pr} \theta_2'' + f\theta_2 \frac{\lambda}{Pr}\left(\varepsilon_2 We + \varepsilon_1 + \frac{k_{Brownian}}{k_f Pr_f \varepsilon_2}\right) f'^2 - \varepsilon_3 \frac{\lambda\beta(\theta_1-\delta)}{Pr}\left[\frac{4f}{(\eta+\alpha)^5} + \frac{2f'}{(\eta+\alpha)^4}\right] = 0 \quad (16)$$
$$+\varepsilon_3 \frac{1}{Pr} \frac{2\lambda\beta}{(\eta+\alpha)^3} f\theta_2 - \delta_e\left(f^2\theta_2'' - 3ff'\theta_2' - 2ff''\theta_2 + 4f'^2\theta_2\right)$$

$$(1-\phi)^{2.5}\frac{1}{Sc}\left(\chi_1'' + 2\chi_2\right) + f\chi_1' - \sigma\chi_1 = 0 \quad (17)$$

$$(1-\phi)^{2.5}\frac{1}{Sc}\chi_2'' + f\chi_2' - 2\chi_2 f' - \sigma\chi_2 = 0 \quad (18)$$

where

$$\varepsilon_1 = \frac{1}{(1-\phi)^{2.5}\left(1-\phi+\phi\frac{\rho_s}{\rho_f}\right)}, \quad \varepsilon_2 = \frac{1}{\left(1-\phi+\phi\frac{\rho_s}{\rho_f}\right)}, \quad \varepsilon_3 = \frac{1}{\left(1-\phi+\phi\frac{(\rho C_p)_s}{(\rho C_p)_f}\right)} \quad (19)$$

Reduced conditions:

$$\left.\begin{array}{l} f(0) = 0, \; f'(0) = 1, \; \theta_1(0) = 1, \; \theta_2(0) = 0, \; \chi_1(0) = 1, \; \chi_2(0) = 0 \\ f'(\infty) \to 0, \; \theta_1(\infty) \to 0, \; \theta_2(\infty) \to 0, \; \chi_1(\infty) \to 0, \; \chi_2(\infty) \to 0 \end{array}\right\} \quad (20)$$

where

$$\alpha = \sqrt{\frac{c}{v_f}}a, \; \beta = \mu_0 K \frac{\gamma\rho_f}{2\pi\mu_f^2}(T_c - T_w), \; We = \frac{1}{\beta_1\epsilon\mu_f}, \; \delta_e = c\lambda_2, \; \delta = \frac{T_c}{(T_c-T_w)}$$
$$\lambda = \frac{c\mu_f^2}{k_f\rho_f(T_c-T_w)}, \; Pr = \frac{\mu_f C_p}{k_f}, \; \sigma = \frac{k_r}{c}, \; Sc = \frac{v_f}{D_f}, \; Re = \frac{cx^2}{v_f} \quad (21)$$

The definitions of the quantities of physical interests are as follows:

$$C_{f_x} = \frac{-2\left(\frac{1}{\beta_e}+\mu_{nf}\right)\left(\frac{\partial^2 u}{\partial x^2}\right)_{y=0}}{\rho(cx)^2}, \quad Nu_x = \frac{-xk_{nf}\left(\frac{\partial T}{\partial y}\right)_{y=0}}{(T_w-T_c)}, \quad (22)$$
$$Sh_x = \frac{-x\left(\frac{\partial C}{\partial y}\right)_{y=0}}{(C_w-C_c)}$$

The quantities of physical interest corresponding to Equations (12) and (13) transform the following equations:

$$\sqrt{Re_x}C_{f_x} = -\frac{2(1+We)}{(1-\phi)^{2.5}}f''(0) \quad (23)$$

$$(Re_x)^{-1/2}Nu_x = -\frac{k_{nf}}{k_f}\left(\theta_1'(0) + \xi^2\theta_2'(0)\right) \quad (24)$$

$$(Re_x)^{-1/2}Sh_x = -(1-\phi)^{2.5}\left(\chi_1'(0) + \xi^2\chi_2'(0)\right) \quad (25)$$

## 3. Solution Method and Details

In order to solve Equations (14–18) under the boundary conditions (19, 20), we use the Homotopy Analysis Method (HAM) with the following procedure. The solutions with the auxiliary parameters $\hbar$ adjust and control the convergence of the solutions.

The initial guesses are selected as follows:

$$f_0(\eta) = (1 - e^{-\eta}), \; \theta_{1,\,0}(\eta) = e^{-\eta}, \; \theta_{2,\,0}(\eta) = \eta e^{-\eta}, \; \chi_{1,\,0}(\eta) = e^{-\eta}, \; \chi_{2,\,0}(\eta) \quad (26)$$

The linear operators are taken as $L_f, \; L_{\theta_1}, \; L_{\theta_2}, \; L_{\chi_1}, \; L_{\chi_2}$

$$L_f(f) = f''' - f', \; L_{\theta_1}(\theta_1) = \theta_1'' - \theta_1, \; L_{\theta_2}(\theta_2) = \theta_2'' - \theta_2$$
$$L_{\chi_1}(\chi_1) = \chi_1'' - \chi_1, \; L_{\chi_2}(\chi_2) = \chi_2'' - \chi_2 \quad (27)$$

which have the following properties:

$$L_f(c_1 + c_2 e^{-\eta} + c_3 e^{\eta}) = 0, \; L_{\theta_1}(c_4 e^{\eta} + c_5 e^{-\eta}) = 0$$
$$L_{\theta_2}(c_6 e^{\eta} + c_7 e^{-\eta}) = 0, \; L_{\chi_1}(c_8 e^{-\eta} + c_9 e^{\eta}) = 0, \; L_{\chi_1}(c_{10} e^{-\eta} + c_{11} e^{\eta}) = 0 \tag{28}$$

where $c_i(i = 1 - 11)$ are the constants in general solution:

The resultant non-linear operatives $N_f$, $N_{\theta_1}$, $N_{\theta_2}$, $N_{\chi_1}$, $N_{\chi_2}$ are given as

$$N_f[f(\eta; p), \theta_1(\eta; p)] = \left(\varepsilon_2 We + \frac{k_{Brownian}}{k_f Pr_f \varepsilon_2}\right)\frac{\partial^3 f(\eta; p)}{\partial \eta^3} - \varepsilon_2 \frac{2\beta}{(\eta + \alpha)^4}\theta_1(\eta; p) - \left(\frac{\partial f(\eta; p)}{\partial \eta}\right)^2 + f(\eta; p)\frac{\partial^2 f(\eta; p)}{\partial \eta^2} \tag{29}$$

$$N_{\theta_1}[f(\eta; p), \theta_1(\eta; p), \theta_2(\eta; p)] = \varepsilon_2 \frac{k_{nf}}{k_f}\frac{1}{Pr}\left(\frac{\partial^2 \theta_1(\eta; p)}{\partial \eta^2} + 2\theta_2(\eta; p)\right) +$$
$$f(\eta; p)\frac{\partial \theta_1(\eta; p)}{\partial \eta} + \varepsilon_2 \frac{1}{Pr}\frac{2\lambda}{(\eta + \alpha)^4}(f(\eta; p)\theta_1(\eta; p) - \varepsilon f(\eta; p)) - \tag{30}$$
$$\delta_e\left((f(\eta; p))^2\frac{\partial^2 \theta_1(\eta; p)}{\partial \eta^2} + f(\eta; p)\frac{\partial f(\eta; p)}{\partial \eta}\frac{\partial \theta_1(\eta; p)}{\partial \eta}\right)$$

$$N_{\theta_2}[f(\eta; p), \theta_1(\eta; p), \theta_2(\eta; p)] = \varepsilon_2 \frac{k_{nf}}{k_f}\frac{1}{Pr}\frac{\partial^2 \theta_2(\eta; p)}{\partial \eta^2} +$$
$$\frac{\lambda}{Pr}\left(\varepsilon_2 We + \varepsilon_1 \frac{k_{Brownian}}{k_f Pr_f \varepsilon_2}\right)f(\eta; p)\left(\frac{\partial f(\eta; p)}{\partial \eta}\right)^2 \theta_1(\eta; p) + \varepsilon_2 \frac{2\lambda}{Pr}\frac{1}{(\eta + \alpha)^3}f(\eta; p)\theta_2(\eta; p)$$
$$-\varepsilon_2 \frac{\lambda\beta}{Pr}\left[\frac{4}{(\eta + \alpha)^5}f(\eta; p) + \frac{2}{(\eta + \alpha)^4}\frac{\partial f(\eta; p)}{\partial \eta}\right] \tag{31}$$
$$\delta_e\left(\begin{array}{c}(f(\eta; p))^2\frac{\partial^2 \theta_2(\eta; p)}{\partial \eta^2} - 3f(\eta; p)\frac{\partial f(\eta; p)}{\partial \eta}\frac{\partial \theta_2(\eta; p)}{\partial \eta} \\ -f(\eta; p)\frac{\partial^2 f(\eta; p)}{\partial \eta^2}\theta_2(\eta; p) + 4\left(\frac{\partial f(\eta; p)}{\partial \eta}\right)^2 \theta_2(\eta; p)\end{array}\right)$$

$$N_{\chi_1}[f(\eta; p), \chi_1(\eta; p), \chi_2(\eta; p)] = (1 - \phi)^{2.5}\frac{1}{Sc}\left(\frac{\partial^2 \chi_1(\eta; p)}{\partial \eta^2} + 2\chi_2(\eta; p)\right) +$$
$$f(\eta; p)\frac{\partial \chi_1(\eta; p)}{\partial \eta} - \sigma\chi_1(\eta; p) \tag{32}$$

$$N_{\chi_2}[f(\eta; p), \chi_2(\eta; p)] = (1 - \phi)^{2.5}\frac{1}{Sc}\frac{\partial^2 \chi_1(\eta; p)}{\partial \eta^2} + f(\eta; p)\frac{\partial \chi_2(\eta; p)}{\partial \eta}$$
$$-2\frac{\partial f(\eta; p)}{\partial \eta}\chi_2(\eta; p) - \sigma\chi_2(\eta; p) \tag{33}$$

The basic idea of the HAM is described in [1–7]; the zeroth-order problems from Equations (14)–(18) are

$$(1 - p)L_f[f(\eta; p) - f_0(\eta)] = p\hbar_f N_f[f(\eta; p), \theta_1(\eta; p)] \tag{34}$$
$$(1 - p)L_{\theta_1}[\theta_1(\eta; p) - \theta_{1,0}(\eta)] = p\hbar_{\theta_1} N_{\theta_1}[f(\eta; p), \theta_1(\eta; p), \theta_2(\eta; p)] \tag{35}$$
$$(1 - p)L_{\theta_2}[\theta_2(\eta; p) - \theta_{2,0}(\eta)] = p\hbar_{\theta_2} N_{\theta_2}[f(\eta; p), \theta_1(\eta; p), \theta_2(\eta; p)] \tag{36}$$
$$(1 - p)L_{\chi_1}[\chi_1(\eta; p) - \chi_{1,0}(\eta)] = p\hbar_{\chi_1} N_{\chi_1}[f(\eta; p), \chi_1(\eta; p), \chi_2(\eta; p)] \tag{37}$$
$$(1 - p)L_{\chi_1}[\chi_2(\eta; p) - \chi_{2,0}(\eta)] = p\hbar_{\chi_2} N_{\chi_2}[f(\eta; p), \chi_2(\eta; p)] \tag{38}$$

The equivalent boundary conditions are

$$f(\eta; p)|_{\eta=0} = 0, \; \left.\frac{\partial f(\eta; p)}{\partial \eta}\right|_{\eta=0} = 1, \; \left.\frac{\partial f(\eta; p)}{\partial \eta}\right|_{\eta\to\infty} = 0$$
$$\begin{aligned}\theta_1(\eta; p)|_{\eta=0} &= 1, \; \theta_1(\eta; p)|_{\eta\to\infty} = 0 \\ \theta_2(\eta; p)|_{\eta=0} &= 0, \; \theta_2(\eta; p)|_{\eta\to\infty} = 0 \\ \chi_1(\eta; p)|_{\eta=0} &= 1, \; \chi_1(\eta; p)|_{\eta\to\infty} = 0 \\ \chi_2(\eta; p)|_{\eta=0} &= 0, \; \chi_2(\eta; p)|_{\eta\to\infty} = 0\end{aligned} \tag{39}$$

where $p \in [0,1]$ is the imbedding parameter; $\hbar_f$, $\hbar_{\theta_1}$, $\hbar_{\theta_2}$, $\hbar_{\chi_1}$, $\hbar_{\chi_2}$ are used to control the convergence of the solution. When $p = 0$ and $p = 1$, we have

$$f(\eta; 1) = f(\eta),\ \theta_1(\eta; 1) = \theta_1(\eta),\ \theta_2(\eta; 1) = \theta_2(\eta),\ \chi_1(\eta; 1) = \chi_1(\eta),\ \chi_2(\eta; 1) = \chi_2(\eta) \tag{40}$$

Expanding $f(\eta; p)$, $\theta_1(\eta; p)$, $\theta_2(\eta; p)$, $\chi_1(\eta; p)$, $\chi_2(\eta; p)$ in Taylor's series about $p = 0$

$$
\begin{aligned}
f(\eta; p) &= f_0(\eta) + \sum_{m=1}^{\infty} f_m(\eta) p^m \\
\theta_1(\eta; p) &= \theta_{1,\,0}(\eta) + \sum_{m=1}^{\infty} \theta_{1,m}(\eta) p^m \\
\theta_2(\eta; p) &= \theta_{2,\,0}(\eta) + \sum_{m=1}^{\infty} \theta_{2,m}(\eta) p^m \\
\chi_1(\eta; p) &= \chi_{1,\,0}(\eta) + \sum_{m=1}^{\infty} \chi_{1,m}(\eta) p^m \\
\chi_2(\eta; p) &= \chi_{2,\,0}(\eta) + \sum_{m=1}^{\infty} \chi_{2,m}(\eta) p^m
\end{aligned}
\tag{41}
$$

where

$$
\begin{aligned}
f_m(\eta) &= \frac{1}{m!} \frac{\partial f(\eta; p)}{\partial \eta}\bigg|_{p=0},\quad \theta_{1,\,m}(\eta) = \frac{1}{m!} \frac{\partial \theta_1(\eta; p)}{\partial \eta}\bigg|_{p=0} \\
\theta_{2,\,m}(\eta) &= \frac{1}{m!} \frac{\partial \theta_2(\eta; p)}{\partial \eta}\bigg|_{p=0},\quad \chi_{1,\,m}(\eta) = \frac{1}{m!} \frac{\partial \chi_1(\eta; p)}{\partial \eta}\bigg|_{p=0},\quad \chi_{2,\,m}(\eta) = \frac{1}{m!} \frac{\partial \chi_2(\eta; p)}{\partial \eta}\bigg|_{p=0}
\end{aligned}
\tag{42}
$$

The secondary constraints $\hbar_f$, $\hbar_{\theta_1}$, $\hbar_{\theta_2}$, $\hbar_{\chi_1}$, $\hbar_{\chi_2}$ are chosen in such a way that the series (40) converges at $p = 1$; switching $p = 1$ in (40), we obtain

$$
\begin{aligned}
f(\eta) &= f_0(\eta) + \sum_{m=1}^{\infty} f_m(\eta) \\
\theta_1(\eta) &= \theta_{1,\,0}(\eta) + \sum_{m=1}^{\infty} \theta_{1,m}(\eta) \\
\theta_2(\eta) &= \theta_{2,\,0}(\eta) + \sum_{m=1}^{\infty} \theta_{2,m}(\eta) \\
\chi_1(\eta) &= \chi_{1,\,0}(\eta) + \sum_{m=1}^{\infty} \chi_{1,m}(\eta) \\
\chi_2(\eta) &= \chi_{2,\,0}(\eta) + \sum_{m=1}^{\infty} \chi_{2,m}(\eta)
\end{aligned}
\tag{43}
$$

The $m^{th}$-order problem satisfies the following:

$$
\begin{aligned}
L_f[f_m(\eta) - \chi_m f_{m-1}(\eta)] &= \hbar_f R_m^f(\eta) \\
L_{\theta_1}[\theta_{1,m}(\eta) - \chi_m \theta_{1,m-1}(\eta)] &= \hbar_{\theta_1} R_m^{\theta_1}(\eta) \\
L_{\theta_2}[\theta_{2,m}(\eta) - \chi_m \theta_{2,m-1}(\eta)] &= \hbar_{\theta_2} R_m^{\theta_2}(\eta) \\
L_{\chi_1}[\chi_{1,m}(\eta) - \chi_m \chi_{1,m-1}(\eta)] &= \hbar_{\chi_1} R_m^{\chi_1}(\eta) \\
L_{\chi_2}[\chi_{2,m}(\eta) - \chi_m \chi_{2,m-1}(\eta)] &= \hbar_{\chi_2} R_m^{\chi_2}(\eta)
\end{aligned}
\tag{44}
$$

The corresponding boundary conditions are as follows:

$$
\begin{aligned}
f_m(0) = f'_m(0) = \theta'_{1,\,m}(0) = \theta'_{2,\,m}(0) = \chi_{1,\,m}(0) = \chi_{2,\,m}(0) &= 0 \\
f'_m(\infty) = \theta_{1,\,m}(\infty) = \theta_{2,\,m}(\infty) = \chi_{1,\,m}(\infty) = \chi_{2,\,m}(\infty) &= 0
\end{aligned}
\tag{45}
$$

Here

$$
R_m^f(\eta) = \left(\varepsilon_2 We + \frac{k_{Brownian}}{k_f Pr_f \varepsilon_2}\right) f'''_{m-1} - \varepsilon_2 \frac{2\beta}{(\eta+\alpha)^4} \theta_{1,\,m-1} - \frac{\partial^3 f(\eta; p)}{\partial \eta^3} -
$$
$$
\varepsilon_2 \frac{2\beta}{(\eta+\alpha)^4} \theta_1(\eta; p) - \sum_{k=0}^{m-1} f'_{m-1} f'_k + \sum_{k=0}^{m-1} f_{m-1-k} f''_k
\tag{46}
$$

$$R_m^{\theta_1}(\eta) = \varepsilon_2 \frac{k_{nf}}{k_f} \frac{1}{Pr} \left( \theta''_{1,\,m-1} + 2\theta_{2,\,m-1} \right) + \varepsilon_2 \frac{1}{Pr} \frac{2\lambda}{(\eta+\alpha)^4} \left( \sum_{k=0}^{m-1} f_{m-1-k}\theta_{1,\,k} - \varepsilon f_{m-1} \right)$$
$$+ \sum_{k=0}^{m-1} f_{m-1-k}\theta'_{1,\,k} - \delta_e \left[ \sum_{k=0}^{m-1} f_{m-1-k} \sum_{i=0}^{k} f_{k-i}\theta''_{1,i} + \sum_{k=0}^{m-1} f_{m-1-k} \sum_{i=0}^{k} f'_{k-i}\theta_{1,\,i} \right] \tag{47}$$

$$R_m^{\theta_2}(\eta) = \varepsilon_2 \frac{k_{nf}}{k_f} \frac{1}{Pr}\theta''_{2,m-1} + \frac{\lambda}{Pr}\left( \varepsilon_2 We + \varepsilon_1 \frac{k_{Brownian}}{k_f Pr_f \varepsilon_2} \right) \sum_{k=0}^{m-1} f_{m-1-k} \sum_{i=0}^{k} f'_k \sum_{p=0}^{i} f'_{i-p}\theta_{1,p}$$
$$- \varepsilon_2 \frac{\lambda\beta}{Pr}\left[ \frac{4}{(\eta+\alpha)^5}f_{m-1} + \frac{2}{(\eta+\alpha)^4}f'_{m-1} \right] + \varepsilon_2 \frac{2\lambda\beta}{Pr}\frac{1}{(\eta+\alpha)^3} \sum_{k=0}^{m-1} f_{m-1-k}\theta_{2,k}$$
$$\delta_e \left[ \begin{array}{l} \sum_{k=0}^{m-1} f_{m-1-k} \sum_{i=0}^{k} f_{k-i}\theta''_{2,i} - 3 \sum_{k=0}^{m-1} f_{m-1-k} \sum_{i=0}^{k} f'_{k-i}\theta_{2,i} - \\ \sum_{k=0}^{m-1} f_{m-1-k} \sum_{i=0}^{k} f''_{k-i}\theta_{2,i} + 4 \sum_{k=0}^{m-1} f'_{m-1-k} \sum_{i=0}^{k} f'_{k-i}\theta_{2,i} \end{array} \right] \tag{48}$$

$$R_m^{\chi_1}(\eta) = (1-\phi)^{2.5} \frac{1}{Sc}\left( \chi''_{1,\,m-1} + 2\chi_{2,\,m-1} \right) + \sum_{k=0}^{m-1} f_{m-1-k}\chi'_{1,k} - \sigma\chi_{1,\,m-1} \tag{49}$$

$$R_m^{\chi_2}(\eta) = (1-\phi)^{2.5} \frac{1}{Sc}\chi''_{2,\,m-1} + \sum_{k=0}^{m-1} f_{m-1-k}\chi'_{2,k} - \sum_{k=0}^{m-1} f'_{m-1-k}\chi_{2,k} - \sigma\chi_{2,\,m-1} \tag{50}$$

where

$$\chi_m = \begin{cases} 0, & if\ p \le 1 \\ 1, & if\ p > 1 \end{cases} \tag{51}$$

*Validation and Comparison*

Table 1 shows the physical properties of nanofluid. A comparison of the validation of the results for the velocity, temperature, and concentration fields using the Homotopy Analysis Method and a numerical (ND-solve) method are shown in Tables 2–4 and in Figures 2–4. From these tables and figures, it can be observed that the results of both methods are in good agreement.

**Table 1.** The base fluids' and nanoparticles' material properties.

| Fluids | $\rho$ (kg/m$^3$) | $C_p$ (J/kg·K) | $k$ (W/m·K) |
|---|---|---|---|
| H$_2$O | 997.1 | 4179 | 0.613 |
| AA7072 | 2810 | 960 | 173 |
| AA7075 | 2720 | 893 | 222 |

**Table 2.** Comparison table for HAM solution and numerical method for velocity field and their results.

| $\eta$ | HAM Solution | Numerical Solution | Absolute Error |
|---|---|---|---|
| 0.0 | 1.000000 | 1.000000 | 0.000000 |
| 0.5 | 0.672325 | 0.672090 | 0.000235 |
| 1.0 | 0.450868 | 0.450368 | 0.000501 |
| 1.5 | 0.300635 | 0.300102 | 0.000532 |
| 2.0 | 0.198485 | 0.198011 | 0.000474 |
| 2.5 | 0.128841 | 0.128458 | 0.000383 |
| 3.0 | 0.081224 | 0.080938 | 0.000286 |
| 3.5 | 0.048575 | 0.048379 | 0.000196 |
| 4.0 | 0.026132 | 0.026014 | 0.000118 |
| 4.5 | 0.010669 | 0.010617 | 0.000053 |
| 5.0 | $8.649750 \times 10^{-8}$ | $1.697710 \times 10^{-8}$ | $6.952040 \times 10^{-8}$ |

**Table 3.** Comparison table for HAM solution and numerical method for temperature field and their results.

| $\eta$ | HAM Solution | Numerical Solution | Absolute Error |
|---|---|---|---|
| 0.0 | 1.000000 | 1.000000 | $2.775560'' \times 10^{-15}$ |
| 0.5 | 1.422530 | 1.422300 | 0.000240 |
| 1.0 | 1.520240 | 1.519810 | 0.000432 |
| 1.5 | 1.422380 | 1.421830 | 0.000551 |
| 2.0 | 1.222520 | 1.221930 | 0.000592 |
| 2.5 | 0.982008 | 0.981442 | 0.000566 |
| 3.0 | 0.737851 | 0.737362 | 0.000489 |
| 3.5 | 0.510493 | 0.510113 | 0.000380 |
| 4.0 | 0.309869 | 0.309615 | 0.000254 |
| 4.5 | 0.139648 | 0.139524 | 0.000124 |
| 5.0 | $-2.16757'' \times 10^{-7}$ | $7.061450'' \times 10^{-8}$ | $2.873720'' \times 10^{-7}$ |

**Table 4.** Comparison table for HAM solution and numerical method for concentration field and their results.

| $\eta$ | HAM Solution | Numerical Solution | Absolute Error |
|---|---|---|---|
| 0.0 | 1.000000 | 1.000000 | 0.000000 |
| 0.5 | 0.662316 | 0.662225 | 0.000090 |
| 1.0 | 0.449273 | 0.449152 | 0.000121 |
| 1.5 | 0.309253 | 0.309134 | 0.000119 |
| 2.0 | 0.213854 | 0.213750 | 0.000104 |
| 2.5 | 0.146837 | 0.146752 | 0.000085 |
| 3.0 | 0.098537 | 0.098472 | 0.000066 |
| 3.5 | 0.062965 | 0.062918 | 0.000047 |
| 4.0 | 0.036264 | 0.036234 | 0.000030 |
| 4.5 | 0.015864 | 0.015850 | 0.000014 |
| 5.0 | $1.648660'' \times 10^{-7}$ | $5.280260'' \times 10^{-8}$ | $1.120640'' \times 10^{-7}$ |

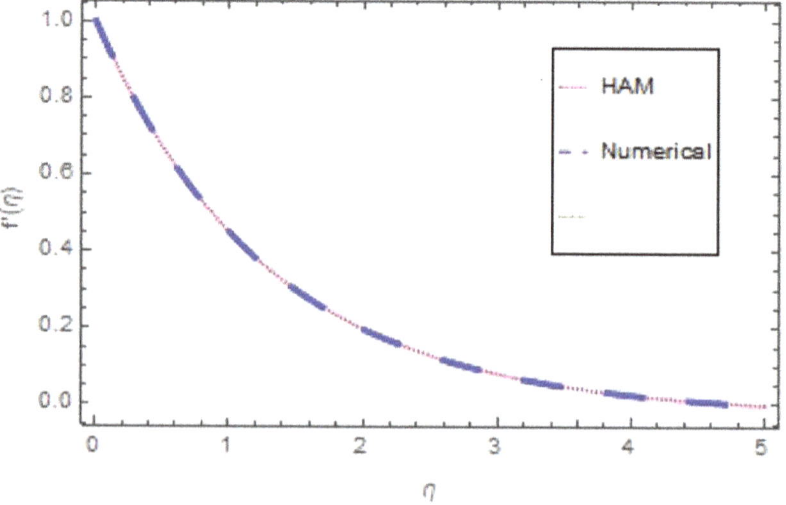


**Figure 2.** Comparison graph for velocity profile.

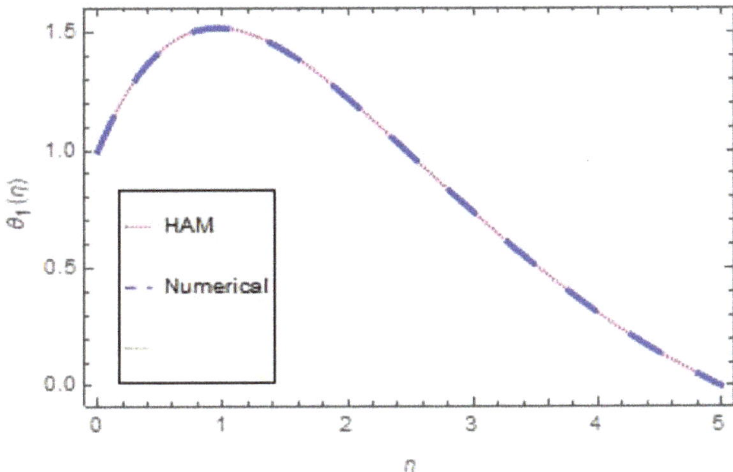

**Figure 3.** Comparison graph for temperature profile.

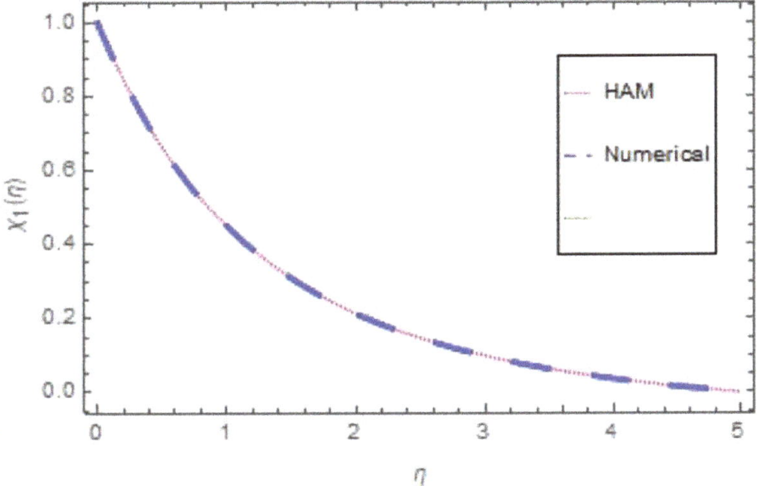

**Figure 4.** Comparison graph for concentration profile.

## 4. Results and Discussion

This section uses plotted figures to discuss the physical aspects of flow, thermal field changes, and concentration profile, and to explain physical interpretations triggered by the dominant dimensionless factors. The HAM is used to solve shortened ODEs numerically. Two diverse cases, namely, the $AA7072$ alloy and the $AA7075$ alloy, are well-thought-out in this analysis. Graphs are used to explain the effects of various specifications on $f'(\eta)$, $\theta_1(\eta)$, and $\chi_1(\eta)$, such as the Ree-Eyring fluid parameter, the ferromagnetic interaction parameter, the Schmidt number, the Prandtl number, and the reaction rate parameter. Additionally, skin friction and the Nusselt number are illustrated graphically.

Figure 5 shows the change in $f'(\eta)$ of both alloys $AA7072$ and $AA7075$ as $\beta$ changes. In this case, increasing $\beta$ lowers the $f'(\eta)$ of both alloys. This means that a large mass flux can reduce the velocity of the liquid on the surface. The occurrence of $\beta$ and Curie temperature in this circumstance is crucial to consider the ferromagnetic stimulus on the flow, which upsurges liquid viscosity and diminishes the velocity gradient. Physically, when the magnetic influence is absent, the fluid velocity upsurges. The magnetic dipole effect, on the other hand, causes the fluid velocity to decrease. Furthermore, when comparing the $AA7072$ alloy to the $AA7075$ alloy, fluid velocity is quite slow. Figure 6 depicts the oscillation in $f'(\eta)$ with various values of $\phi$ for both alloys. The increase in $\phi$ lowers the $f'(\eta)$. The velocity $f'(\eta)$ of both alloys decreases as the volume fraction rises. Furthermore, when compared to the $AA7075$ alloy, the velocity of the $AA7072$ alloy is strongly motivated by the volume fraction and falls faster. Figure 7 depicts the behaviour of $f'(\eta)$ in relation to the Weissenberg number $We$. The velocity of the liquid is observed to be reduced across the entire domain as the Weissenberg number rises. Furthermore, when the Weissenberg number increases, the velocity layer thickness decreases. Mathematically, the Weissenberg number $We$ is used in the investigation of viscoelastic flows. It is the ratio of viscous and elastic forces. As a result, as the Weissenberg number rises, the viscous forces diminish, and the velocity profile rises. Figure 8 indicates the effects of $\beta$ on $\theta_1(\eta)$ for both alloys. This indicates that an increase in $\beta$ values significantly improves the temperature profile $\theta_1(\eta)$. This is due to the fact that as the tension between the fluid particles boosts, too much heat is produced, resulting in higher fluid temperatures. Furthermore, for both $AA7072$ and $AA7075$ alloys, the inter-relevance thickness of the thermal layer is increased. Additionally, in $AA7075$ and when treated with $AA7072$ alloy, the closeness of the thermal layer further improves. Figure 9 shows that as $Pr$ increases, so does the temperature of the fluid $\theta_1(\eta)$. According to the observations, the thickness of the boundary layer appears to decrease as $Pr$ increases. As a result, as the Prandtl number upsurges, so does the rate of thermal conductivity. $Pr$ is the ratio of thermal diffusivity and momentum diffusivity. As a result, with a higher $Pr$, heat will dissipate from the sheet more quickly. Fluids with a higher $Pr$ have a lower thermal conduction value. As a result, the $Pr$ attempts to improve the cooling behaviour of the flows. The effect of $\lambda$ on the $\theta_1(\eta)$ profile is portrayed in Figure 10. It shows that as the value of $\lambda$ increases, the temperature field decays. Additionally, for booming values of $\lambda$, the inter-relevance thickness is reduced for both alloys. Furthermore, heat abatement is enhanced in $AA7072$ alloy when treated with $AA7075$ alloy. The fluctuation in the thermal gradient for various values of $\delta_e$ for both alloys is shown in Figure 11. The thermal distribution is enhanced when the values of the thermal relaxation parameters are increased. The heat flow relaxation time causes this parameter to emerge physically. The higher the $\delta_e$ value, the longer it takes for the liquid particles to exchange heat with nearby particles, resulting in a decrease in temperature but an improvement in the temperature gradient. Figure 12 describes the outcome of volume fraction $\phi$ on heat transport in both alloys. The heat transmission of both alloys is improved as the volume fraction increases. Furthermore, in $AA7075$ and when treated with $AA7072$ alloy, the closeness of the thermal layer further improves. Figure 13 depicts the effect of $\sigma$ on $\chi_1(\eta)$ in both alloys. This figure confirms that $\chi_1(\eta)$ has a decreasing nature for various $\sigma$ values, and an increase in the reaction rate parameter $\sigma$ diminishes the concentration of the liquids. In fact, as the reaction rate parameter values increase, the concentration field and related boundary layer thickness decreases. According to Figure 14, a higher Schmidt number corresponds to a lower solute diffusivity, allowing for a shallower penetration of the solute effect. As a result, as $Sc$ rises, $\chi_1(\eta)$ falls. Thus, with lower concentrations of $Sc$, the solute boundary layer is thicker, and vice versa.

Figure 15 depicts the variants in surface drag force $C_{f_x}$ versus $We$ for various $\phi$ values for both alloys. It has been discovered that significantly greater values of $\phi$ enhance the surface drag force, whereas contrasting actions are observed for growing values of $\beta$; see Figure 16. Figure 17 shows the outcome of $\delta_e$ on the rate of heat transfer versus $We$ for both alloys $AA7072$ and $AA7075$. In both $AA7072$ and $AA7075$ alloys, an increase in $\delta_e$ degrades the Nusselt number. Figure 18 illustrates the importance of $\phi$ on $Re_x^{-1/2}Nu_x$ versus $We$ for both $AA7072$ and $AA7075$ alloys. For both alloys, boosting the $\phi$ values improves the heat transmission rate. Figure 19 depicts the variation in $Re_x^{-1/2}Nu_x$ versus $We$ for various $\beta$ values. It can be observed that significantly higher values of $\beta$ enhance the heat transmission rate. Figure 20 depicts the variation in $Re_x^{-1/2}Sh_x$ versus $Sc$ for various $\phi$ values. It can also be observed that significantly higher values of $\phi$ enhance the concentration rate. A comparison between previous and present works for the validation of the results for skin friction is presented in Table 5.

**Table 5.** Comparison of $-f''(0)$ with literature.

| Published Papers | $-f''(0)$ |
| --- | --- |
| Zeeshan and Majeed [36] | 0.6058427 |
| B.C Prsannakumara [37] | 0.6069352 |
| Present results | 0.603457 |

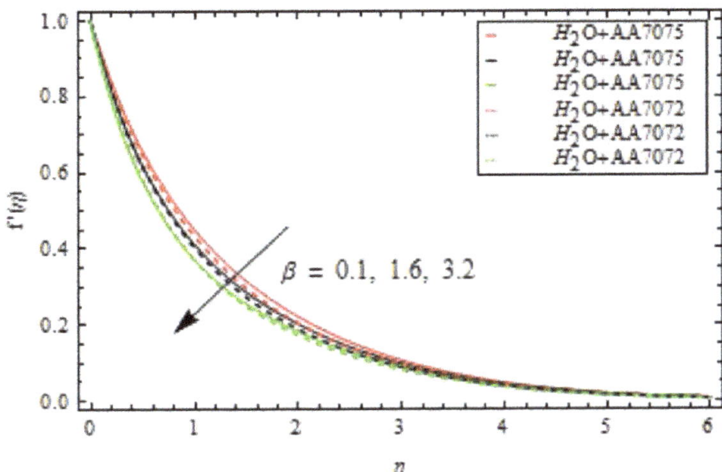

Figure 5. Influence of ferromagnetic interaction parameter $\beta$ on velocity profile.

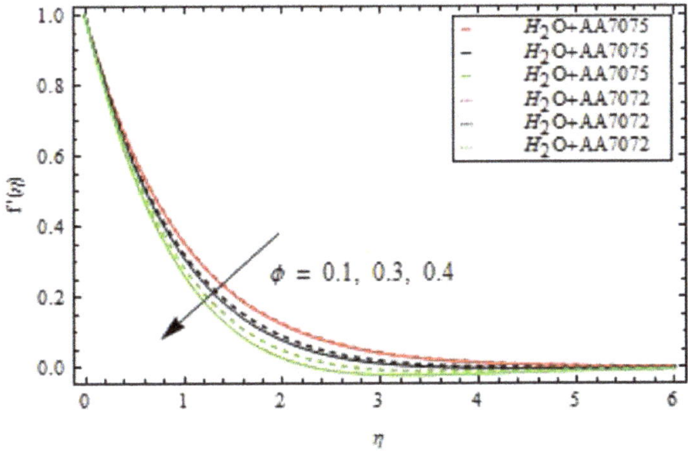

Figure 6. Influence of volume fraction $\phi$ on velocity profile.

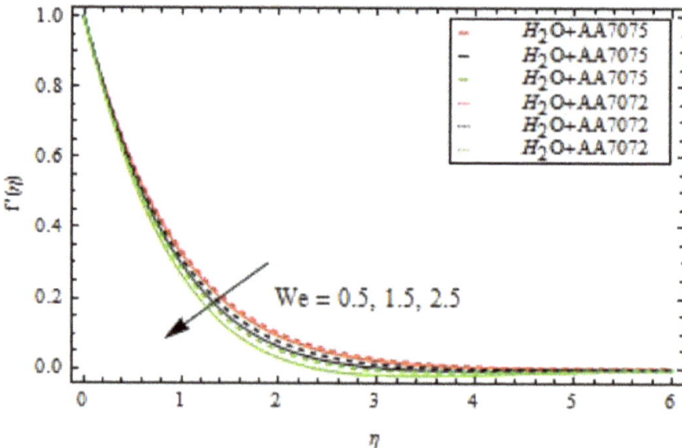


**Figure 7.** Influence of Weissenberg number We on velocity profile.

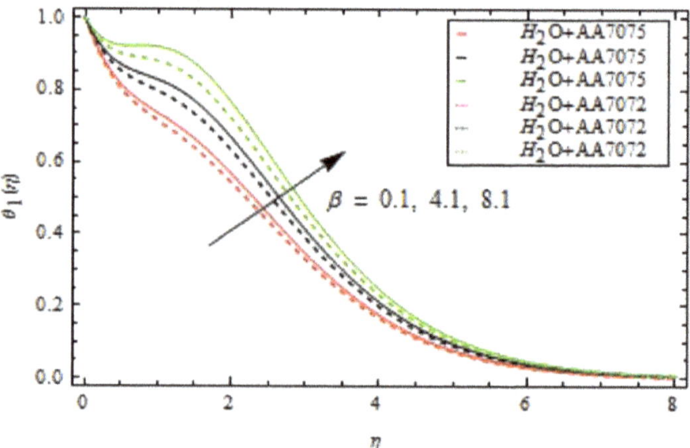


**Figure 8.** Influence of ferromagnetic interaction parameter $\beta$ on temperature profile.

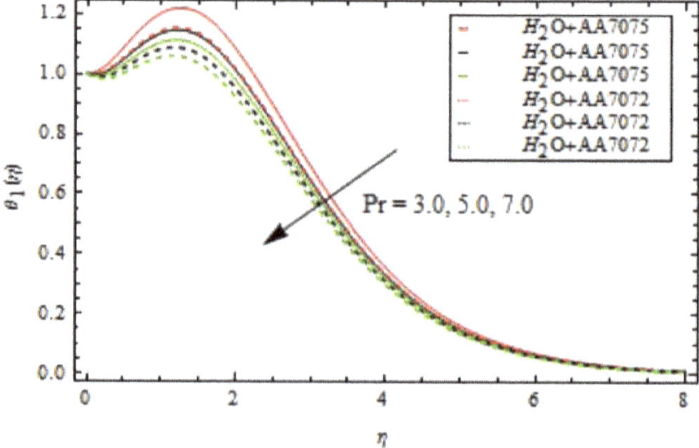


**Figure 9.** Influence of Prandtl number Pr on temperature profile.

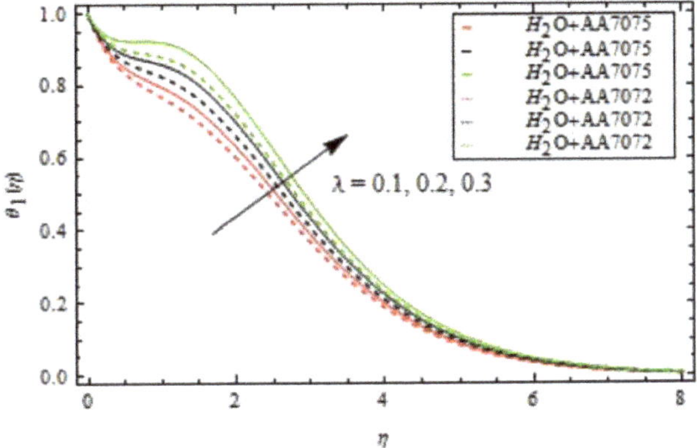

**Figure 10.** Influence of viscous dissipation parameter $\lambda$ on temperature profile.

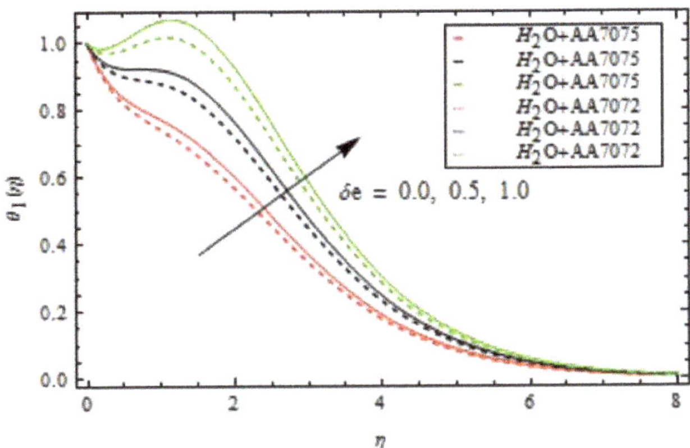

**Figure 11.** Influence of thermal relaxation parameter $\delta_e$ on temperature profile.

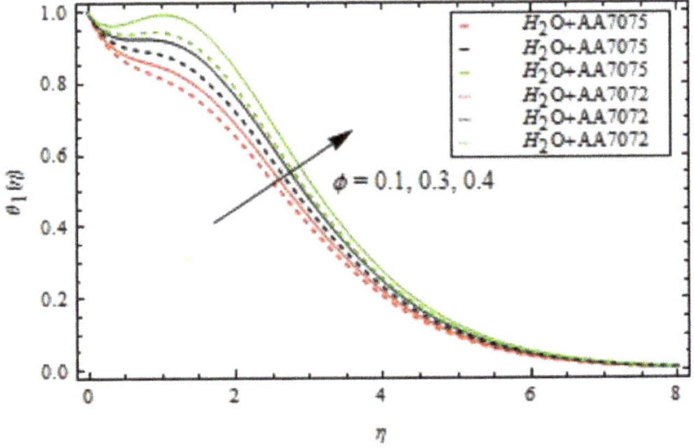

**Figure 12.** Influence of volume fraction $\phi$ on temperature profile.

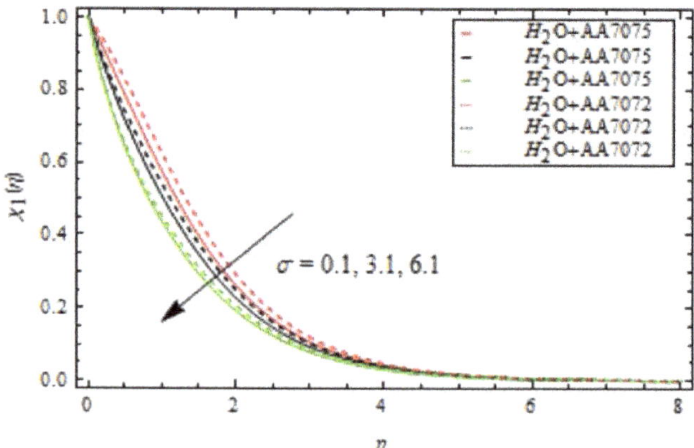

Figure 13. Influence of reaction rate parameter $\sigma$ on concentration profile.

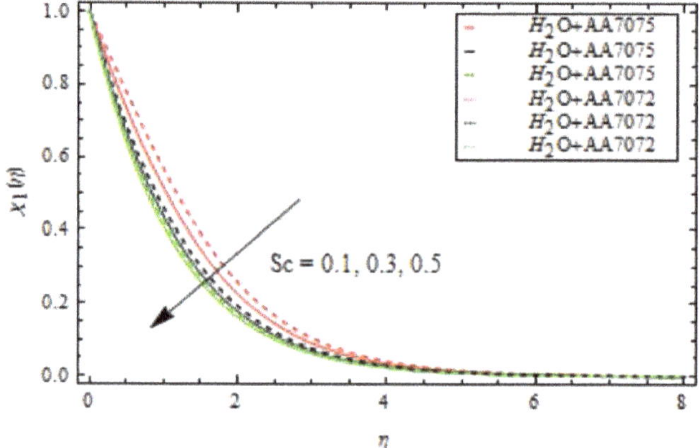

Figure 14. Influence of Schmidt number Sc on concentration profile.

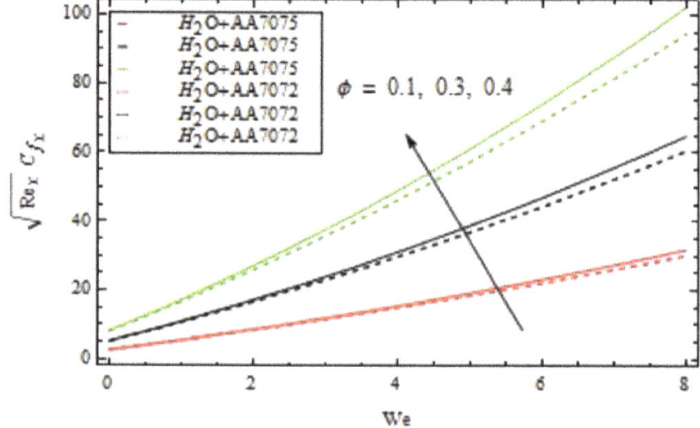

Figure 15. Various values of $\phi$ versus We for skin friction.

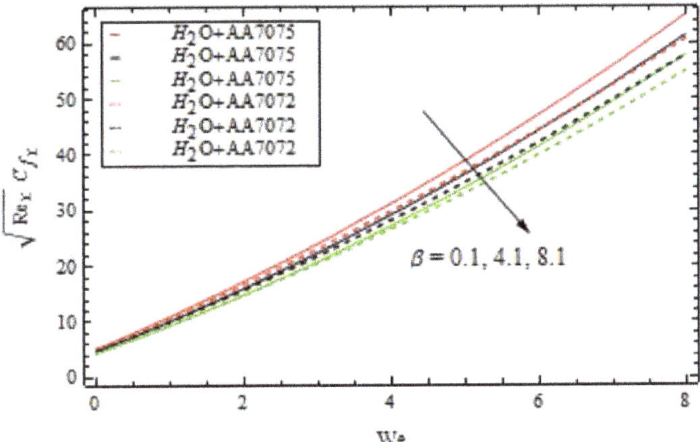



**Figure 16.** Various values of $\beta$ versus We for skin friction.

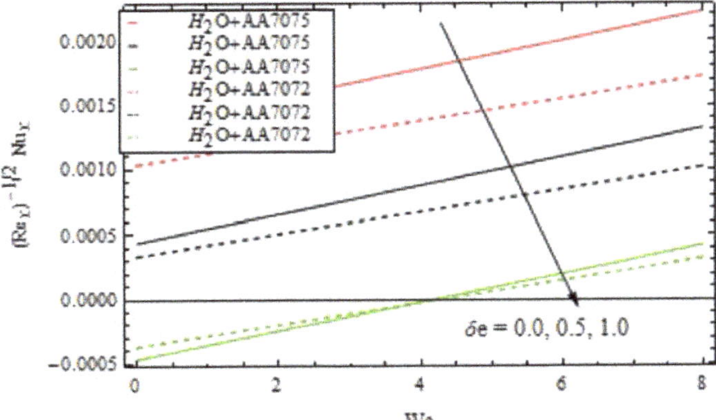



**Figure 17.** Various values of $\delta_e$ versus We.

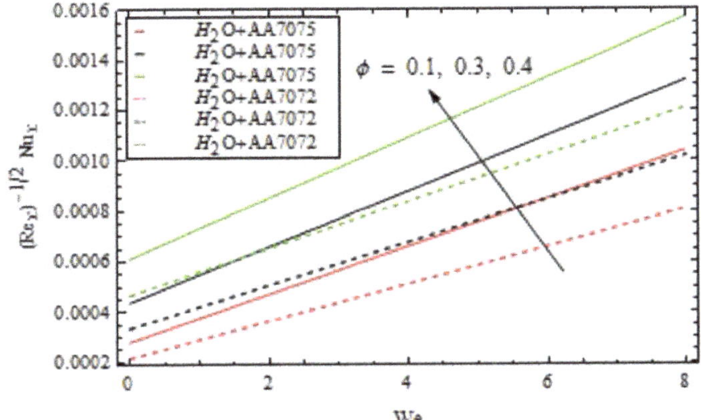



**Figure 18.** Various values of $\phi$ versus We for nusselt number.

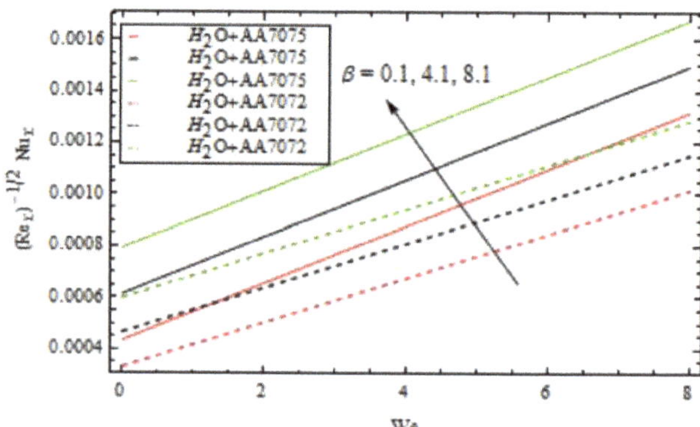



**Figure 19.** Various values of $\beta$ versus We for nusselt number.

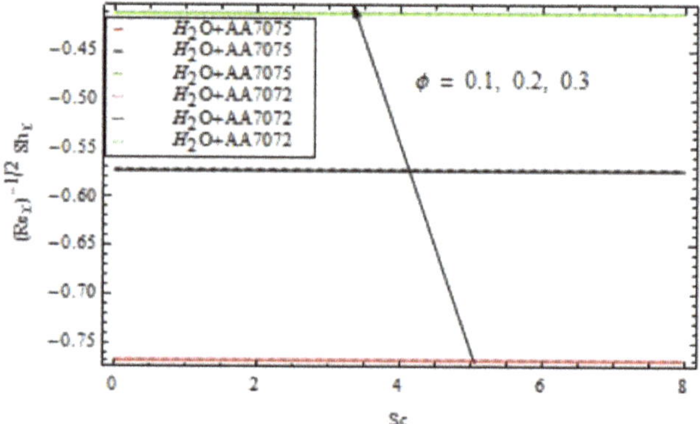



**Figure 20.** Various values of $\phi$ versus Sc.

## 5. Conclusions

Using the influence of magnetic dipoles and the Koo-Kleinstreuer model, the momentum, heat transfer, and mass transfer behavioru of Ree-Eyring nanoliquids through a stretching surface are investigated in this research. Moreover, the heat transmission is described by the Cattaneo-Christov heat flux model, and viscous dissipation is taken into account. Finally, the constructed governing equations related to the momentum, thermal, and mass distributions are converted to ODEs and solved with the HAM. The following are the results of the present analysis:

1. An escalation in volume fraction, Weissenberg number, and the ferromagnetic interaction parameter affects the velocity gradient. Furthermore, all these parameters negatively influence the velocity gradient of alloy $AA7075$, which falls quicker than the velocity gradient of alloy $AA7072$.

2. As the ferromagnetic interaction, viscous dissipation parameter, thermal relaxation parameter, and volume fraction grow, the temperature gradient of both alloys increases, whereas contrasting behaviour is revealed for the Prandtl number. Moreover, in $AA7075$ and when treated with $AA7072$ alloy, the closeness of the thermal layer further improves.

3. A growth in the reaction rate parameter and the Schmidt number brings down the concentration profile. Similarly, all parameters negatively influence the concentration profile of alloy $AA7075$, which drops quicker than the concentration profile of alloy $AA7072$.

4. An improvement in the volume fraction enhances the surface drag force; however, an improvement in the ferromagnetic interaction decreases the surface drag force.

5. The Nusselt number rises as the volume fraction and ferromagnetic interaction grow; however it falls as the thermal relaxation parameter rises.

**Author Contributions:** Conceptualization, Z.S.; Data curation, M.R., W.D. and M.S.; Formal analysis, Z.S., M.R. and W.D.; Funding acquisition, N.V.; Investigation, Z.S.; Methodology, Z.S., M.R. and M.S.; Project administration, Z.S. and M.S.; Resources, Z.S. and N.V.; Software, Z.S., M.R., W.D. and M.S.; Supervision, Z.S.; Validation, Z.S., N.V., W.D. and M.S.; Visualization, Z.S., N.V., M.R. and M.S.; Writing—original draft, Z.S. and W.D.; Writing—review & editing, Z.S. and N.V. All authors have read and agreed to the published version of the manuscript.

**Funding:** The project was financed by Lucian Blaga University of Sibiu and Hasso Plattner Foundation research Grants LBUS-IRG-2021-07.

**Institutional Review Board Statement:** Not applicable.

**Informed Consent Statement:** Not applicable.

**Data Availability Statement:** The data that support the findings of this study are available from the corresponding author upon reasonable request.

**Conflicts of Interest:** The authors declare no conflict of interest.

## Nomenclature

| | |
|---|---|
| $a$ | Distance |
| $c$ | Constant |
| $C_p$ | Specific heat transfer $\left(\text{J·kg}^{-1}\text{·K}^{-1}\right)$ |
| $C$ | Concentration |
| $k$ | Thermal conductivity $\left(\text{W·m}^{-1}\text{·K}^{-1}\right)$ |
| $K$ | Constant |
| $H$ | Magnetic field intensity $\left(\text{A·m}^{-1}\right)$ |
| $M$ | Magnetisation $\left(\text{A·m}^{-1}\right)$ |
| $Pr$ | Prandtl number |
| $Re$ | Local Reynolds number |
| $Sc$ | Schmidt number |
| $T$ | Temperature of fluid |
| $u, v$ | Velocity components $\left(\text{m·s}^{-1}\right)$ |
| $We$ | Weissenberg number |
| $x, y$ | Coordinates axis $\left(\text{m}\right)$ |
| Greek Letter | |
| $\alpha$ | Dimensionless distance |
| $\beta$ | Ferromagnetic interaction parameter |
| $\gamma$ | Constant |
| $\beta_1, \epsilon$ | Material constant of the fluid |
| $\delta$ | Dimensionless Curie temperature |
| $\delta_e$ | Thermal relaxation parameter |
| $\eta, \xi$ | Independent coordinate |
| $\theta_1(\eta), \theta_2(\eta)$ | Dimensionless temperature profile |
| $\lambda$ | Viscous dissipation parameter |
| $\lambda_2$ | Thermal relaxation time |
| $\mu$ | Dynamic viscosity |
| $\mu_0$ | Magnetic permeability |
| $v$ | Kinematic viscosity |
| $\rho$ | Density $\left(\text{kg·m}^{-3}\right)$ |
| $\rho C_p$ | Heat capacitance |
| $\sigma$ | Reaction rate parameter |

| | |
|---|---|
| $\phi_1$ | Scalar potential |
| $\phi$ | Volume fraction |
| $\chi_1(\eta)$, $\chi_2(\eta)$ | Dimensionless concentration profile |
| $\psi$ | Stream function $(m^2 \cdot s^{-1})$ |
| Subscript | |
| $f$ | Fluid |
| $nf$ | Nanofluid |
| $c$ | Curie |
| $w$ | Surface |
| $s$ | Solid particle |

## References


1. Choi, S.U.; Eastman, J.A. Enhancing thermal conductivity of fluids with nanoparticles. *ASME Int. Mech. Eng. Congr. Expo.* **1995**, *10*, 12–17.
2. Kishan, N.; Deepa, G. Viscous Dissipation Effects on Stagnation Point Flow and Heat Transfer of a Micropolar Fluid with Uniform Suction or Blowing. *Adv. Appl. Sci. Res.* **2012**, *3*, 430–439.
3. Alim, M.; Alam, M.; Mamun, A.A.; Hossain, B. Combined effect of viscous dissipation and joule heating on the coupling of conduction and free convection along a vertical flat plate. *Int. Commun. Heat Mass Transf.* **2008**, *35*, 338–346. [CrossRef]
4. Sheikholeslami, M.; Hatami, M.; Ganji, D. Nanofluid flow and heat transfer in a rotating system in the presence of a magnetic field. *J. Mol. Liq.* **2014**, *190*, 112–120. [CrossRef]
5. He, W.; Mashayekhi, R.; Toghraie, D.; Akbari, O.A.; Li, Z.; Tlili, I. Hydrothermal performance of nanofluid flow in a sinusoidal double layer microchannel in order to geometric optimization. *Int. Commun. Heat Mass Transf.* **2020**, *117*, 104700. [CrossRef]
6. Mahdavi, M.; Sharifpur, M.; Meyer, J.P. Fluid flow and heat transfer analysis of nanofluid jet cooling on a hot surface with various roughness. *Int. Commun. Heat Mass Transf.* **2020**, *118*, 104842. [CrossRef]
7. Abdelsalam, S.I.; Mekheimer, K.; Zaher, A. Alterations in blood stream by electroosmotic forces of hybrid nanofluid through diseased artery: Aneurysmal/stenosed segment. *Chin. J. Phys.* **2020**, *67*, 314–329. [CrossRef]
8. Nadeem, S.; Israr-Ur-Rehman, M.; Saleem, S.; Bonyah, E. Dual solutions in MHD stagnation point flow of nanofluid induced by porous stretching/shrinking sheet with anisotropic slip. *AIP Adv.* **2020**, *10*, 065207. [CrossRef]
9. Khan, U.; Waini, I.; Ishak, A.; Pop, I. Unsteady hybrid nanofluid flow over a radially permeable shrinking/stretching surface. *J. Mol. Liq.* **2021**, *331*, 115752. [CrossRef]
10. Abbas, N.; Nadeem, S.; Saleem, S.; Issakhov, A. Transportation of modified nanofluid flow with time dependent viscosity over a Riga plate: Exponentially stretching. *Ain Shams Eng. J.* **2021**, *12*, 3967–3973. [CrossRef]
11. Awais, M.; Awan, S.E.; Raja, M.; Nawaz, M.; Khan, W.; Malik, M.Y.; He, Y. Heat Transfer in Nanomaterial Suspension ($CuO$ and $Al_2O_3$) Using KKL Model. *Coatings* **2021**, *11*, 417. [CrossRef]
12. Sandeep, N.; Animasaun, I. Heat transfer in wall jet flow of magnetic-nanofluids with variable magnetic field. *Alex. Eng. J.* **2017**, *56*, 263–269. [CrossRef]
13. Kandasamy, R.; Adnan, N.A.B.; Abbood, J.A.A.; Kamarulzaki, M.; Saifullah, M. Electric field strength on water based aluminum alloys nanofluids flow up a non-linear inclined sheet. *Eng. Sci. Technol. Int. J.* **2018**, *22*, 229–236. [CrossRef]
14. Tlili, I.; Nabwey, H.A.; Ashwinkumar, G.; Sandeep, N. 3-D magnetohydrodynamic AA7072-AA7075/methanol hybrid nanofluid flow above an uneven thickness surface with slip effect. *Sci. Rep.* **2020**, *10*, 4265. [CrossRef]
15. Ahmed, J.; Shahzad, A.; Khan, M.; Ali, R. A note on convective heat transfer of an MHD Jeffrey fluid over a stretching sheet. *AIP Adv.* **2015**, *5*, 117117. [CrossRef]
16. Masood, K.H.A.N.; Ramzan, A.L.I.; Shahzad, A. MHD Falkner-Skan Flow with Mixed Convection and Convective Boundary Conditions. *Walailak J. Sci. Technol.* **2013**, *10*, 517–529. [CrossRef]
17. Malik, M.; Salahuddin, T.; Hussain, A.; Bilal, S. MHD flow of tangent hyperbolic fluid over a stretching cylinder: Using Keller box method. *J. Magn. Magn. Mater.* **2015**, *395*, 271–276. [CrossRef]
18. Sheikholeslami, M.; Rashidi, M.; Hayat, T.; Ganji, D. Free convection of magnetic nanofluid considering MFD viscosity effect. *J. Mol. Liq.* **2016**, *218*, 393–399. [CrossRef]
19. Crane, L.J. Flow past a stretching plate. *Z. Für Angew. Math. Und Phys.* **1970**, *21*, 645–647. [CrossRef]
20. Brady, J.F.; Acrivos, A. Steady flow in a channel or tube with an accelerating surface velocity. An exact solution to the Navier—Stokes equations with reverse flow. *J. Fluid Mech.* **1981**, *112*, 127–150. [CrossRef]
21. Wang, C.Y. Fluid flow due to a stretching cylinder. *Phys. Fluids* **1988**, *31*, 466. [CrossRef]
22. Elbashbeshy, E.M.; Aldawody, D.A. Heat transfer over an unsteady stretching surface with variable heat flux in the presence of a heat source or sink. *Comput. Math. Appl.* **2010**, *60*, 2806–2811. [CrossRef]
23. Nadeem, S.; Hussain, A.; Khan, M. HAM solutions for boundary layer flow in the region of the stagnation point towards a stretching sheet. *Commun. Nonlinear Sci. Numer. Simul.* **2010**, *15*, 475–481. [CrossRef]
24. Awan, A.U.; Abid, S.; Ullah, N.; Nadeem, S. Magnetohydrodynamic oblique stagnation point flow of second grade fluid over an oscillatory stretching surface. *Results Phys.* **2020**, *18*, 103233. [CrossRef]

25. Jawad, M.; Saeed, A.; Gul, T. Entropy Generation for MHD Maxwell Nanofluid Flow Past a Porous and Stretching Surface with Dufour and Soret Effects. *Braz. J. Phys.* **2021**, *51*, 469–480. [CrossRef]
26. Bestman, A.R. Natural convection boundary layer with suction and mass transfer in a porous medium. *Int. J. Energy Res.* **1990**, *14*, 389–396. [CrossRef]
27. Mustafaa, M.; Hayat, T.; Obaidat, S. Boundary layer flow of a nanofluid over an exponentially stretching sheet with convective boundary conditions. *Int. J. Numer. Methods Heat Fluid Flow* **2013**, *23*, 945–959. [CrossRef]
28. Mohyud-Din, S.T.; Khan, U.; Ahmed, N.; Bin-Mohsin, B. Heat and mass transfer analysis for MHD flow of nanofluid inconvergent/divergent channels with stretchable walls using Buongiorno's model. *Neural Comput. Appl.* **2016**, *28*, 4079–4092. [CrossRef]
29. Aleem, M.; Asjad, M.I.; Shaheen, A.; Khan, I. MHD Influence on different water based nanofluids (TiO$_2$, Al$_2$O$_3$, CuO) in porous medium with chemical reaction and Newtonian heating. *Chaos Solitons Fractals* **2019**, *130*, 109437. [CrossRef]
30. Hayat, T.; Khan, S.A.; Khan, M.I.; Alsaedi, A. Theoretical investigation of Ree–Eyring nanofluid flow with entropy optimization and Arrhenius activation energy between two rotating disks. *Comput. Methods Programs Biomed.* **2019**, *177*, 57–68. [CrossRef]
31. Tanveer, A.; Malik, M. Slip and porosity effects on peristalsis of MHD Ree-Eyring nanofluid in curved geometry. *Ain Shams Eng. J.* **2020**, *12*, 955–968. [CrossRef]
32. Khan, M.I.; Kadry, S.; Chu, Y.-M.; Khan, W.A.; Kumar, A. Exploration of Lorentz force on a paraboloid stretched surface in flow of Ree-Eyring nanomaterial. *J. Mater. Res. Technol.* **2020**, *9*, 10265–10275. [CrossRef]
33. Al-Mdallal, Q.M.; Renuka, A.; Muthtamilselvan, M.; Abdalla, B. Ree-Eyring fluid flow of Cu-water nanofluid between infinite spinning disks with an effect of thermal radiation. *Ain Shams Eng. J.* **2021**, *12*, 2947–2956. [CrossRef]
34. Rao, D.P.C.; Thiagarajan, S.; Kumar, V.S. Darcy-Forchheimer flow of Ree–Eyring fluid over an inclined plate with chemical reaction: A statistical approach. *Heat Transf.* **2021**, *50*, 7120–7138. [CrossRef]
35. Zeeshan, A.; Majeed, A. Effect of Magnetic Dipole on Radiative Non-Darcian Mixed Convective Flow over a Stretching Sheet in Porous Medium. *J. Nanofluids* **2016**, *5*, 617–626. [CrossRef]
36. Prasannakumara, B. Numerical simulation of heat transport in Maxwell nanofluid flow over a stretching sheet considering magnetic dipole effect. *Partial Differ. Equations Appl. Math.* **2021**, *4*, 100064. [CrossRef]
37. Souayeh, B.; Yasin, E.; Alam, M.W.; Hussain, S.G. Numerical Simulation of Magnetic Dipole Flow Over a Stretching Sheet in the Presence of Non-Uniform Heat Source/Sink. *Front. Energy Res.* **2021**, *9*, 824. [CrossRef]
38. García-Merino, J.A.; Torres-Torres, D.; Carrillo-Delgado, C.; Trejo-Valdez, M.; Torres-Torres, C. Trejo-Valdez and C. Torres-Torres Optofluidic and strain measurements induced by polarization-resolved nanosecond pulses in gold-based nanofluids. *Optik* **2019**, *182*, 443–451. [CrossRef]

*Article*

# Microscopy-Assisted Digital Image Analysis with Trainable Weka Segmentation (TWS) for Emulsion Droplet Size Determination

Pelin Salum [1], Onur Güven [2,*], Levent Yurdaer Aydemir [1] and Zafer Erbay [1]

[1] Department of Food Engineering, Faculty of Engineering, Adana Alparslan Türkeş Science and Technology University, Adana 01250, Turkey; pelinsalum@ymail.com (P.S.); lyaydemir@atu.edu.tr (L.Y.A.); zerbay@atu.edu.tr (Z.E.)

[2] Department of Mining Engineering, Faculty of Engineering, Adana Alparslan Türkeş Science and Technology University, Adana 01250, Turkey

* Correspondence: oguven@atu.edu.tr; Tel.: +90-322-4550000 (ext. 2094)

**Abstract:** The size distribution of droplets in emulsions is very important for adjusting the effects of many indices on their quality. In addition to other methods for the determination of the size distribution of droplets, the usage of machine learning during microscopic analyses can enhance the reliability of the measurements and decrease the measurement cost at the same time. Considering its role in emulsion characteristics, in this study, the droplet size distributions of emulsions prepared with different oil/water phase ratios and homogenization times were measured with both a microscopy-assisted digital image analysis technique and a well-known laser diffraction method. The relationships between the droplet size and the physical properties of emulsions (turbidity and viscosity) were also investigated. The results showed that microscopic measurements yielded slightly higher values for the D(90), D[3,2], and D[4,3] of emulsions compared to the laser diffraction method for all oil/water phase ratios. When using this method, the droplet size had a meaningful correlation with the turbidity and viscosity values of emulsions at different oil/water phase ratios. From this point of view, the usage of the optical microscopy method with machine learning can be useful for the determination of the size distribution in emulsions.

**Keywords:** emulsion; droplet size; microscopy-assisted; image analysis; laser diffraction; turbidity; viscosity

**Citation:** Salum, P.; Güven, O.; Aydemir, L.Y.; Erbay, Z. Microscopy-Assisted Digital Image Analysis with Trainable Weka Segmentation (TWS) for Emulsion Droplet Size Determination. *Coatings* **2022**, *12*, 364. https://doi.org/10.3390/coatings12030364

Academic Editor: Eduardo Guzmán

Received: 16 February 2022
Accepted: 7 March 2022
Published: 9 March 2022

**Publisher's Note:** MDPI stays neutral with regard to jurisdictional claims in published maps and institutional affiliations.

## 1. Introduction

An emulsion can be defined as a system consisting of two immiscible liquids, in which one of the liquids is dispersed as small spherical droplets in the other liquid [1]. The size and distribution of droplets depend upon the energy input and temperature during homogenization, the characteristics and ratios of the two phases (dispersed and continuous), and the type and concentration of the emulsifier [2]. It is well known that the size and distribution of droplets have a great impact on emulsion stability, optical properties, rheology, and sensorial characteristics [1]. The distribution of droplets in emulsion systems can be monodispersed or polydispersed. If all droplets in an emulsion are the same size, the emulsion is referred to as "monodisperse" and can be characterized by the size of a single droplet (the radius or diameter of the droplet). However, the vast majority of emulsions, such as food emulsions, are polydisperse systems containing droplets with a range of different sizes. Therefore, they should be characterized by the particle or droplet size distribution, which represents the concentration of droplets in different size classes [1]. The droplet size distribution of an emulsion is one of the important factors that control aggregation, coalescence, and resistance to sedimentation or creaming. The size distribution can also be used as a representative of stability if measured as a function of time. It is well known that the smaller and more uniform the droplets, the more stable the emulsions, provided that all other conditions are the same [3]. Therefore, determining the size distribution of the droplets in the continuous phase in a precise and accurate manner is essential for studies in emulsion science.

In the literature, considerable work has been carried out for the development of analytical techniques to obtain information about the droplet size distribution, such as light scattering, electrical conductivity, acoustics/electroacoustics, near-infrared spectroscopy, nuclear magnetic resonance, and various kinds of microscopic measurements (optical microscopy, transmission electron microscopy, and scanning electron microscopy) [4–7]. Several particle size analyzers have been designed and are commercially available for the determination of particle size distribution based on several physical principles, such as the scattering of light, the velocity of particles in a field, scattering, or absorption of ultrasonic waves [1,4]. In general, these analyzers are automated, rapid, and reliable systems with high installation costs. On the other hand, among all of these measurement techniques, microscopic measurement methods differ from other techniques, as they rely directly on the visual measurement of droplets [3]. Optical microscopy stands out in particular, as it is an inexpensive, relatively easy-to-use instrument available in most laboratories. However, droplet size determination by optical microscopy is generally time-consuming and laborious [3,4,6,8,9]. These weaknesses of microscopic analysis techniques become especially apparent when considering the necessity of observing thousands of droplets and quantifying their sizes to obtain meaningful results in droplet size distribution analysis.

Optical microscopes can be coupled with a digital camera, and in this way, images can be recorded and digitalized. It is possible to utilize these digitalized data with the aid of image processing techniques to reduce the time and workload required for the analysis [10]. Microscopy and image processing techniques have been used in combination to determine the droplet size distribution of emulsions [2,3,10–12]. Jokela et al. (1990) used a computerized microscope image analysis technique to determine the droplet size distribution of an oil-in-water emulsion [10]. The threshold method was used for the discrimination of droplets from the background. They found that the computerized microscope image analysis results were satisfactory and in agreement with "Coulter counting" and "laser diffraction" methods. Moradi et al. (2011) used optical microscopy and image analysis to determine the droplet size distribution of water-in-crude oil emulsions by using image enhancement techniques. They noted that applying general enhancement techniques such as brightness and contrast adjustment, sharpening, and open filters improved the detection of droplets [3]. Maaref and Ayatollahi (2018) also utilized some of the general enhancement techniques, including brightness, smoothing, and sharpening, for distinguishing emulsion droplets from the surroundings to evaluate the droplet size distribution of water-in-oil emulsions [11].

Digital image analysis with the assistance of microscopy consists of four basic steps: (i) image acquisition, (ii) image restoration, (iii) segmentation and filtering, and (iv) measurement [13]. First, the appropriate image is transferred from the microscope to the computer via the image transmitter and a Charge-Coupled Device (CCD) or Complementary Metal-Oxide-Semiconductor (CMOS) camera. In this step, proper focusing is crucial, as the droplets should not have overlapping structures and should not cause any disturbances during the analysis [3]. After image acquisition, imaging defects, noise, or disturbances can be reduced, and the brightness and contrast of the images can also be adjusted. Then, segmentation and filtering processes are performed [13]. After the segmentation is completed and the images are converted to binary form, particle size analysis is performed. In addition, obtaining reproducible, highly accurate results that reflect the sample requires creating a protocol to perform as much automated or semi-automated particle size analysis as possible. Moreover, an adequate number of droplets representing the system are necessary to conduct a meaningful statistical analysis. Moradi et al. (2011) reported that reliable results that guarantee convergence of the distribution were produced with 2000 or more droplets [3].

The other challenge is the poor contrast between components of emulsions. Hu et al. (2018) stated that this could be due to the close refractive indices of water and oil, which are the two major components of the emulsions, under an optical microscope. This can cause difficulties in segmentation during the image analysis of emulsions [8]. Although visual separation of water and oil phases is possible with the use of dyes, there are limitations, as they show interfacial activity and change the physical properties (pH, electrical conductivity, density, etc.) that have an impact on the emulsion character [1,14]. Most traditional segmentation methods rely on the density and spatial relationships of pixels or constrained patterns, such as pixel-based, edge-based, texture-based, or region-based methods. Each of these methods has various advantages and disadvantages. Compliance with the subsequent processing and the obtained dataset can only be achieved by determining the threshold appropriately. As a result, the application of these techniques is not suitable for all situations. However, machine learning techniques overcome the problem based on the manual calibration of parameters by applying optimization techniques to a given set of training images [15]. Therefore, in recent years, the use of trainable machine learning methods, which enable more dynamic and accurate results to be obtained, has come to the fore [16,17]. The Trainable Weka Segmentation (TWS) is one of the plugins of Fiji, which is an open-source image processing package based on ImageJ [16,18]. It uses Waikato Environment for Knowledge Analysis, which was developed at Waikato University in New Zealand, for data mining tasks [16,19]. It is a combination of image segmentation and machine learning algorithms [17].

This study aimed to develop a protocol for the determination of droplet size of an emulsion with a microscopy-assisted digital image analysis technique using TWS for the segmentation step. For this purpose, emulsions (O/W) were prepared with different oil/water phase ratios and homogenization times, and their droplet size parameters were determined. For verification of the method, instrumental measurements of the same emulsion samples were also performed with the laser diffraction method and the results were compared. Moreover, the relationships between the droplet size and the physical properties of emulsions (turbidity and viscosity) were investigated. To the best of the authors' knowledge, this is the first time in the literature that TWS was used for the segmentation of emulsion oil droplets from the background in digital image analysis of emulsion micrographs.

## 2. Materials and Methods

### 2.1. Emulsion Preparation

Oil-in-water (O/W) emulsions were prepared in the present study. Commercial sunflower seed oil was purchased and used as the oil phase. Polysorbate 80 (Crillet 4™) was kindly provided by Croda (Croda International Plc, Snaith, UK) and used as an emulsifier at a constant ratio of 2% (*w/w*) in all emulsions. The experiment was designed with the aim of obtaining emulsions with different droplet sizes. For this purpose, the oil/water phase ratios of emulsions (5%, 10%, 20%, 30%, 40%, and 50%) and homogenization times (3, 6, and 9 min) were varied and are presented in Table 1.

In the emulsion preparation, the total weight of each emulsion was kept constant at 30 g. Firstly, Crillet 4™ was dissolved in distilled water in a 50 mL Falcon tube. Then, sunflower oil was added to the tube and mixed. The mixture was homogenized with Ultra-Turrax (T-18 Digital Package, IKA, Staufen, Germany) at 10,000 rpm. The preparations were carried out in two repetitions, and the measurements were performed in triplicate.

**Table 1.** Oil/water phase ratios and homogenization times of the prepared emulsions.

| Run | Oil (%) | Homogenization (min) |
|---|---|---|
| 1 | 5 | 3 |
| 2 * | | 3 |
| 3 * | 10 | 6 |
| 4 * | | 9 |
| 5 | 20 | 3 |
| 6 * | | 3 |
| 7 * | 30 | 6 |
| 8 * | | 9 |
| 9 | 40 | 3 |
| 10 | 50 | 3 |

* The samples were also analyzed after one day of storage.

## 2.2. Determination of Particle Size Distribution

The particle size distributions of the emulsions were measured with two different techniques, using a laser diffraction particle size analyzer and microscopy-assisted digital image analysis. For this purpose, D(10), D(50), D(90), span, D[3,2], and D[4,3] values were determined with both methods. D(10), D(50), and D(90) are the equivalent volume diameters at 10%, 50%, and 90% cumulative volume, respectively, while D[3,2] and D[4,3] values represent the area- and volume-weighted mean diameters, respectively. The span value is a measure of distribution width:

$$D[3,2] = \frac{\sum n_i d_i^3}{\sum n_i d_i^2} \tag{1}$$

$$D[4,3] = \frac{\sum n_i d_i^4}{\sum n_i d_i^3} \tag{2}$$

$$\text{Span} = \frac{D(90) - D(10)}{D(50)} \tag{3}$$

where $n_i$ is the number of particles of diameter $d_i$.

### 2.2.1. Determination of Particle Size Distribution by Laser Diffraction

A Mastersizer (Mastersizer 3000, Malvern Instruments Ltd., Malvern, Worcestershire, UK) was used as the laser diffraction particle size analyzer. The emulsions were prepared according to the formulation listed in Table 1, and they were cooled to room temperature before measurements. The refractive index ratios of sunflower oil and distilled water were assumed to be 1.472 and 1.333, respectively. The stirring speed was adjusted to 2100 rpm, and emulsions (approximately 50–100 μL) were added to 500 mL of aqueous dispersant (distilled water) until a laser obscuration value of 10–14% was obtained [20–22].

### 2.2.2. Determination of Particle Size Distribution by Microscopy-Assisted Digital Image Analysis

A total of 50 μL of distilled water was dropped onto the microscope slide, and 5 μL of the emulsion was spread over the slide with a syringe needle. It was carefully covered with a coverslip, and then immersion oil was dripped onto the coverslip. For imaging, 100×/1.25 oil, 160/0.17 objective lenses were used. Micrographs were obtained by a compound microscope (M83EZ, OMAX Microscopes, Kent, WA, USA) combined with a 5-megapixel CMOS camera (A3550U, OMAX Microscopes, Kent, WA, USA). The camera was located at the trinocular head, and micrographs were taken under transparent light conditions. At least 100 photographs were taken from 5 separate slides for each sample. The analysis was carried out immediately after the emulsion preparation and simultaneously with the other analyses. The light intensity, field, and condenser diaphragm apertures of the microscope were adjusted carefully for reproducible results. No pre-correction was applied to the images. ImageJ/Fiji (ver.1.53c.) was used as the software, and Trainable Weka Segmentation (TWS) (ver.3.2.35) was used for the segmentation of the images.

For the training of TWS, 10 images with different droplet densities were selected. As these images were used in the training, they were not included in the droplet particle size analysis. Since the processing of individual images slows the training process, they were converted into a stack of a single image and then used in the training (Image > Stacks > Images to Stack). Then, 800 × 800 sized regions were duplicated from the center of these images (Figure 1). Afterwards, TWS was activated from the plugins (Plugins > Segmentation > Trainable Weka Segmentation). Droplets and background were drawn manually on each image and defined as different "Classes". After the first training, minor corrections were made to the images based on the responses obtained from TWS, and the training process was repeated until the results for the images were obtained from TWS (Figure 2). After training of TWS, segmentations of the micrographs of the samples were performed using the trained classifier on TWS. The images obtained as a result of TWS segmentation were turned into a single stack, and further operations were performed on this stack. Firstly, this stack was converted to 8 bits and then transformed into the binary format (Image > Type > 8-Bit, Process > Binary > Make Binary). After the empty droplets were filled with the "Fill holes" command, the "Open" command was applied to clear the pixels from the droplets (Process > Binary > Fill Holes-Open). Eventually, the particle sizes of these images were calculated (Analyze > Analyze particles).

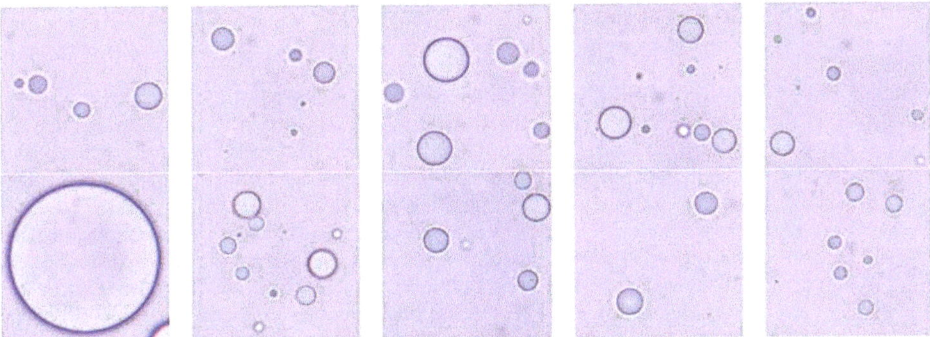

**Figure 1.** The micrograms used in training of TWS.

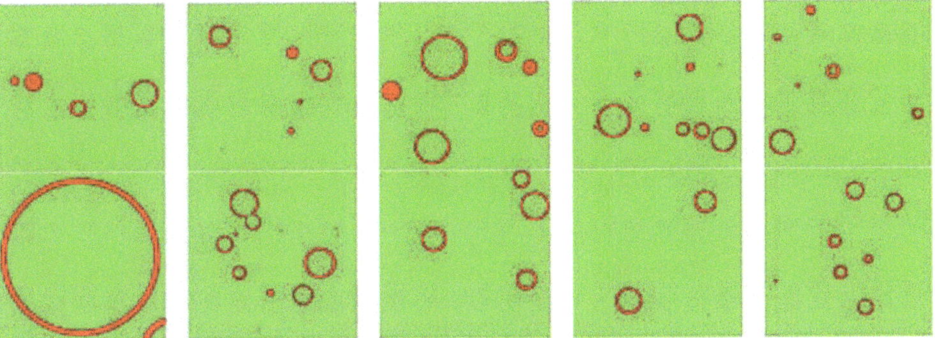

**Figure 2.** The classified images after the training of TWS.

The calculation was applied to objects with a circularity value greater than 0.85 to exclude composite or half-droplet images, which are illustrated in the regions marked in red circles in Figure 3. A scale of 1 mm (100 × 10 µm) was used for the size calibration. The top of the scale was covered with a coverslip, and then immersion oil was dripped onto it. The number of pixels corresponding to the distance between two points with known distances was determined on the micrograph of the calibration slide. The measured distance was defined as global scale in ImageJ/Fiji (Analyze > Set Scale).

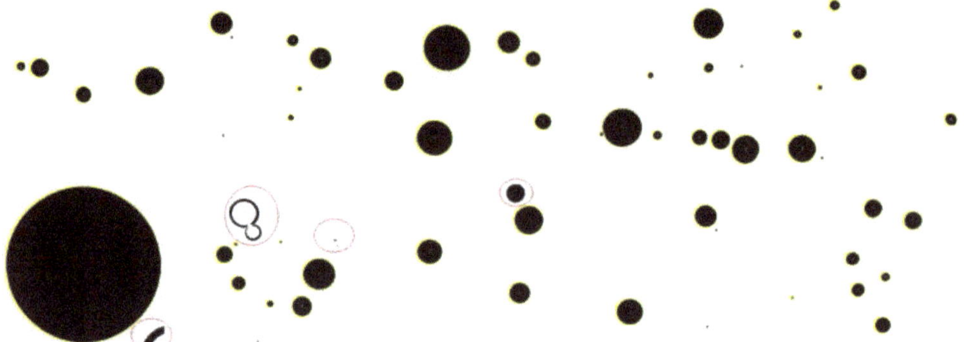

**Figure 3.** An example of droplet size analysis with classified images. While the particles circled in yellow are defined as droplets and used in the droplet size distribution calculations, those in the red circles, which are have a roundness value below 0.85, are not included in calculations.

*2.3. Determination of Turbidity*

The turbidity of emulsions was measured with both a portable turbidity meter (MI 415, Milwaukee Instruments, Rocky Mount, NC, USA) and UV spectrophotometer (Carry 60, Agilent Technologies, Santa Clara, CA, USA) at different wavelengths, such as 450 and 850 nm. A 150 µL emulsion sample was diluted with 15 mL of distilled water, and the mixture was homogenized by shaking before measurements. The method of measurement with the turbidity meter is based on the detection of scattered light according to ISO 7027 [23]. Before analysis, calibration of the device was performed with the 10 and 500 Formazin Nephelometric Unit (FNU) standards provided by the manufacturer. The wavelengths were set to 450 and 800 nm for the spectrophotometric determination of turbidity.

*2.4. Determination of Viscosity*

Flow properties of the emulsions were determined at 35 °C by using a viscometer (DV-II+ Pro Viscometer, Brookfield Engineering, Middleborough, MA, USA) and an SC4-18 spindle. A small sample adapter (SSA-13RD, Brookfield Engineering, Middleborough, MA, USA) was used in the measurements, and the temperatures of the samples were adjusted by a water circulation system (ICC Basic Eco 8, IKA, Staufen, Germany). Samples were analyzed immediately after the emulsion preparation. The sample cup was filled with 6.7 mL of the emulsion, and temperature equilibration was obtained in approximately 15 min. Experiments were duplicated and the average results were determined.

*2.5. Statistical Analysis*

The linear relations between the results of the microscopic measurements, laser diffraction particle size measurements, and other emulsion stability tests were investigated with bivariate correlation. For this purpose, SPSS statistical package program (SPSS ver. 13.0 for Windows, SPSS Inc., Chicago, IL, USA) was used, and Pearson's correlation coefficients were calculated. It was assumed that when the variables were continuous and approximately normally distributed without significant outliers, there was a linear relationship between the variables. The linear correlation was assumed to be significant at $p < 0.05$, and a high correlation was determined at $p < 0.01$.

### 3. Results and Discussion

In this study, 10 emulsions with different formulations and/or homogenization times were prepared. Immediately after the production of these emulsions, analyses of the droplet size distribution were carried out with the laser diffraction technique, and simultaneously, microscope images of the emulsions were taken. The microscope images were processed with the help of TWS. In addition, the same analyses were repeated for six of the emulsions after one day of storage. The results of analyses performed with two different methods for the droplet size distribution are shown in Table 2. On the other hand, it is known that emulsion stability and viscosity properties are related to emulsion droplet sizes in emulsions [24,25]. Therefore, the stability and viscosity of the produced emulsions were analyzed, and the results are presented in Table 3.

**Table 2.** The particle size distributions of emulsions measured with two different techniques.

| Oil (%) | Homogenization (min) | Storage (Day) | Microscopic Measurements | | | Laser Diffraction | | |
|---|---|---|---|---|---|---|---|---|
| | | | D(90) (μm) | D[3,2] (μm) | D[4,3] (μm) | D(90) (μm) | D[3,2] (μm) | D[4,3] (μm) |
| 5 | 3 | 0 | 29.67 | 7.20 | 15.12 | 21.10 | 2.69 | 8.46 |
| 10 | 3 | 0 | 32.83 | 7.78 | 16.70 | 27.80 | 3.87 | 13.60 |
| | | 1 | 36.01 | 11.51 | 19.51 | 27.00 | 3.79 | 13.10 |
| | 6 | 0 | 31.95 | 7.38 | 12.12 | 20.90 | 3.12 | 9.43 |
| | | 1 | 32.18 | 10.38 | 18.71 | 23.80 | 3.54 | 11.40 |
| | 9 | 0 | 32.40 | 7.48 | 13.25 | 22.20 | 3.49 | 10.50 |
| | | 1 | 31.51 | 10.18 | 17.86 | 24.00 | 3.69 | 11.50 |
| 20 | 3 | 0 | 34.22 | 12.81 | 21.71 | 32.30 | 6.00 | 18.60 |
| 30 | 3 | 0 | 37.41 | 12.85 | 22.92 | 34.10 | 7.23 | 20.70 |
| | | 1 | 35.25 | 14.97 | 23.44 | 33.70 | 7.07 | 20.20 |
| | 6 | 0 | 37.37 | 10.82 | 21.63 | 31.40 | 6.68 | 18.80 |
| | | 1 | 33.11 | 11.85 | 19.78 | 31.70 | 6.61 | 18.80 |
| | 9 | 0 | 37.16 | 9.75 | 18.73 | 27.80 | 5.76 | 16.00 |
| | | 1 | 33.97 | 11.93 | 20.62 | 28.00 | 5.65 | 16.00 |
| 40 | 3 | 0 | 37.30 | 16.68 | 24.44 | 31.70 | 7.22 | 19.40 |
| 50 | 3 | 0 | 37.84 | 18.50 | 24.83 | 26.00 | 6.35 | 15.80 |

For a realistic evaluation of droplet size distribution in emulsions, the largest possible number of droplets must be analyzed. It is stated in the literature that at least 2000 droplets should be analyzed to obtain reliable results in studies carried out with microscopic measurements [3]. The present study was conducted to determine the droplet size distribution with microscopy-assisted digital image analysis, and more than 2000 droplets (droplets between 3836–7850) were analyzed for each emulsion sample. Moreover, the Pearson correlation test was used to determine the relationship between the results obtained (Table 4).

**Table 3.** Turbidity and viscosity values of the emulsions.

| Oil (%) | Homogenization (min) | Storage (Day) | Turbidity (FNU) | Absorbance | | Viscosity (cP) |
|---|---|---|---|---|---|---|
| | | | | **450 nm** | **850 nm** | |
| 5 | 3 | 0 | $188 \pm 2$ [a] | $0.461 \pm 0.002$ [a] | $0.340 \pm 0.003$ [a] | $1.57 \pm 0.01$ [a] |
| 10 | 3 | 0 | $283 \pm 4$ [b,y] | $0.581 \pm 0.016$ [b,x] | $0.491 \pm 0.035$ [b,y] | $1.82 \pm 0.08$ [a,x] |
| | | 1 | $313 \pm 1$ [y] | $0.605 \pm 0.015$ [x] | $0.514 \pm 0.014$ [y] | $1.81 \pm 0.05$ [x] |
| | 6 | 0 | $290 \pm 1$ [y] | $0.641 \pm 0.011$ [y] | $0.503 \pm 0.005$ [y] | $1.74 \pm 0.08$ [x] |
| | | 1 | $311 \pm 9$ [y] | $0.659 \pm 0.010$ [y] | $0.513 \pm 0.014$ [y] | $1.97 \pm 0.09$ [x] |
| | 9 | 0 | $262 \pm 1$ [x] | $0.574 \pm 0.014$ [x] | $0.449 \pm 0.011$ [x] | $1.73 \pm 0.06$ [x] |
| | | 1 | $281 \pm 1$ [x] | $0.596 \pm 0.007$ [x] | $0.462 \pm 0.001$ [x] | $2.00 \pm 0.03$ [x] |
| 20 | 3 | 0 | $471 \pm 3$ [c] | $0.804 \pm 0.010$ [c] | $0.758 \pm 0.006$ [c] | $2.92 \pm 0.34$ [a] |
| 30 | 3 | 0 | $623 \pm 7$ [d,x] | $0.882 \pm 0.023$ [d,x] | $0.869 \pm 0.012$ [d,x] | $5.19 \pm 0.55$ [b,x] |
| | | 1 | $600 \pm 18$ [x] | $0.892 \pm 0.018$ [x] | $0.885 \pm 0.013$ [x] | $4.54 \pm 0.44$ [x] |
| | 6 | 0 | $692 \pm 21$ [y] | $0.976 \pm 0.018$ [y] | $0.943 \pm 0.005$ [y] | $3.92 \pm 0.06$ [x] |
| | | 1 | $656 \pm 30$ [x] | $1.043 \pm 0.009$ [y] | $0.956 \pm 0.022$ [y] | $3.82 \pm 0.18$ [x] |
| | 9 | 0 | $816 \pm 21$ [z] | $1.164 \pm 0.014$ [z] | $1.078 \pm 0.005$ [z] | $3.79 \pm 0.09$ [x] |
| | | 1 | $782 \pm 1$ [y] | $1.183 \pm 0.006$ [z] | $1.082 \pm 0.003$ [z] | $3.68 \pm 0.08$ [x] |
| 40 | 3 | 0 | $984 \pm 13$ [e] | $1.298 \pm 0.009$ [e] | $1.218 \pm 0.011$ [e] | $6.45 \pm 0.52$ [b] |
| 50 | 3 | 0 | +1000 | $1.560 \pm 0.019$ [f] | $1.494 \pm 0.023$ [f] | $12.48 \pm 1.66$ [c] |

[a–f] Different superscript letters indicate significant differences between samples prepared with 3 min of homogenization time with the same storage period and different oil/water ratios ($p > 0.05$). [x–z] Different superscript letters indicate significant differences between samples prepared with a constant oil/water ratio with the same storage period and different homogenization times ($p > 0.05$).

The results of both droplet size distribution analyses showed similar trends (Tables 2 and 4). However, the nominal values calculated by microscopy-assisted digital image analysis were slightly higher than the values obtained from the laser diffraction particle sizer. While D(90) and D[4,3] values calculated by microscopic measurements were in the range of 29.67–40.74 μm and 12.12–24.06 μm, the same parameters obtained by the laser diffraction technique were in the range of 20.90–34.10 μm and 8.46–20.70 μm, respectively (Table 2). It is known that measurement conditions (sample dilution, mixing speed, ultrasound application, etc.) have important effects on the nominal values obtained from measurements taken with a laser diffraction particle size analyzer.

The dilution and stirring processes may lead to floc disruption and result in smaller particle size values [13,26]. In this study, ultrasound was not applied during the sample injection, whereas a mixing speed of 2100 rpm was used. Although this may be the reason for the difference in nominal values between the microscopic and laser diffraction measurement results, especially for the stored emulsions, this effect seemed to be insignificant, as the other emulsion analyses showed similar trends to those observed with particle size analysis (Table 3). In general, the measured values of D(90) and D[4,3] slightly fluctuated with the storage time. According to the results, droplet size data determined by microscopy-assisted digital image analysis and droplet size analysis determined by laser diffraction were positively correlated in all parameters ($p < 0.05$). Notably, D[4,3], D[3,2], and D(90) values had higher positive correlations ($p < 0.01$). It is known that D(90) is less affected by the obscuration ratio than other particle size parameters, such as D(10) and D(50), in the laser diffraction technique and yields more stable results. It is reported in the literature that D[4,3] is more sensitive to the presence of large particles in an emulsion system than the other parameters, and the differences between D[4,3] and D[3,2] generally indicate a broad or multimodal particle size distribution [1]. In the present study, the difference between D[4,3] and D[3,2] was significant according to the results of both measurement techniques and varied in the range of 4.7–10.8 μm and 5.8–13.5 μm for microscopic and laser diffraction measurements, respectively (Table 2).

**Table 4.** The results of statistical analysis for the relations between microscopic measurements, laser diffraction, and other emulsion tests.

| | Property | Statistical Evaluation | Microscopic Measurements | | | | |
| --- | --- | --- | --- | --- | --- | --- | --- |
| | | | D(10) (μm) | D(90) (μm) | Span | D[3,2] (μm) | D[4,3] (μm) |
| Laser Diffraction | D(10) (μm) | Pearson correlation coefficient | 0.676 ** | 0.767 ** | −0.705 ** | 0.769 ** | 0.847 ** |
| | | Significance (2-tailed) | 0.004 | 0.001 | 0.002 | 0.000 | 0.000 |
| | D(90) (μm) | Pearson correlation coefficient | 0.440 | 0.624 ** | −0.663 ** | 0.593 * | 0.784 ** |
| | | Significance (2-tailed) | 0.088 | 0.010 | 0.005 | 0.015 | 0.000 |
| | Span | Pearson correlation coefficient | −0.535 * | −0.731 ** | 0.535 * | −0.635 ** | −0.691 ** |
| | | Significance (2-tailed) | 0.033 | 0.001 | 0.033 | 0.008 | 0.003 |
| | D[3,2] (μm) | Pearson correlation coefficient | 0.663 ** | 0.763 ** | −0.714 ** | 0.766 ** | 0.858 ** |
| | | Significance (2-tailed) | 0.005 | 0.001 | 0.002 | 0.001 | 0.000 |
| | D[4,3] (μm) | Pearson correlation coefficient | 0.560 * | 0.714 ** | −0.701 ** | 0.693 ** | 0.840 ** |
| | | Significance (2-tailed) | 0.024 | 0.002 | 0.002 | 0.003 | 0.000 |
| Other Tests | Turbidity | Pearson correlation coefficient | 0.820 ** | 0.779 ** | −0.593 * | 0.817 ** | 0.763 ** |
| | | Significance (2-tailed) | 0.000 | 0.000 | 0.015 | 0.000 | 0.001 |
| | Spectrophotometric Measurement (450 nm) | Pearson correlation coefficient | 0.793 ** | 0.732 ** | −0.574 * | 0.788 ** | 0.727 ** |
| | | Significance (2-tailed) | 0.000 | 0.001 | 0.020 | 0.000 | 0.001 |
| | Spectrophotometric Measurement (850 nm) | Pearson correlation coefficient | 0.803 ** | 0.776 ** | −0.612 * | 0.815 ** | 0.778 ** |
| | | Significance (2-tailed) | 0.000 | 0.000 | 0.012 | 0.000 | 0.000 |
| | Viscosity (cP) | Pearson correlation coefficient | 0.887 ** | 0.672 ** | −0.585 * | 0.849 ** | 0.717 ** |
| | | Significance (2-tailed) | 0.000 | 0.004 | 0.023 | 0.000 | 0.002 |

* Correlation is significant at the 0.05 level (2-tailed). ** Correlation is significant at the 0.01 level (2-tailed).

Moreover, all of the parameters calculated from the microscopic measurements were highly correlated with the emulsion stability and viscosity results, and these relationships were even stronger than those between laser diffraction measurements and emulsion properties (Table 4). It was observed that the viscosity of the emulsions did not change significantly with storage ($p > 0.05$) (Table 3). When the effect of homogenization time was evaluated, the smallest droplet size was achieved with a 6 min process at a 10% oil ratio and with a 9 min process for an emulsion with a 30% oil ratio. These results are in agreement with the turbidity results (measurements made with both the turbidimeter and spectrophotometer), and turbidity values are high in samples with a small droplet size [2].

With the increase in the oil ratio in the produced emulsion formulation, increases in viscosity, turbidity, and droplet size values, especially those measured with the microscope (mainly D[3,2] and D[4,3] parameters), were observed (Tables 2 and 3). While no significant changes were detected in the laser diffraction measurements of turbidity, viscosity, or droplet size values of emulsions after storage, some fluctuations in the variation of droplet sizes measured with the microscope were observed (Tables 2 and 3).

## 4. Conclusions

The results of the present study showed that the same trends and similar particle size values were obtained by microscopy-assisted digital image analysis and laser diffraction particle size analysis for the determination of particle size in emulsions. Considering the installation costs of devices for measuring laser diffraction, microscopy-assisted digital image analysis seems to be a very useful and effective alternative. On the other hand, if sample characterization is to be performed, it should be noted that there are certain differences in the measured values, and slightly larger droplet sizes are calculated with the microscopy-assisted digital image analysis technique. In addition, while the results can be obtained in about 5 min when taking measurements with laser diffraction devices (total time for analysis), the analysis and calculation process may take one or several days per sample with the procedure developed in the present study, unless working with high-capacity workstations. It was observed that comparable results were obtained for D(90), D[3,2], and D[4,3], which were correlated with the physical properties of the emulsions.

In summary, the microscopy-assisted digital image analysis procedure developed in this study can provide very reasonable results when it is impossible to obtain/use laser diffraction particle size analyzers. However, it should be noted that performing this method is labor- and time-intensive.

**Author Contributions:** P.S.: formal analysis, validation, investigation, data curation, and writing—original draft. O.G.: conceptualization, methodology, and data curation. L.Y.A.: writing—review and editing and supervision. Z.E.: conceptualization, methodology, writing—original draft, and project administration. All authors have read and agreed to the published version of the manuscript.

**Funding:** This work was supported by the Scientific and Technological Research Council of Turkey (TUBITAK) (Grant No. 120O763). The authors are also grateful to Croda for providing the emulsifiers.

**Institutional Review Board Statement:** Not applicable.

**Informed Consent Statement:** Not applicable.

**Data Availability Statement:** Not applicable.

**Conflicts of Interest:** The authors declare no conflict of interest.

## References

1.  McClements, D.J. Critical review of techniques and methodologies for characterization of emulsion stability. *Crit. Rev. Food Sci. Nutr.* **2007**, *47*, 611–649. [CrossRef] [PubMed]
2.  Alade, O.S.; Mahmoud, M.; Al Shehri, D.A.; Sultan, A.S. Rapid Determination of Emulsion Stability Using Turbidity Measurement Incorporating Artificial Neural Network (ANN): Experimental Validation Using Video/Optical Microscopy and Kinetic Modeling. *ACS Omega* **2021**, *6*, 5910–5920. [CrossRef] [PubMed]
3.  Moradi, M.; Alvarado, V.; Huzurbazar, S. Effect of salinity on water-in-crude oil emulsion: Evaluation through drop-size distribution proxy. *Energy Fuels* **2011**, *25*, 260–268. [CrossRef]
4.  McClements, D.J. Principles of ultrasonic droplet size determination in emulsions. *Langmuir* **1996**, *12*, 3454–3461. [CrossRef]
5.  Li, M.; Wilkinson, D.; Patchigolla, K. Comparison of particle size distributions measured using different techniques. *Part. Sci. Technol.* **2005**, *23*, 265–284. [CrossRef]
6.  Fieber, W.; Hafner, V.; Normand, V. Oil droplet size determination in complex flavor delivery systems by diffusion NMR spectroscopy. *J. Colloid Interface Sci.* **2011**, *356*, 422–428. [CrossRef] [PubMed]
7.  Richter, A.; Voigt, T.; Ripperger, S. Ultrasonic attenuation spectroscopy of emulsions with droplet sizes greater than 10 $\mu$m. *J. Colloid Interface Sci.* **2007**, *315*, 482–492. [CrossRef]
8.  Hu, Y.T.; Ting, Y.; Hu, J.Y.; Hsieh, S.C. Techniques and methods to study functional characteristics of emulsion systems. *J. Food Drug Anal.* **2017**, *25*, 16–26. [CrossRef]
9.  Tyuftin, A.A.; Mohammed, H.; Kerry, J.P.; O'Sullivan, M.G.; Hamill, R.; Kilcawley, K. Microscopy-Assisted Digital Photography as an Economical Analytical Tool for Assessment of Food Particles and Their Distribution Through The use of the ImageJ Program. *Adv. Nutr. Food Sci.* **2021**, *2021*. [CrossRef]
10. Jokela, P.; Fletcher, P.D.I.; Aveyard, R.; Lu, J.R. The use of computerized microscopic image analysis to determine emulsion droplet size distributions. *J. Colloid Interface Sci.* **1990**, *134*, 417–426. [CrossRef]
11. Maaref, S.; Ayatollahi, S. The effect of brine salinity on water-in-oil emulsion stability through droplet size distribution analysis: A case study. *J. Dispers. Sci. Technol.* **2018**, *39*, 721–733. [CrossRef]
12. Tripathi, S.; Bhattacharya, A.; Singh, R.; Tabor, R.F. Rheological behavior of high internal phase water-in-oil emulsions: Effects of droplet size, phase mass fractions, salt concentration, and aging. *Chem. Eng. Sci.* **2017**, *174*, 290–301. [CrossRef]
13. Merkus, H.G. *Particle Size Measurements: Fundamentals, Practice, Quality*; Merkus, H.G., Ed.; Springer: New York, NY, USA, 2009; ISBN 9781402090158.
14. McClements, D.J. *Food Emulsions Principles, Practices, and Techniques*; CRC Press: Boca Raton, FL, USA, 2016; ISBN 9781498726696.
15. Wesemeyer, T.; Jauer, M.L.; Deserno, T.M. Annotation quality vs. quantity for deep-learned medical image segmentation. In *Medical Imaging 2021: Imaging Informatics for Healthcare, Research, and Applicationsi*; International Society for Optics and Photonics: Bellingham, WA, USA, 2021; Volume 11601, p. 116010C.

16. Arganda-Carreras, I.; Kaynig, V.; Rueden, C.; Eliceiri, K.W.; Schindelin, J.; Cardona, A.; Seung, H.S. Trainable Weka Segmentation: A machine learning tool for microscopy pixel classification. *Bioinformatics* **2017**, *33*, 2424–2426. [CrossRef]
17. Lormand, C.; Zellmer, G.F.; Németh, K.; Kilgour, G.; Mead, S.; Palmer, A.S.; Sakamoto, N.; Yurimoto, H.; Moebis, A. Weka trainable segmentation plugin in ImageJ: A semi-automatic tool applied to crystal size distributions of microlites in volcanic rocks. *Microsc. Microanal.* **2018**, *24*, 667–675. [CrossRef] [PubMed]
18. Schindelin, J.; Arganda-Carreras, I.; Frise, E.; Kaynig, V.; Longair, M.; Pietzsch, T.; Preibisch, S.; Rueden, C.; Saalfeld, S.; Schmid, B.; et al. Fiji: An open-source platform for biological-image analysis. *Nat. Methods* **2012**, *9*, 676–682. [CrossRef]
19. Hall, M.; Frank, E.; Holmes, G.; Pfahringer, B.; Reutemann, P.; Witten, I.H. The WEKA data mining software: An update. *ACM SIGKDD Explor. Newsl.* **2009**, *11*, 10–18. [CrossRef]
20. Himmetagaoglu, A.B.; Erbay, Z.; Cam, M. Production of microencapsulated cream: Impact of wall materials and their ratio. *Int. Dairy J.* **2018**, *83*, 20–27. [CrossRef]
21. O'Dwyer, S.P.; O'Beirne, D.; Eidhin, D.N.; O'Kennedy, B.T. Effects of emulsification and microencapsulation on the oxidative stability of camelina and sunflower oils. *J. Microencapsul.* **2013**, *30*, 451–459. [CrossRef]
22. Meyer, S.; Berrut, S.; Goodenough, T.I.J.; Rajendram, V.S.; Pinfield, V.J.; Povey, M.J.W. A comparative study of ultrasound and laser light diffraction techniques for particle size determination in dairy beverages. *Meas. Sci. Technol.* **2006**, *17*, 289–297. [CrossRef]
23. ISO 7027-1:2016 Water Quality—Determination of Turbidity—Part 1: Quantitative Methods 2016. Available online: https://www.iso.org/standard/62801.html (accessed on 16 February 2022).
24. Pal, R. Effect of droplet size on the rheology of emulsions. *AIChE J.* **1996**, *42*, 3181–3190. [CrossRef]
25. Castaño, E.P.M.; Leite, R.H.T.; de Souza Mendes, P.R. Microscopic phenomena inferred from the rheological analysis of an emulsion. *Phys. Fluids* **2021**, *33*, 073102. [CrossRef]
26. Kim, H.J.; Decker, E.A.; McClements, D.J. Comparison of droplet flocculation in hexadecane oil-in-water emulsions stabilized by β-lactoglobulin at pH 3 and 7. *Langmuir* **2004**, *20*, 5753–5758. [CrossRef] [PubMed]

*Article*

# Patterning Configuration of Surface Hydrophilicity by Graphene Nanosheet towards the Inhibition of Ice Nucleation and Growth

**Biao Jiang** [1]**, Yizhou Shen** [1,*]**, Jie Tao** [1,*]**, Yangjiangshan Xu** [1]**, Haifeng Chen** [2]**, Senyun Liu** [3]**, Weilan Liu** [4] **and Xinyu Xie** [1]

1   College of Materials Science and Technology, Nanjing University of Aeronautics and Astronautics, 29 Yudao Steet, Nanjing 210016, China; jiangbiao@nuaa.edu.cn (B.J.); xyjshan@nuaa.edu.cn (Y.X.); xiexinyu@nuaa.edu.cn (X.X.)
2   Department of Materials Chemistry, Qiuzhen School, Huzhou University, 759# East 2nd Road, Huzhou 313000, China; headder@zjhu.edu.cn
3   Key Laboratory of Icing and Anti/De-Icing, China Aerodynamics Research and Development Center, Mianyang 621000, China; liusenyun@cardc.cn
4   Institute of Advanced Materials, Nanjing Tech University, 30 Puzhu South Road, Nanjing 210009, China; iamwlliu@njtech.edu.cn
*   Correspondence: shenyizhou@nuaa.edu.cn (Y.S.); taojie@nuaa.edu.cn (J.T.)

**Citation:** Jiang, B.; Shen, Y.; Tao, J.; Xu, Y.; Chen, H.; Liu, S.; Liu, W.; Xie, X. Patterning Configuration of Surface Hydrophilicity by Graphene Nanosheet towards the Inhibition of Ice Nucleation and Growth. *Coatings* **2022**, *12*, 52. https://doi.org/ 10.3390/coatings12010052

Academic Editors: Eduardo Guzmán and Alicia de Andrés

Received: 28 October 2021
Accepted: 29 December 2021
Published: 2 January 2022

**Publisher's Note:** MDPI stays neutral with regard to jurisdictional claims in published maps and institutional affiliations.

**Abstract:** Freezing of liquid water occurs in many natural phenomena and affects countless human activities. The freezing process mainly involves ice nucleation and continuous growth, which are determined by the energy and structure fluctuation in supercooled water. Herein, considering the surface hydrophilicity and crystal structure differences between metal and graphene, we proposed a kind of surface configuration design, which was realized by graphene nanosheets being alternately anchored on a metal substrate. Ice nucleation and growth were investigated by molecular dynamics simulations. The surface configuration could induce ice nucleation to occur preferentially on the metal substrate where the surface hydrophilicity was higher than the lateral graphene nanosheet. However, ice nucleation could be delayed to a certain extent under the hindering effect of the interfacial water layer formed by the high surface hydrophilicity of the metal substrate. Furthermore, the graphene nanosheets restricted lateral expansion of the ice nucleus at the clearance, leading to the formation of a curved surface of the ice nucleus as it grew. As a result, ice growth was suppressed effectively due to the Gibbs–Thomson effect, and the growth rate decreased by 71.08% compared to the pure metal surface. Meanwhile, boundary misorientation between ice crystals was an important issue, which also prejudiced the growth of the ice crystal. The present results reveal the microscopic details of ice nucleation and growth inhibition of the special surface configuration and provide guidelines for the rational design of an anti-icing surface.

**Keywords:** surface hydrophilicity; graphene; ice formation; clearance; molecular dynamic simulation

## 1. Introduction

Understanding and regulating the crystallization of supercooled water on surfaces is essential in both basic research and engineering applications [1,2]. The freezing of water on surfaces is, in fact, a complicated phenomenon that requires collective understanding of nucleation, crystal growth, surface science and thermodynamics [3–5], and plenty of research has been done theoretically and experimentally [6,7]. According to previous studies, ice nucleation can be affected by many factors, including surface morphology [8,9], wettability [10,11], shear flow [12], ions and contamination particles [13–15], etc. For example, Yue et al. and Wang et al. found that micro-hierarchical structures or patterns on a silicon surface would delay ice nucleation due to enhanced free energy for nucleation [16,17]. He et al. reported the effect of counter ions on heterogeneous ice nucleation on a polyelectrolyte brush and discovered that a distinct efficiency of ions in turning ice a freezing temperature follows a certain sequence [18].Their study revealed that counter ions have a profound ion-specific effect on the relaxation of the hydrogen bond and the formation rate of interfacial water molecules.

Over the past few years, it has been investigated that graphene and its derivatives could influence ice nucleation and growth [19–21]. The two-dimensional material, graphene, has become one of the most promising materials in recent decades due to its potential applications in high performance electronics, sensors and energy storage devices [22–27]. It can be made into fibers, membranes and can be drop-casted onto various substrates [28–30]. Using a molecular dynamics simulation, Lupi and Molinero studied the heterogeneous ice nucleation of liquid water in contact with graphitic surfaces of various dimensions and curvatures, which were reported in experimental characterizations of soot [31,32]. Their results indicated that the ordering of interfacial liquid water on a graphene surface was the main reason for the facilitated heterogeneous nucleation [33]. Not only a promoting material, graphene and its derivatives can also restrain ice formation under specific conditions [22,34]. The exceptional Joule's heating effect and electrothermal effect of graphene-based composites have received much attention in the design of anti-icing systems [35,36]. The lightweight graphene-based material is expected to be an ideal heater to prevent ice freezing. Specifically, there are numerous studies taking advantage of these properties for anti-icing and de-icing applications [37,38].

Recently, graphene oxide (GO) was used to mimic antifreeze proteins (AFPs) and considered to be an advanced icing inhibitor [39–42]. The repeated hexagonal carbon ring structure arranges the functional groups on the basal plane of GO to match with an ice crystal lattice, leading to the preferred adsorption of GO on existing ice in liquid water. Thus, the growth of the ice crystal is suppressed owing to the Gibbs–Thomson effect; that is, the curved surface lowers the freezing temperature [43,44]. The curved surface of the ice crystal derives from the special surface configuration, which involves different surface hydrophilicity and a crystal structure between the two materials. By using molecular dynamics simulations, Zhang and Chen studied ice nucleation on graphene surfaces functionalized by several kind of ions and methane molecules [45]. Their results indicated that the ice nucleation ability of the functionalized surfaces was weakened compared with that of the smooth graphene surface, depending on the type and the number of functional groups. Akhtar et al. presented an anti-icing coating based on fluorinated graphene, which could strikingly delay ice formation in a high humidity environment [46]. The anti-icing performance of fluorinated graphene was attributed to a robust liquid layer arising from the interface confinement effect that increases the ice-water contact angle and viscosity of water molecules near the surface. Additionally, the coupling of surface crystallinity and surface hydrophilicity was found to be a controlling factor for heterogeneous ice nucleation [7]. With an appropriate hydrophilicity, the arrangement of the water layer in contact with crystalline graphene can be changed; thus, the ice nucleation rate decreases consequently. The interplay between surface morphology and hydrophobicity on heterogeneous ice nucleation was also studied by Martin [4]. They showed that lattice mismatch of the surface with respect to ice is desirable for a good ice nucleating agent. Hence, it is interesting to investigate the synergistic effect of a surface hydrophilicity discrepancy and the clearance configuration created by an anchored graphene sheet on ice nucleation and growth.

In this work, a graphene nanosheet was introduced to design a special surface configuration of surface hydrophilicity on a metal surface. The characteristics of ice nucleation and growth on the surface were investigated using a molecular dynamics (MD) simulation. We constructed a series of surface configurations with various surface hydrophilicity discrepancies between a graphene nanosheet and metal substrate and explored the initiation of ice nucleation and growth processes under specific surface conditions. Besides, we studied the growth process of an ice nucleus and computed the growth rate of the ice nucleus. A different surface configuration resulted in varying levels of ice growth inhibition. We also discussed the misorientation between grain boundaries of the ice crystal [47].

## 2. System and Simulations

The simulation system contained four graphene nanosheets anchored on a metal substrate, which were cleaved in the (100) surface of a face-centred cubic (fcc) crystal. The lattice parameter $a_{fcc}$ was 4.0495 Å, and 6189 water molecules were covered on this surface, as shown in Figure 1a. The simulation box had dimensions of $L_x$ = 9.758 nm, $L_y$ = 9.744 nm and $L_z$ = 10.000 nm. A void space was set in the simulation box to avoid the influence of a periodic boundary condition in the z direction. Four graphene nanosheets with a size of 2.85 nm × 3.27 nm were fixed atop of the metal substrate with a distance of 6 Å [33,48], resulting in a 2-nm-wide cross-shaped clearance between the graphene nanosheets, as shown in Figure S1.

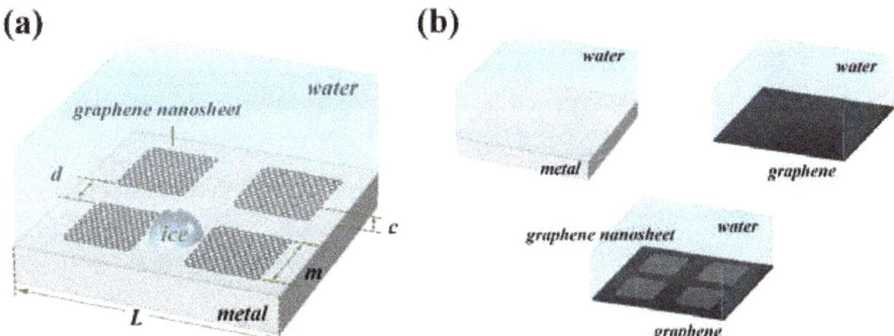


**Figure 1.** Illustrations of simulation systems. (**a**) Model of the metal–graphene nanosheet surface that was covered with a box of liquid water. A typical ice nucleus formation is illustrated on the surface. (**b**) Models of three other simulation systems. Pure metal, pure graphene and graphene–graphene nanosheet surface systems.

The system described above is defined as a metal–graphene nanosheet system. Additionally, we constructed three other simulation systems for comparison, which were pure metal, pure graphene and graphene–graphene nanosheet surface systems. Details are shown in Figure 1b and Table S1.

The coarse-grained monatomic water (mW) model was employed in this paper to describe the interaction between water molecules [49]. This specific water model has excellent structural properties and a melting point close to the experiment. The mW model treats water molecules as a single particle that interacts through short-range two-body and three-body interactions [50]. Since it is monatomic, it exhibits faster dynamics, which allows for the alleviation of computational costs of our simulations for the ice freezing process. Detailed information about the mW water model is provided in Supplementary Note 1. The Lennard-Jones (LJ) potential was used to model the interaction between water molecules and substrate atoms. Length and energy parameters, $\sigma_{w\text{-}g}$ and $\varepsilon_{w\text{-}g}$, were set to 2.488 Å and 0.13 kcal/mol, respectively, for each water molecule interacting with carbon atoms [31]. Otherwise, for interactions between water molecules and metal atoms, $\varepsilon_{w\text{-}m}$ was tuned from 0.13 kcal/mol to 14.0 kcal/mol in each simulation to obtain a different surface hydrophilicity of the metal substrate, while $\sigma_{w\text{-}m}$ was fixed in 2.8798 Å [51]. As a result, the difference between $\varepsilon_{w\text{-}g}$ and $\varepsilon_{w\text{-}m}$ is defined as surface hydrophilicity discrepancy $D_\varepsilon$. There is no need to define the metal–metal and carbon–carbon interaction potentials, because they are fixed in the simulation system. Periodic boundary conditions were applied in three dimensions, and the integration time-step for the velocity Verlet algorithm was set to 5 fs. After 0.2 ps of relaxation at temperature 290 K, the whole system was cooled down to 200 K gradually with a cooling rate of 0.9 K/ns in the canonical ensemble (NVT), and the process of water freezing was studied during the quenching period. To obtain statistically reasonable results, we performed 5 repetitions of each cooling simulation for every system. All MD simulations were performed using the LAMMPS simulation package. Phase and structure identification was carried out by the Identify Diamond Structure modifier in OVITO software (version 3.6.0) during the freezing process of liquid water on each surface configuration [52]. The initiation and ending times of the icing process were recorded through the entire simulation. The initiation time of icing was identified as the moment when the number of water molecules in the ice nucleus ($N_{ice}$) started to increase rapidly. Conversely, the ending time of icing was identified as the moment when $N_{ice}$ tended to be stable, which is marked in Figure 2a.

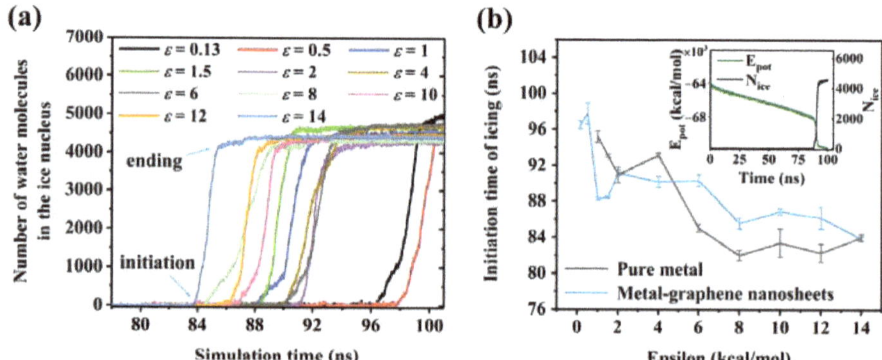

**Figure 2.** Freezing of liquid water in different surface conditions. (**a**) Number of water molecules in the ice cluster on metal–graphene nanosheet surfaces with different surface hydrophilicities. The interaction strength between the metal and water molecule $\varepsilon_{w\text{-}m}$ varied from 0.13 to 14.0 kcal/mol. (**b**) Initiation time of icing with different surface hydrophilicities of a metal substrate. Inset shows total potential energy $E_{pot}$ of the system and the number of water molecules in the ice nucleus, $N_{ice}$, during the whole simulation time. The data refer to the metal–graphene nanosheet system, and $\varepsilon_{w\text{-}m}$ was fixed to 1.0 kcal/mol in the inset of (**b**).

## 3. Results and Discussion

### 3.1. Surface Hydrophilicity and Ice Nucleation

The role of surface hydrophilicity [53,54] has been a central argument to determine whether ice nucleation and the subsequent growth process can happen [55–57]. In order to understand the impact of surface clearance on ice formation where hydrophilicity is different from peripheral graphene sheets, we performed a series of MD simulations using the interaction strength $\varepsilon_{w\text{-}m}$ as the collective variate to explore a desirable freezing delay surface, as shown in Figure 1a.

Ice formation can be obviously observed on every metal–graphene nanosheet surface as the system cooled down gradually, and the details are shown in Figure S3. The freezing process of liquid water on each metal–graphene nanosheet surface configuration is shown in Figure 2a. Curves with different colors represent various surface hydrophilicities of the metal substrate $\varepsilon_{w\text{-}m}$. Generally, lower interaction strength between a solid atom and water molecule results in later ice nucleation, which has been demonstrated in other research [54]. However, there was no simple trend for the triggering time of ice nucleation for $\varepsilon_{w\text{-}m}$ ranging from 1.0 to 6.0 kcal/mol, as shown in Figure 2b. Water molecules in the simulation system could interact with substrate atoms freely when $\varepsilon_{w\text{-}m}$ was smaller than 1.0 kcal/mol, as shown in Figure 3a. Note that a layer of water molecules was arranged in an fcc structure gradually with the increasing $\varepsilon_{w\text{-}m}$, as shown in Figure 3b. This interfacial water layer was formed adjacent to the substrate with a distance of about 1.88 Å, and it could be detected by the count of water molecules along the z direction, as shown in Figure 3b.

On one hand, water molecules in this layer were closely packed in an fcc structure that differed from the structure of an ice crystal, which is not beneficial for ice nucleation. On the other hand, other free water molecules were separated from the metal substrate by the interfacial water layer, which decreased the interaction between free water molecules and the metal substrate. Therefore, the interfacial water layer acts as a barrier and hinders the nucleation of an ice nucleus [7,58]. Meanwhile, the surrounding conditions of the graphene nanosheets were not beneficial to ice nucleation, owing to the extremely low surface hydrophilicity of 0.13 kcal/mol. Therefore, ice nucleation was delayed for $\varepsilon_{w-m}$ ranging from 1.0 to 6.0 kcal/mol because of the hindering effect of the interfacial water layer formed on the metal substrate.

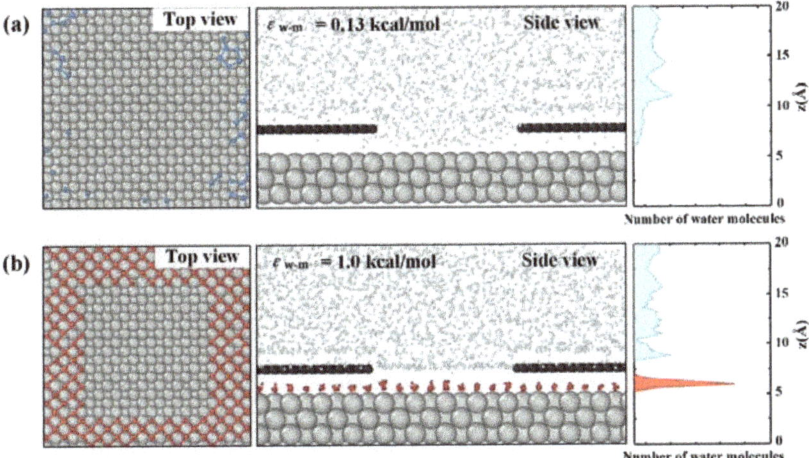


**Figure 3.** The formation of an interfacial water layer on a metal–graphene nanosheet surface. (a) Snapshot of the interface between a metal substrate and liquid water. The dense interfacial water layer was not formed at $\varepsilon_{w-m}$ = 0.13 kcal/mol. (b) Snapshot of the interfacial water layer viewed from the top and the side. The right panel is the number of water molecules along the z direction before water freezing. In all cases, the interfacial water layer is colored red for ease of visualization.

However, with the $\varepsilon_{w-m}$ increasement, a higher surface hydrophilicity of a metal substrate forced more water molecules to arrange in an orderly manner, which was convenient for the formation of an ice structure during the $\varepsilon_{w-m}$ value range from 6.0 to 14.0 kcal/mol. The competition of both effects lead to the nonlinear dependence, as shown in Figure 2b. Therefore, a proper surface hydrophilicity discrepancy between a graphene nanosheet and metal substrate is considered as a desirable agent for the delayed initiation of ice nucleation. The inset shows the total potential energy of a system changed with simulation time, together with the number of water molecules in the ice nucleus $N_{ice}$ [59]. As a plunge of the potential energy $E_{pot}$ of the system occurred, $N_{ice}$ increased suddenly, indicating the nucleation of an ice embryo. When $\varepsilon_{w-m}$ was set to 6.0 kcal/mol, i.e., a discrepancy of the surface hydrophilicity $D_\varepsilon$ of 5.87 kcal/mol, the ice nucleation on the metal–graphene nanosheet surface was delayed by 5.25 ns, which was a maximum value compared to that on a pure metal surface, as shown in Figure 2b. We noticed that a higher extent of hydrophilicity could induce a greater thickness of the interfacial water layer. The hindering effect of the interfacial water layer prevailed again, which well explained the jump of the triggering time of ice nucleation for $\varepsilon_{w-m}$ values between 6.0 to 8.0 kcal/mol.

It should be mentioned that small ice nuclei form and expand constantly under the action of structure and energy undulations in supercooled water [60]. Additionally, these unstable ice nuclei prefer to form at the edge of graphene sheets, where the energy barrier for ice nucleation is lower compared with those in other positions [19,61]. The edge of the graphene sheet is a phase contact area where water molecules interact with carbon and metal atoms simultaneously.

### 3.2. Priority of Ice Nucleation

We noticed that the position of initial ice nucleation on the metal–graphene nanosheet surface depended on the surface hydrophilicity of the metal substrate. For $\varepsilon_{\text{w-m}}$ lower than 1.0 kcal/mol, ice nucleation tended to happen at the top of the graphene nanosheets, as shown in Figure 4e and Figure S4. On this condition, ice nucleation was triggered by the repeated hexagonal carbon ring structure of the graphene nanosheet, which matched better with an ice structure compared to the fcc crystal structure of a metal substrate. However, ice nucleation generally happened at the clearance between the graphene nanosheets, as shown in Figure 4a and Figure S5, when $\varepsilon_{\text{w-m}}$ was higher than the critical value mentioned above, i.e., 1.0 kcal/mol. In this case, a more hydrophilic metal substrate, rather than the hexagonal structure on the basal plane of the graphene nanosheet, dominates the position that a stable ice nucleus would form initially [60].

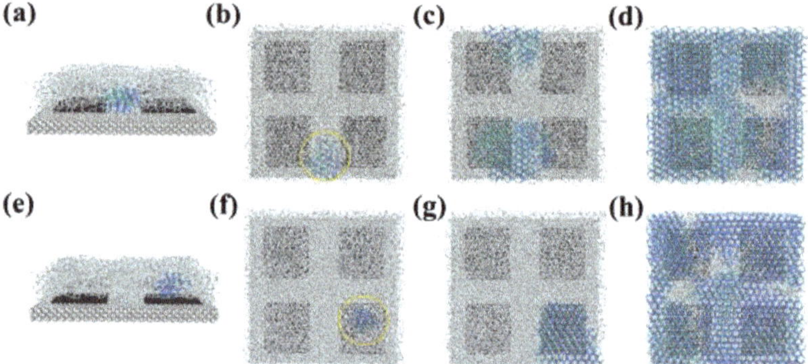

**Figure 4.** Snapshots of the preferential site of ice nucleation and subsequent growth of ice crystals on a metal–graphene surface. (**a**) Ice nucleation at the clearance of the surface, $\varepsilon_{\text{w-m}}$ = 4.0 kcal/mol. (**b–d**) Representative growth trajectories of an ice nucleus that appeared at the clearance, which is indicated by a yellow circle (from the top view). (**e**) Ice nucleation at the top of the graphene nanosheet, $\varepsilon_{\text{w-m}}$ = 1.0 kcal/mol. (**f–h**) Representative growth trajectories of an ice nucleus that appeared at the top of the graphene nanosheet. Silver and black spheres represent metal and carbon atoms. White sticks covered on them represent liquid water molecules, and blue and green sticks connect pairs of ice molecules with hexagonal and cubic structure orders, respectively.

Therefore, in this work, we draw the conclusion that an ice nucleus tends to form at the clearance of a surface with surface hydrophilicity discrepancy $D_\varepsilon$ larger than 0.87 kcal/mol. On the contrary, an ice nucleus tended to form at the top of the graphene nanosheets when the $D_\varepsilon$ was lower than 0.87 kcal/mol. Either at the clearance or at the top of the graphene nanosheets, an ice nucleus could grow continuously as the system temperature cooled down, as shown in Figure 4b–d or f–h. Nearly all liquid water molecules on the surface arranged in the regular order finally to form the ice structures. In addition, the growth rate of the ice crystal that nucleated initially at the clearance was much lower than that of the ice crystal that nucleated at the top of the graphene sheets, and the lowest growth rate of an ice crystal on the metal–graphene nanosheet surface could be found under the condition of the surface hydrophilicity discrepancy $D_\varepsilon$ being set to 7.87 kcal/mol.

### 3.3. Ice Growth and Stunting Effect of Boundary Misorientation

We also compared the mean growth rate of an ice nucleus under several surface conditions, including metal–graphene nanosheets, pure metal, pure graphene and graphene–graphene nanosheet systems, as shown in Figure 5. Different surface conditions and their models are illustrated in Figure 1b. The ice growth rate was calculated by the method described in Figure S6.

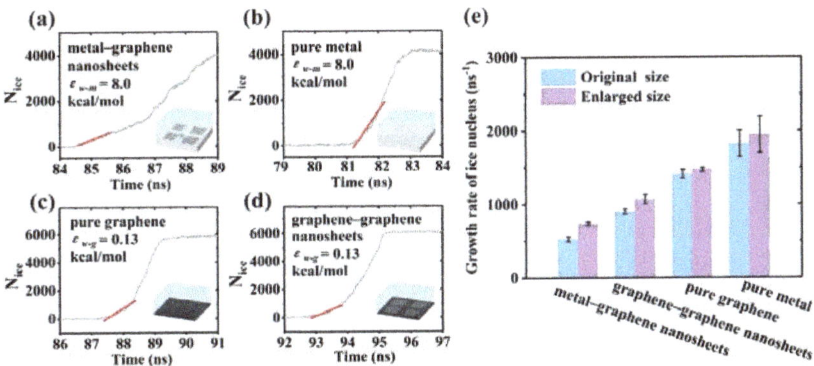

**Figure 5.** Growth rate of an ice nucleus on different surface configurations. (**a**) Number of water molecules in the ice nucleus, $N_{ice}$, changed with simulation time on the metal–graphene nanosheet surface. (**b–d**) $N_{ice}$ on a pure metal, pure graphene and graphene–graphene nanosheet surface, respectively. The data refer to $\varepsilon_{w-m}$ = 8.0 kcal/mol both in (**a,b**). The red line in each figure is the linear fitting of $N_{ice}$ within the first 1 ns after the initiation of icing. (**e**) Comparison of the ice growth rate on different surface configurations. Blue and purple bars represent the original and enlarged size of the simulation box.

The lowest growth rate of an ice crystal was on the metal–graphene nanosheet surface when $\varepsilon_{w-m}$ was set to 8.0 kcal/mol, i.e., 7.87 kcal/mol surface hydrophilicity discrepancy for the clearance configuration. Compared to the pure metal surface, this ice growth rate was decreased by 71.08%, as shown in Figure 5e. Similarly, for the graphene–graphene nanosheet surface, the growth rate of the ice nucleus decreased by 36.07% compared to that on the pure graphene surface. Ice growth was suppressed on both the metal and graphene substrate, while graphene nanosheets were anchored on them. These alternately distributed graphene nanosheets can restrict the growth of an ice nucleus in the direction parallel to the substrate [42]. An ice nucleus at the clearance tends to grow along the z axis, leading to the curved surface of the ice nucleus. According to the Gibbs–Thomson effect [43,62], the surface curvature of ice would cause additional pressure on the ice–water interface, resulting in an external resistance for the further growth of the ice nucleus. [42,63]. Hence, the free energy barrier for ice growth increases, and the freezing point of liquid water is decreased. However, the suppression effect was more prominent on the metal–graphene nanosheet surface configuration, owing to the proper surface hydrophilicity discrepancy and the rearranged interfacial water layer existing at the clearance configuration. Additionally, we found that ice nucleation occurred at a random position of the graphene–graphene nanosheet surface. The ice growth was suppressed only when the ice nucleus was formed at the clearance; otherwise, the ice nucleus grew on the graphene nanosheet without the suppression effect of the clearance. These results indicate that the suppression effect of the clearance generated by graphene nanosheets does occur both on metal and graphene substrates, and the optimal freezing delay ability of the surface can be achieved in the metal–graphene nanosheet surface system with a proper surface hydrophilicity discrepancy. The suppression effect of these surfaces followed the sequence of metal–graphene nanosheets > graphene–graphene nanosheets > pure graphene, as illustrated in Figure 5e.

On the other hand, we also constructed models with an enlarged size of the simulation box, which means an enlarged clearance width on these surfaces, and performed simulations to investigate the impact of clearance width on the growth of the ice nucleus. Details of the models are shown in Table S2. The simulation parameters of the enlarged systems were the same to that of the systems with the original size for each surface configuration. The initiation of ice nucleation on the metal–graphene nanosheet surface was greatly advanced (10–20 ns) by merely a 1 nm increase of the clearance width, and the ice growth rate increased noticeably by 40.02%, as shown in Figure S7 and Figure 5e. In the enlarged metal–graphene nanosheet system, the ice embryo could still nucleate at the enlarged clearance configuration. However, a stable ice nucleus formed before it contacted the lateral graphene nanosheets. Ice growth would not be restricted and the curved surface of an ice nucleus would not form. Consequently, the suppression effect of the clearance configuration disappeared, when the clearance width was larger than 2 nm on the surface both in metal–graphene nanosheet and graphene–graphene nanosheet systems. Therefore, our investigation leads to the conclusion that a pronounced ice inhibition will exhibit on the metal–graphene nanosheet surface configuration only when the clearance between graphene nanosheets is narrow enough. Further research needs to be developed for a critical size of the clearance configuration. In contrast, the enlarged size of the simulation box had no obvious influence on ice growth on the pure metal and pure graphene surfaces. For the graphene–graphene nanosheet surface, ice nucleation occurred at a random position on the surface in both the original and enlarged systems. When the clearance width was enlarged, ice growth would not be restricted by the graphene nanosheets, even if the ice nucleus was formed at the clearance. The growth rate of the ice nucleus on the graphene–graphene nanosheet surface slightly increased by 18.46% due to the enlarged clearance width, as shown in Figure 5e.

Figure 6a suggests that a stunting effect exists during the growth process of an ice crystal on a metal–graphene nanosheet surface. This is due to the boundary misorientation between ice crystals [47], as illustrated in Figure 6b. Obviously, the ice cluster marked with a red circle has a different orientation with the ice cluster on the right side. Generally, two growing ice nuclei contact each other, and the boundary misorientation forms. With the decreasing entropy and potential energy of a system, ice clusters with a boundary misorientation tended to transform into a consistent crystal, and the disorder boundary structure disappeared gradually, which suspended the growth of the ice crystal on the surface about 0.5–1.0 ns. This stunting effect could take a lot of time during the rapid growth of the ice crystal, as shown in Figure 6a. We also found that the duration of the stunting effect extended by increasing the contact area between ice clusters with different orientations, finally slowing down the growth rate of the ice crystal.

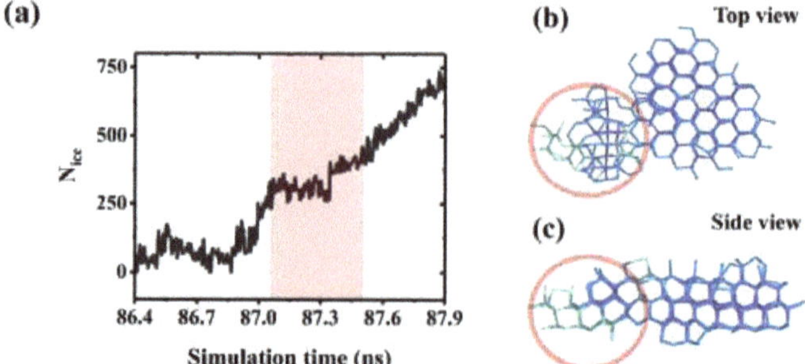



**Figure 6.** Stunting effect of a boundary misorientation during ice growth. (**a**) The inset shows the stunting period (shaded red) in the ice growth process caused by boundary misorientation between ice crystals. (**b**) Top view snapshots of boundary misorientation between ice crystals. (**c**) Side view of the boundary misorientation between ice crystals.

## 4. Conclusions

In summary, we carried out molecular dynamics simulations to investigate the freezing delay capability of metal–graphene nanosheet surfaces and gained a microscopic understanding of ice nucleation and growth, which were restricted at the clearance configuration. The interfacial water layer would form when the surface hydrophilicity of the metal substrate $\varepsilon_{w\text{-}m}$ was higher than 1.0 kcal/mol, separating other free water molecules from interacting with the metal substrate. This hindering effect of interfacial water layer competed with the increasing interaction strength between water molecules and the metal substrate. Ice nucleation was delayed when $\varepsilon_{w\text{-}m}$ ranged from 4.0 to 14.0 kcal/mol and the delayed time peaks to 5.25 ns with the surface hydrophilicity discrepancy $D_\varepsilon$ of 5.87 kcal/mol.

The nucleation position of an ice embryo changed from the top of graphene nanosheets to the clearance since the surface hydrophilicity of the metal substrate $\varepsilon_{w\text{-}m}$ surpassed 1.0 kcal/mol. The growth of the ice nucleus at the clearance was restricted by surrounding graphene nanosheets both on metal and graphene substrates. The ice nucleus grew with the lowest rate when $\varepsilon_{w\text{-}m}$ was 8.0 kcal/mol. An enlarged width of the clearance weakened the suppression effect remarkably, which is an important element in the design of anti-icing nanomaterials. Furthermore, the boundary misorientation between ice crystals can also suppress ice growth because of the stunting effect, which is proportional to the contact area between ice crystals. We believe these findings provide a simple model to explore the mechanism of nano-sized graphene and its derivatives on ice nucleation and growth, as well as their potential application for anti-icing materials.

**Supplementary Materials:** The following supporting information can be downloaded at: https://www.mdpi.com/article/10.3390/coatings12010052/s1, Figure S1: Morphology of the simulation system. (a) A typical metal–graphene nanosheets surface, containing four single graphene nanosheets anchored alternately on a metal substrate. (b) Entire simulation box with a water slab covered on the surface. Atoms of metal substrate and graphene nanosheets are colored by silver and black respectively, hydrogen bonds between liquid water molecules are represented by white sticks with a cut off distance of 3.2 Å, which is realized by the Create Bonds modifier in OVITO software; Figure S2: Example of a simulation box used in ice freezing. Periodic boundary conditions (PBC) were applied in x, y and z directions. A 7-nm-thick void region was incorporated in the simulation box to avoid the undesired effect of metal substrate in the adjacent simulation box caused by the PBC in the z direction; Figure S3: Ice formation on the metal–graphene nanosheets surface. (a) Stacking disordered ice formed on the surface ($\varepsilon_{w\text{-}m}$ = 1.0 kcal/mol), from a cross section view. Random layers of ice $I_h$ and ice $I_c$ were colored in bule and green respectively. (b) Count of water molecules above the surface along z direction, which depicts the regular order of ice; Figure S4: Ice growth at the top the graphene nanosheets with a surface hydrophilicity of metal substrate of 0.13 kcal/mol, which is equal to the surface hydrophilicity of graphene $\varepsilon_{w\text{-}g}$. (a) Ice embryo nucleating at the top the graphene nanosheets owing to the repeated hexagonal carbon ring structure on graphene. (b–d) Growth of the ice crystal on the graphene nanosheets; Figure S5: Ice growth at the clearance, between the peripheral graphene nanosheets, with a surface hydrophilicity of metal substrate of 4.0 kcal/mol. (a) Ice embryo nucleated at the clearance configuration. (b–d) Restricted growth of the ice crystal on the metal–graphene nanosheets surface; Figure S6: Number of water molecules in ice nucleus changed with simulation time on metal–graphene nanosheets surface with different surface hydrophilicity of metal substrate; Figure S7: Initiation time of icing on metal–graphene nanosheets with different size of simulation box. The clearance width is about 2.0 nm in original sized system, whereas it is 3.0 nm in enlarged sized system; Table S1: The size of simulation box; Table S2: The interaction parameters for the Model Systems in this work.

**Author Contributions:** Conceptualization, Y.S.; methodology, Y.X.; validation, B.J., Y.S., J.T., Y.X., H.C., and S.L.; writing—original draft preparation, B.J. and Y.S.; writing—review and editing, J.T. and Y.S.; visualization, W.L. and X.X.; funding acquisition, J.T. and Y.S. All authors have read and agreed to the published version of the manuscript.

**Funding:** This work was supported by the National Natural Science Foundation of China (Nos. 52075246, 12002364, and U1937206), the Natural Science Foundation of Jiangsu Province (No. BK20211568), the Project Funded by China Postdoctoral Science Foundation (No. 2019M661826), the

Open Fund of Key Laboratory of Icing and Anti/De-icing (Nos. IADL20190202, IADL20200407) and the NUAA Innovation Program for Graduate Education (kfjj20200613, kfjj20200605).

**Institutional Review Board Statement:** Not applicable.

**Informed Consent Statement:** Not applicable.

**Data Availability Statement:** The data presented in this study are available within the article.

**Conflicts of Interest:** The authors declare no conflict of interest.

## References


1. Liu, B.; Zhang, K.; Tao, C.; Zhao, Y.; Li, X.; Zhu, K.; Yuan, X. Strategies for anti-icing: Low surface energy or liquid-infused? *RSC Adv.* **2016**, *6*, 70251–70260. [CrossRef]
2. Shen, Y.; Wu, X.; Tao, J.; Zhu, C.; Lai, Y.; Chen, Z. Icephobic materials: Fundamentals, performance evaluation, and applications. *Prog. Mater. Sci.* **2019**, *103*, 509–557. [CrossRef]
3. Brini, E.; Fennell, C.J.; Fernandez-Serra, M.; Hribar-Lee, B.; Luksic, M.; Dill, K.A. How water's properties are encoded in its molecular structure and energies. *Chem. Rev.* **2017**, *117*, 12385–12414.
4. Fitzner, M.; Sosso, G.C.; Cox, S.J.; Michaelides, A. The many faces of heterogeneous ice nucleation: Interplay between surface morphology and hydrophobicity. *J. Am. Chem. Soc.* **2015**, *137*, 13658–13669. [CrossRef]
5. Bayer, I.S. Mechanisms of surface icing and deicing technologies. *Ice Adhes. Mech. Meas. Mitig.* **2020**, 325–359. [CrossRef]
6. Ma, R.; Cao, D.; Zhu, C.; Tian, Y.; Peng, J.; Guo, J.; Chen, J.; Li, X.Z.; Francisco, J.S.; Zeng, X.C.; et al. Atomic imaging of the edge structure and growth of a two-dimensional hexagonal ice. *Nature* **2020**, *577*, 60–63. [CrossRef]
7. Bi, Y.; Cabriolu, R.; Li, T. Heterogeneous Ice Nucleation controlled by the coupling of surface crystallinity and surface hydrophilicity. *J. Phys. Chem. C* **2016**, *120*, 1507–1514. [CrossRef]
8. Jiang, J.; Li, G.X.; Sheng, Q.; Tang, G.H. Microscopic mechanism of ice nucleation: The effects of surface rough structure and wettability. *Appl. Surf. Sci.* **2020**, *510*, 145520. [CrossRef]
9. Lin, Y.; Chen, H.; Wang, G.; Liu, A. Recent progress in preparation and anti-icing applications of superhydrophobic coatings. *Coatings* **2018**, *8*, 208. [CrossRef]
10. Li, C.; Gao, X.; Li, Z. Roles of surface energy and temperature in heterogeneous ice nucleation. *J. Phys. Chem. C* **2017**, *121*, 11552–11559. [CrossRef]
11. Li, C.; Gao, X.; Li, Z. Surface energy-mediated multistep pathways for heterogeneous ice nucleation. *J. Phys. Chem. C* **2018**, *122*, 9474–9479. [CrossRef]
12. Luo, S.; Wang, J.; Li, Z. Homogeneous ice nucleation under shear. *J. Phys. Chem. B* **2020**, *124*, 3701–3708. [CrossRef] [PubMed]
13. Metya, A.K.; Singh, J.K. Ice Nucleation on a graphite surface in the presence of nanoparticles. *J. Phys. Chem. C* **2018**, *122*, 19056–19066. [CrossRef]
14. Matsuno, R.; Kokubo, Y.; Kumagai, S.; Takamatsu, S.; Hashimoto, K.; Takahara, A. Molecular design and characterization of ionic monomers with varying ion pair interaction energies. *Macromolecules* **2020**, *53*, 1629–1637. [CrossRef]
15. Zhuo, Y.; Xiao, S.; Håkonsen, V.; He, J.; Zhang, Z. Anti-icing ionogel surfaces: Inhibiting ice nucleation, growth, and adhesion. *ACS Mater. Lett.* **2020**, *2*, 616–623. [CrossRef]
16. Yue, X.; Liu, W.; Wang, Y. Effects of black silicon surface structures on wetting behaviors, single water droplet icing and frosting under natural convection conditions. *Surf. Coat. Technol.* **2016**, *307*, 278–286. [CrossRef]
17. Wang, Y.; Wang, Z.-g. Sessile droplet freezing on polished and micro-micro-hierarchical silicon surfaces. *Appl. Therm. Eng.* **2018**, *137*, 66–73. [CrossRef]
18. He, Z.Y.; Xie, W.J.; Liu, Z.Q.; Liu, G.M.; Wang, Z.W.; Gao, Y.Q.; Wang, J.J. Tuning ice nucleation with counterions on polyelectrolyte brush surfaces. *Sci. Adv.* **2016**, *2*, e1600345. [CrossRef]
19. Kyrkjebø, S.; Cassidy, A.; Akhtar, N.; Balog, R.; Scheffler, M.; Hornekær, L.; Holst, B.; Flatabø, R. Graphene and graphene oxide on Ir(111) are transparent to wetting but not to icing. *Carbon* **2021**, *174*, 396–403. [CrossRef]
20. Joghataei, M.; Ostovari, F.; Atabakhsh, S.; Tobeiha, N. Heterogeneous ice nucleation by graphene nanoparticles. *Sci. Rep.* **2020**, *10*, 9723. [CrossRef]
21. Xue, H.; Lu, Y.; Geng, H.; Dong, B.; Wu, S.; Fan, Q.; Zhang, Z.; Li, X.; Zhou, X.; Wang, J. hydroxyl groups on the graphene surfaces facilitate ice nucleation. *J. Phys. Chem. Lett.* **2019**, *10*, 2458–2462. [CrossRef]
22. Wang, Y.-R.; Xu, J.-Y.; Ma, C.-K.; Shi, M.-X.; Tu, Y.-B.; Sun, K.; Meng, S.; Wang, J.-Z. Ice II-like monolayer ice grown on graphite surface. *J. Phys. Chem. C* **2019**, *123*, 20297–20303. [CrossRef]
23. Nasser, J.; Lin, J.; Zhang, L.; Sodano, H.A. Laser induced graphene printing of spatially controlled super-hydrophobic/hydrophilic surfaces. *Carbon* **2020**, *162*, 570–578. [CrossRef]
24. Peng, W.J.; Li, H.Q.; Liu, Y.Y.; Song, S.X. A review on heavy metal ions adsorption from water by graphene oxide and its composites. *J. Mol. Liq.* **2017**, *230*, 496–504. [CrossRef]
25. Babaahmadi, V.; Montazer, M.; Gao, W. Low temperature welding of graphene on PET with silver nanoparticles producing higher durable electro-conductive fabric. *Carbon* **2017**, *118*, 443–451. [CrossRef]

26. Naumis, G.G.; Barraza-Lopez, S.; Oliva-Leyva, M.; Terrones, H. Electronic and optical properties of strained graphene and other strained 2D materials: A review. *Rep. Prog. Phys.* **2017**, *80*, 1–62. [CrossRef] [PubMed]
27. Shao, Y.Y.; Wang, J.; Wu, H.; Liu, J.; Aksay, I.A.; Lin, Y.H. Graphene based electrochemical sensors and biosensors: A review. *Electroanalysis* **2010**, *22*, 1027–1036. [CrossRef]
28. Akinwande, D.; Brennan, C.J.; Bunch, J.S.; Egberts, P.; Felts, J.R.; Gao, H.J.; Huang, R.; Kim, J.S.; Li, T.; Li, Y.; et al. A review on mechanics and mechanical properties of 2D materials-Graphene and beyond. *Extrem. Mech. Lett.* **2017**, *13*, 42–77. [CrossRef]
29. Papageorgiou, D.G.; Kinloch, I.A.; Young, R.J. Mechanical properties of graphene and graphene-based nanocomposites. *Prog. Mater. Sci.* **2017**, *90*, 75–127. [CrossRef]
30. Wu, J.B.; Lin, M.L.; Cong, X.; Liu, H.N.; Tan, P.H. Raman spectroscopy of graphene-based materials and its applications in related devices. *Chem. Soc. Rev.* **2018**, *47*, 1822–1873. [CrossRef] [PubMed]
31. Lupi, L.; Hudait, A.; Molinero, V. Heterogeneous nucleation of ice on carbon surfaces. *J. Am. Chem. Soc.* **2014**, *136*, 3156–3164. [CrossRef] [PubMed]
32. Lupi, L.; Kastelowitz, N.; Molinero, V. Vapor deposition of water on graphitic surfaces: Formation of amorphous ice, bilayer ice, ice I, and liquid water. *J. Chem. Phys.* **2014**, *141*, 18C508. [CrossRef] [PubMed]
33. Carrasco, J.; Hodgson, A.; Michaelides, A. A molecular perspective of water at metal interfaces. *Nat. Mater.* **2012**, *11*, 667–674. [CrossRef] [PubMed]
34. Zokaie, M.; Foroutan, M. Confinement effects of graphene oxide nanosheets on liquid–solid phase transition of water. *RSC Adv.* **2015**, *5*, 97446–97457. [CrossRef]
35. Redondo, O.; Prolongo, S.G.; Campo, M.; Sbarufatti, C.; Giglio, M. Anti-icing and de-icing coatings based Joule's heating of graphene nanoplatelets. *Compos. Sci. Technol.* **2018**, *164*, 65–73. [CrossRef]
36. Vertuccio, L.; De Santis, F.; Pantani, R.; Lafdi, K.; Guadagno, L. Effective de-icing skin using graphene-based flexible heater. *Compos. Part B Eng.* **2019**, *162*, 600–610. [CrossRef]
37. Karim, N.; Zhang, M.; Afroj, S.; Koncherry, V.; Potluri, P.; Novoselov, K.S. Graphene-based surface heater for de-icing applications. *RSC Adv.* **2018**, *8*, 16815–16823. [CrossRef]
38. Chen, L.; Zhang, Y.; Wu, Q. Effect of graphene coating on the heat transfer performance of a composite anti-/deicing component. *Coatings* **2017**, *7*, 158. [CrossRef]
39. Liu, K.; Wang, C.; Ma, J.; Shi, G.; Yao, X.; Fang, H.; Song, Y.; Wang, J. Janus effect of antifreeze proteins on ice nucleation. *Proc. Natl. Acad. Sci. USA* **2016**, *113*, 14739–14744. [CrossRef]
40. He, Z.; Liu, K.; Wang, J. Bioinspired materials for controlling ice nucleation, growth, and recrystallization. *Acc. Chem. Res.* **2018**, *51*, 1082–1091. [CrossRef]
41. Hudait, A.; Odendahl, N.; Qiu, Y.; Paesani, F.; Molinero, V. Ice-nucleating and antifreeze proteins recognize ice through a diversity of anchored clathrate and ice-like motifs. *J. Am. Chem. Soc.* **2018**, *140*, 4905–4912. [CrossRef]
42. Geng, H.; Liu, X.; Shi, G.; Bai, G.; Ma, J.; Chen, J.; Wu, Z.; Song, Y.; Fang, H.; Wang, J. Graphene oxide restricts growth and recrystallization of ice crystals. *Angew Chem. Int. Ed. Engl.* **2017**, *56*, 997–1001. [CrossRef]
43. Knight, C.A. Structural biology—Adding to the antifreeze agenda. *Nature* **2000**, *406*, 249–251. [CrossRef]
44. Bai, G.; Gao, D.; Liu, Z.; Zhou, X.; Wang, J. Probing the critical nucleus size for ice formation with graphene oxide nanosheets. *Nature* **2019**, *576*, 437–441. [CrossRef]
45. Zhang, X.-X.; Chen, M. Icephobicity of functionalized graphene surfaces. *J. Nanomater.* **2016**, *2016*, e1600345. [CrossRef]
46. Akhtar, N.; Anemone, G.; Farias, D.; Holst, B. Fluorinated graphene provides long lasting ice inhibition in high humidity. *Carbon* **2019**, *141*, 451–456. [CrossRef]
47. Ribeiro, I.d.A.; Koning, M.d. Grain-boundary sliding in ice ih: Tribology and rheology at the nanoscale. *J. Phys. Chem. C* **2021**, *125*, 627–634. [CrossRef]
48. Liu, X.; Geng, H.; Sheng, N.; Wang, J.; Shi, G. Suppressing ice growth by integrating the dual characteristics of antifreeze proteins into biomimetic two-dimensional graphene derivatives. *J. Mater. Chem. A* **2020**, *8*, 23555–23562. [CrossRef]
49. Molinero, V.; Moore, E.B. Water modeled as an intermediate element between carbon and silicon. *J. Phys. Chem. B* **2009**, *113*, 4008–4016. [CrossRef] [PubMed]
50. Malkin, T.L.; Murray, B.J.; Salzmann, C.G.; Molinero, V.; Pickering, S.J.; Whale, T.F. Stacking disorder in ice I. *Phys. Chem. Chem. Phys.* **2015**, *17*, 60–76. [CrossRef]
51. Huang, H.S.; Roy, A.K.; Varshney, V.; Wohlwend, J.L.; Putnam, S.A. Temperature dependence of thermal conductance between aluminum and water. *Int. J. Therm. Sci.* **2012**, *59*, 17–20. [CrossRef]
52. Maras, E.; Trushin, O.; Stukowski, A.; Ala-Nissila, T.; Jónsson, H. Global transition path search for dislocation formation in Ge on Si(001). *Comput. Phys. Commun.* **2016**, *205*, 13–21. [CrossRef]
53. Metya, A.K.; Singh, J.K.; Muller-Plathe, F. Ice nucleation on nanotextured surfaces: The influence of surface fraction, pillar height and wetting states. *Phys. Chem. Chem. Phys.* **2016**, *18*, 26796–26806. [CrossRef] [PubMed]
54. Qiu, H.; Guo, W. phase diagram of nanoscale water on solid surfaces with various wettabilities. *J. Phys. Chem. Lett.* **2019**, *10*, 6316–6323. [CrossRef]
55. Kreder, M.J.; Alvarenga, J.; Kim, P.; Aizenberg, J. Design of anti-icing surfaces: Smooth, textured or slippery? *Nat. Rev. Mater.* **2016**, *1*, 15003. [CrossRef]

56. Guo, P.; Zheng, Y.M.; Wen, M.X.; Song, C.; Lin, Y.C.; Jiang, L. Icephobic/anti-icing properties of micro/nanostructured surfaces. *Adv. Mater.* **2012**, *24*, 2642–2648. [CrossRef] [PubMed]
57. Liu, J.; Zhu, C.; Liu, K.; Jiang, Y.; Song, Y.; Francisco, J.S.; Zeng, X.C.; Wang, J. Distinct ice patterns on solid surfaces with various wettabilities. *Proc. Natl. Acad. Sci. USA* **2017**, *114*, 11285–11290. [CrossRef]
58. Cox, S.J.; Kathmann, S.M.; Slater, B.; Michaelides, A. Molecular simulations of heterogeneous ice nucleation. II. Peeling back the layers. *J. Chem. Phys.* **2015**, *142*, 184705. [PubMed]
59. Matsumoto, M.; Saito, S.; Ohmine, I. Molecular dynamics simulation of the ice nucleation and growth process leading to water freezing. *Nature* **2002**, *416*, 409–413. [CrossRef]
60. Moore, E.B.; Molinero, V. Structural transformation in supercooled water controls the crystallization rate of ice. *Nature* **2011**, *479*, 506–508. [CrossRef]
61. Lu, N.; Yin, D.; Li, Z.; Yang, J. Structure of graphene oxide: Thermodynamics versus kinetics. *J. Phys. Chem. C* **2011**, *115*, 11991–11995. [CrossRef]
62. Perez, M. Gibbs-Thomson effects in phase transformations. *Scr. Mater.* **2005**, *52*, 709–712. [CrossRef]
63. Wu, T.; Chen, Y.Z. Analytical studies of Gibbs-Thomson effect on the diffusion controlled spherical phase growth in a subcooled medium. *Heat Mass Transf.* **2003**, *39*, 665–674. [CrossRef]

*coatings*

*Article*

# Study of 3-D Prandtl Nanofluid Flow over a Convectively Heated Sheet: A Stochastic Intelligent Technique

Muhammad Shoaib [1], Ghania Zubair [1], Muhammad Asif Zahoor Raja [2,*], Kottakkaran Sooppy Nisar [3,*], Abdel-Haleem Abdel-Aty [4,5] and I. S. Yahia [6,7,8]

[1]  Department of Mathematics, COMSATS University Islamabad, Attock Campus, Attock 43600, Pakistan; dr.shoaib@cuiatk.edu.pk (M.S.); sp20-rmt-002@cuiatk.edu.pk (G.Z.)
[2]  Future Technology Research Center, National Yunlin University of Science and Technology, 123 University Road, Section 3, Douliou 64002, Taiwan
[3]  Department of Mathematics, College of Arts and Science, Prince Sattam bin Abdulaziz University, Wadi Aldawaser 11991, Saudi Arabia
[4]  Department of Physics, College of Sciences, University of Bisha, P.O. Box 344, Bisha 61922, Saudi Arabia; amabdelaty@ub.edu.sa
[5]  Physics Department, Faculty of Science, Al-Azhar University, Assiut 71524, Egypt
[6]  Laboratory of Nano-Smart Materials for Science and Technology (LNSMST), Department of Physics, Faculty of Science, King Khalid University, P.O. Box 9004, Abha 61413, Saudi Arabia; isyahia@gmail.com
[7]  Research Center for Advanced Materials Science (RCAMS), King Khalid University, P.O. Box 9004, Abha 61413, Saudi Arabia
[8]  Nanoscience Laboratory for Environmental and Biomedical Applications (NLEBA), Semiconductor Lab., Department of Physics, Faculty of Education, Ain Shams University, Roxy, Cairo 11757, Egypt
*  Correspondence: rajamaz@yuntech.edu.tw (M.A.Z.R.); n.sooppy@psau.edu.sa or ksnisar1@gmail.com (K.S.N.)

**Citation:** Shoaib, M.; Zubair, G.; Raja, M.A.Z.; Nisar, K.S.; Abdel-Aty, A.-H.; Yahia, I.S. Study of 3-D Prandtl Nanofluid Flow over a Convectively Heated Sheet: A Stochastic Intelligent Technique. *Coatings* 2022, 12, 24. https://doi.org/10.3390/coatings12010024

Academic Editors: Eduardo Guzmán and Rahmat Ellahi

Received: 8 October 2021
Accepted: 1 December 2021
Published: 28 December 2021

**Publisher's Note:** MDPI stays neutral with regard to jurisdictional claims in published maps and institutional affiliations.

**Abstract:** In this article, we examine the three-dimensional Prandtl nanofluid flow model (TD-PNFM) by utilizing the technique of Levenberg Marquardt with backpropagated artificial neural network (TLM-BANN). The flow is generated by stretched sheet. The electro conductive Prandtl nanofluid is taken through magnetic field. The PDEs representing the TD-PNFM are converted to system of ordinary differential equations, then the obtained ODEs are solved through Adam numerical solver to compute the reference dataset with the variations of Prandtl fluid number, flexible number, ratio parameter, Prandtl number, Biot number and thermophoresis number. The correctness and the validation of the proposed TD-PNFM are examined by training, testing and validation process of TLM-BANN. Regression analysis, error histogram and results of mean square error (MSE), validates the performance analysis of designed TLM-BANN. The performance is ranges $10^{-10}$, $10^{-10}$, $10^{-10}$, $10^{-11}$, $10^{-10}$ and $10^{-10}$ with epochs 204, 192, 143, 20, 183 and 176, as depicted through mean square error. Temperature profile decreases whenever there is an increase in Prandtl fluid number, flexible number, ratio parameter and Prandtl number, but temperature profile shows an increasing behavior with the increase in Biot number and thermophoresis number. The absolute error values by varying the parameters for temperature profile are $10^{-8}$ to $10^{-3}$, $10^{-8}$ to $10^{-3}$, $10^{-7}$ to $10^{-3}$, $10^{-7}$ to $10^{-3}$, $10^{-7}$ to $10^{-4}$ and $10^{-8}$ to $10^{-3}$. Similarly, the increase in Prandtl fluid number, flexible number and ratio parameter leads to a decrease in the concentration profile, whereas the increase in thermophoresis parameter increases the concentration distribution. The absolute error values by varying the parameters for concentration profile are $10^{-8}$ to $10^{-3}$, $10^{-7}$ to $10^{-3}$, $10^{-7}$ to $10^{-3}$ and $10^{-8}$ to $10^{-3}$. Velocity distribution shows an increasing trend for the upsurge in the values of Prandtl fluid parameter and flexible parameter. Skin friction coefficient declines for the increase in Hartmann number and ratio parameter Nusselt number falls for the rising values of thermophoresis parameter against $N_b$.

**Keywords:** Prandtl nanofluid flow; convectively heated surface; stochastic intelligent technique; Levenberg Marquardt method; backpropagated network; artificial neural network; Adam numerical solver

## 1. Introduction

Nanofluid is a mixture of base liquid and nanosize particles, and the size of these nanoparticles is between 1 to 100 nanometer. There are two types of fluids; first one is Newtonian fluid and the second is non-Newtonian fluid. Nanofluids consist of rods, fibers and nanometer sized particles suspended in base fluids [1]. The base fluid is usually water, ethylene glycol or oil. There are some articles published, that investigate the progress of nanofluids [2–4]. Some researchers study the applications of nanofluids in heat exchange [5], car radiators [6], medical appliances [7] and solar collectors [8].

Heat generation is a phenomenon of generation, conversion, use and exchange of thermal energy between the two objects. Conduction, convection and Radiation are the different modes of heat transfer. Heat transfer of nanofluids is also examined by research workers [9–11]. Three-dimensional flows mean that the flow is describing in three space coordinates. Any physical flow is three dimensional. Three-dimensional flow over a stretched sheet was studied by Wang [12], Ariel [13], Xu et al. [14] and Liu et al. [15].

The main purpose of this article is to study the three-dimensional flow of Prandtl nanofluid containing nanoparticles. Hayat [16,17] examined the peristaltic flow of Prandtl nanoliquid and Kumar [18] examined the impact of mass transfer in Prandtl liquid flow. Hayat [19] also analyzed the Prandtl liquid flow with Cattaneo-Christov double diffusion. Nadeem et al. [20] studied the Prandtl liquid model in an endoscope. Some researchers have carried out tremendous work using the Prandtl nanofluid model, i.e., Akbar [21,22], Sooppy Nisar [23] and Hamid [24], examined the flow of Prandtl fluid flow in their research models. Over a deformable surface, Soomro et al. [25] examined the passive control of nanoparticles in Prandtl nanofluid flow. Nilankush [26,27] examined the Spectral quasi linearization simulation of radiative nanofluidic transport over a permeable inclined disk and a bended surface in his two research articles. Sabu [28] explored the role of nanoparticle form and thermo-hydrodynamic slip limitations in Magnetohydrodynamic alumina-water nanofluid flows across a rotating hot surface. Virmani [29] reviewed the nanostructured materials for exterior panel elements in automotive.

There are many models examined by the scientist over a convectively heated surface. Uddin [30] examined the mixed convective Prandtl-Eyring flow over a surface. Zaka Ullah [31] and Patil [32] demonstrated the flow of Prandtl fluid over a convectively heated surface. Over a convectively heated sheet, Hosseinzadeh [33] analyzed the flow of Maxwell liquid. Ahmed [34] explore the chemically reacting fluid flow through a convectively heated sheet. Alamri et al. [35] investigated the novel viewpoint of Cattaneo-Christove heat flux model. Yausif et al. [36] implemented the numerical technique for heat transfer analysis subjected to fluid flow system under the impacts of thermal radiation and internal heat source/sink. Ellahi et al. [37,38] analyzed the heat transfer impacts on bi-phase flow coatings. The authors also studied the entropy optimized fluid flow system under the influence of heat transfer and magneto hydrodynamics. Moreover, heat transfer generation, heat transfer consumption and thermal radiation for non-Newtonian fluid flow are studies by Saeed et al. [39].

The three-dimensional flow of Prandtl nanofluid is examined by using different numerical and analytical method, but stochastic numerical methods are well known due to their effectiveness, robustness and worth. Research workers already applied stochastic numerical technique on their research problems [40–44]. Some research models examined by implementing the artificial intelligence techniques are Ree-Eyring fluid model [45], third-grade fluid model [46], MHD boundary layer flow model [47] and Maxwell nanofluid model [48].

To analyze the Prandtl fluid models based on differential equations, many scholars employed various numerical simulations. However, no one has applied the solution method which is based on the Levenberg-Marquardt approach in artificial neural networks to improve the solver technique's computing power and precision level. Due to their usefulness, efficiency, and reliability, stochastic numerical approaches are effective and reliable to investigate the Prandtl fluid flow related problems. All of these motivating factors encourage authors to use a precise and consistent AI algorithm-based mathematical simulation framework for mathematical solution of Prandtl fluid flow over a convectively heated surface by performing numerical and graphical research to analyze the influence of all variations on velocity, concentration, and temperature distributions, which is the novelty of this study.

In this article, Mathematica (version 12) and MATLAB (version R2019b) software are used for numerical treatment.

The innovative contributions of computing procedure are as follows:

- The numerical computation has been designed through the technique of Levenberg Marquardt with backpropagated artificial neural network (TLM-BANN) for the comparative study of three dimensional Prandtl nanofluid flow model (TD-PNFM) with convectively heated surface.
- The TLM-BANN coupled PDEs representing TD-PNFM are transformed into system of ODEs by utilizing suitable transformation.
- The Mathematica software command 'NDSolve' is used to compute the dataset for designed TLM-BANN for the variation of Prandtl fluid number, flexible number, ratio parameter, Prandtl number, Biot number and thermophoresis number.
- The MATLAB software is used to interpret the solution and the AE analysis plots of TD-PNFM.
- The correctness and the validation of the proposed TD-PNFM is examined by training, testing and validation process of TLM-BANN.
- Regression analysis, error histogram and results of mean square error (MSE), validates the performance analysis of designed TLM-BANN.

## 2. Mathematical Modeling

We examine the 3D flow of Prandtl nanofluid with a convectively heated surface. Thermal convection and zero nanoparticles mass flux are the two boundary conditions discussed in this model. Thermophoretic and Brownian motion impacts are also examined. The fluid is electro conductive with the magnetic field applied in the direction of $z$-axis, shown by the geometrical interpretation represented through Figure 1. The figure shows that the sheet is extended in $x$ and $y$- direction, and the fluid is moving at that extended sheet. The model is discussed in Cartesian coordinates system. The velocity $U_w$ is along the $x$-axis and the velocity $V_w$ is along the $y$-axis at $z = 0$. Along $z$-axis, magnetic field $B_0$ is applied. The current hall and magnetic field impacts are ignored for a small value of Reynold number. Then the resulting ODEs along with the boundary conditions after solving the system of PDEs [49] are:

$$\beta_1 f''' + (f+g)f'' - f'^2 + \beta_2 f''^2 f''' - (Ha)^2 f' = 0, \tag{1}$$

$$\beta_1 g''' + (f+g)g'' - g'^2 + \beta_2 g''^2 g''' - (Ha)^2 g' = 0, \tag{2}$$

$$\theta'' + \Pr\left((f+g)\theta' + N_b\theta'\phi' + N_t\theta'^2\right) = 0, \tag{3}$$

$$\phi'' + Sc(f+g)\phi' + \frac{N_t}{N_b}\theta'' = 0, \tag{4}$$

And the BCs are;

$$\begin{aligned}
f(0) &= g(0) = 0, f'(0) = 1, g'(0) = \alpha, \\
\theta'(0) &= -\gamma(1 - \theta(0)), N_b\phi'(0) + N_t\theta'(0) = 0,
\end{aligned} \tag{5}$$

$$f'(\infty) \to 0, g'(\infty) \to 0, \theta(\infty) \to 0, \phi(\infty) \to 0, \tag{6}$$

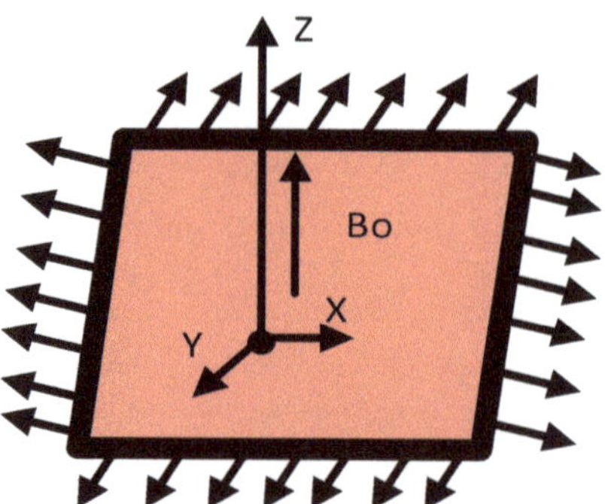



**Figure 1.** Flow Diagram.

### 3. Solution Methodology

The TLM-BANN coupled with PDEs representing TD-PNFM are converted into system of ODEs by applying suitable transformation. Adam numerical method is used to compute the reference dataset for all the six scenarios of TD-PNFM. The MATLAB command 'nftool' is used to execute the technique of Levenberg Marquardt with backpropagated artificial neural network (TLM-BANN) for the study of TD-PNFM.

The neural network figure for LMT-BANN is given below as Figure 2 and the flow chart is shown in Figure 3.

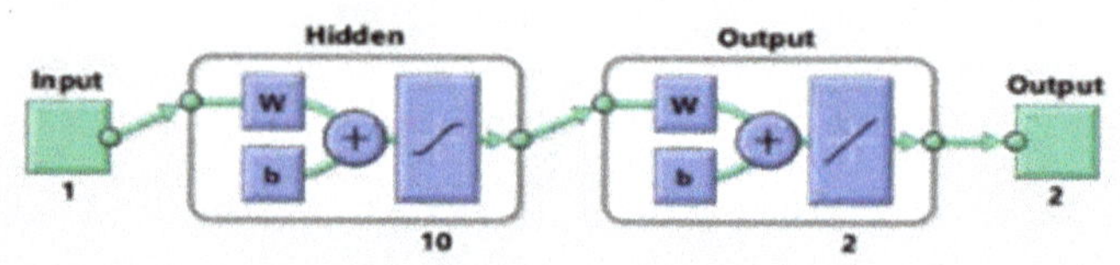



**Figure 2.** Neural Network diagram for TD-PNFM.

The current model is discussed for six scenarios and the scenarios consist of the variations of Prandtl fluid number, flexible number, ratio parameter, Prandtl number, Biot number and thermophoresis parameter. Each scenario has further four cases and the values for all the cases are written below in Table 1. With $Ha = 0.4$, $N_b = 0.9$ and $Sc = 1.0$. We can easily see the impact of variations of physical parameters on temperature profile and concentration profile. There are 10 hidden neurons with the input lies between 0 and 8 and the step size is 0.08. The dataset is computed for 101 points in which 81 points are for training, 10 points for testing and 10 points for validation.

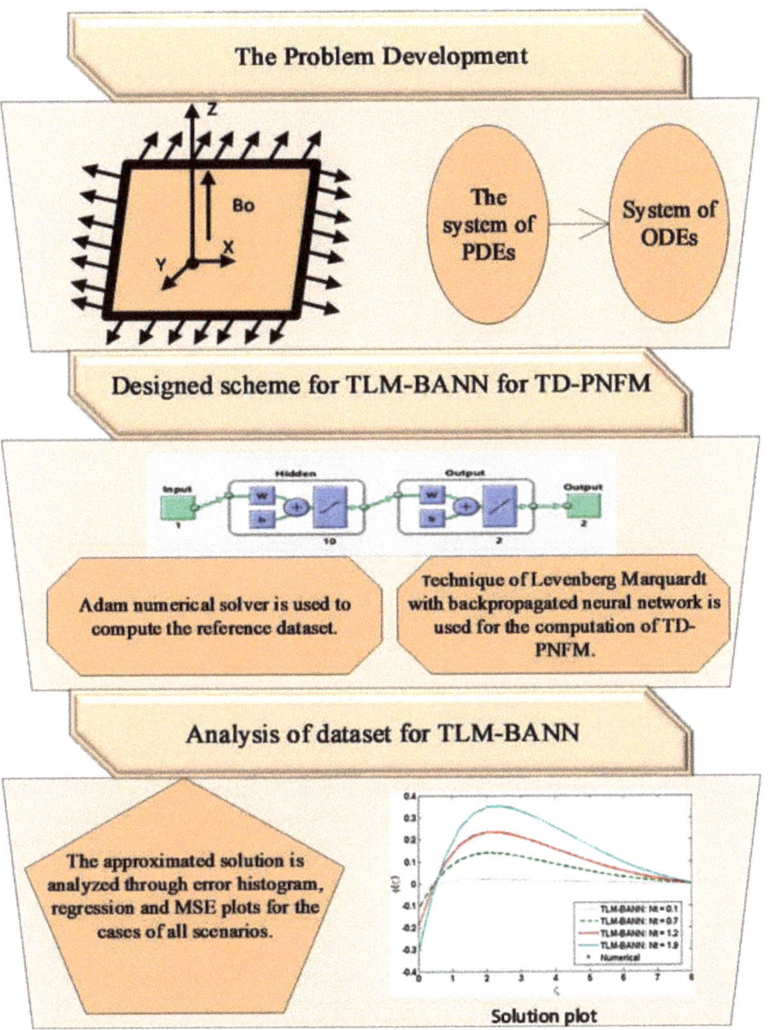



**Figure 3.** Flow chart for TD-PNFM.

**Table 1.** Variation of parameters of TD-PNFM.

| Scenarios | Cases | Physical Quantities | | | | | |
| --- | --- | --- | --- | --- | --- | --- | --- |
| | | $\beta_1$ | $\beta_2$ | $\alpha$ | $\gamma$ | Pr | $N_t$ |
| 01 | 1 | 0.2 | 0.6 | 0.2 | 0.8 | 1.1 | 0.8 |
| | 2 | 0.4 | 0.6 | 0.2 | 0.8 | 1.1 | 0.8 |
| | 3 | 0.6 | 0.6 | 0.2 | 0.8 | 1.1 | 0.8 |
| | 4 | 0.8 | 0.6 | 0.2 | 0.8 | 1.1 | 0.8 |
| 02 | 1 | 0.2 | 0.0 | 0.2 | 0.8 | 1.1 | 0.8 |
| | 2 | 0.2 | 0.4 | 0.2 | 0.8 | 1.1 | 0.8 |
| | 3 | 0.2 | 0.8 | 0.2 | 0.8 | 1.1 | 0.8 |
| | 4 | 0.2 | 1.2 | 0.2 | 0.8 | 1.1 | 0.8 |
| 03 | 1 | 0.2 | 0.6 | 0.0 | 0.8 | 1.1 | 0.8 |
| | 2 | 0.2 | 0.6 | 0.15 | 0.8 | 1.1 | 0.8 |
| | 3 | 0.2 | 0.6 | 0.30 | 0.8 | 1.1 | 0.8 |
| | 4 | 0.2 | 0.6 | 0.45 | 0.8 | 1.1 | 0.8 |
| 04 | 1 | 0.2 | 0.6 | 0.2 | 0.1 | 1.1 | 0.8 |
| | 2 | 0.2 | 0.6 | 0.2 | 0.5 | 1.1 | 0.8 |
| | 3 | 0.2 | 0.6 | 0.2 | 1.0 | 1.1 | 0.8 |
| | 4 | 0.2 | 0.6 | 0.2 | 1.5 | 1.1 | 0.8 |
| 05 | 1 | 0.2 | 0.6 | 0.2 | 0.8 | 0.75 | 0.8 |
| | 2 | 0.2 | 0.6 | 0.2 | 0.8 | 1.0 | 0.8 |
| | 3 | 0.2 | 0.6 | 0.2 | 0.8 | 1.25 | 0.8 |
| | 4 | 0.2 | 0.6 | 0.2 | 0.8 | 1.50 | 0.8 |
| 06 | 1 | 0.2 | 0.6 | 0.2 | 0.8 | 1.1 | 0.1 |
| | 2 | 0.2 | 0.6 | 0.2 | 0.8 | 1.1 | 0.7 |
| | 3 | 0.2 | 0.6 | 0.2 | 0.8 | 1.1 | 1.2 |
| | 4 | 0.2 | 0.6 | 0.2 | 0.8 | 1.1 | 1.9 |

## 4. Discussion of Results

The six scenarios of TD-PNFM by the variation of Prandtl fluid number, flexible number, ratio parameter, Prandtl number, Biot number and thermophoresis parameter are formulated for four different cases for both, temperature and concentration profile of three-dimensional Prandtl nanofluid flow model as elaborated in Table 1.

The Prandtl number described as the ratio of momentum to thermal diffusivity. As the Prandtl number increases, the temperature decreases. As $Pr$ increases, thermal diffusivity reduces, resulting in a drop in temperature. The Biot number represents the ratio of inner to outside thermal resistance for a solid item that transfers heat to a liquid flow. This ratio indicates if the inside temperature of a body fluctuate considerably in space when it heated or cools over time due to a thermal gradient given to its sheet. A rise in the Biot number induces intense convection, resulting in an increase in thermal profile. Thermophoresis is a force caused by the difference of temperature among the cold wall and hot gas that causes particulate particles to migrate more toward the cold wall. This variable is caused by nanoparticles. Nanoparticles increased the nanoliquids thermal conductivity. The thermal conductivity of nanoliquids enhances with temperature. As a result, an increase in temperature is noticed for a better estimation of $N_t$.

The dataset for the designed TLM-BANN is computed with the help of Adam numerical method, for which, the input lies between 0 and 8 with 0.08 step size for all the four cases of six different scenarios of TLM-BANN of TD-PNFM. The solution for three-dimensional Prandtl nanofluid flow is determined by using the command 'nftool' in MATLAB. The dataset is generated for 101 points in which 81 points are for training, 10 points for testing and 10 points for validation of designed TLM-BANN.

The performance, state transition, fitness, error histogram and regression for the case I of the scenarios I, II, III, IV, V and VI are shown in Figures 4–9, respectively. The MSE based performance reflects the difference between observation and simulation; the lower the MSE value, the higher the performance. The performance is $10^{-10}$, $10^{-10}$, $10^{-10}$, $10^{-11}$, $10^{-10}$ and $10^{-10}$ with epochs 204, 192, 143, 20, 183 and 176, as depicted in the performance graphs. The histogram displays the technique's dependability by displaying the difference between the anticipated and targeted results after neural network training. A histogram plot has twenty vertical bars, which are referred to as bins. The data is distributed evenly between positive and negative components is shown by the zero line, which is the red vertical line in the histogram plot. The regression plots demonstrate the output and the target relationship. Regression R measures that how accurately measured values fit a straight line or curve, and the straight line reflecting the perfect fit. If R = 1, the outputs and the targets relation is accurate. The fitness graph depicts the relation of training, testing, and validation of output and training, where overlap of all three indicating that algorithm is properly trained and provides an accurate answer. Inside the training state, there is a portion of error plot that represents the error related to output. Moreover, the data for MSE, performance, epochs, Mu and time taken is given in Table 2. The Mu and gradient values corresponding to epoch reflect whether the convergence is gradual or rapid; as the epoch advances, the values of gradient and Mu fall, indicating the convergence is quick using the Levenberg-Marquardt solver.

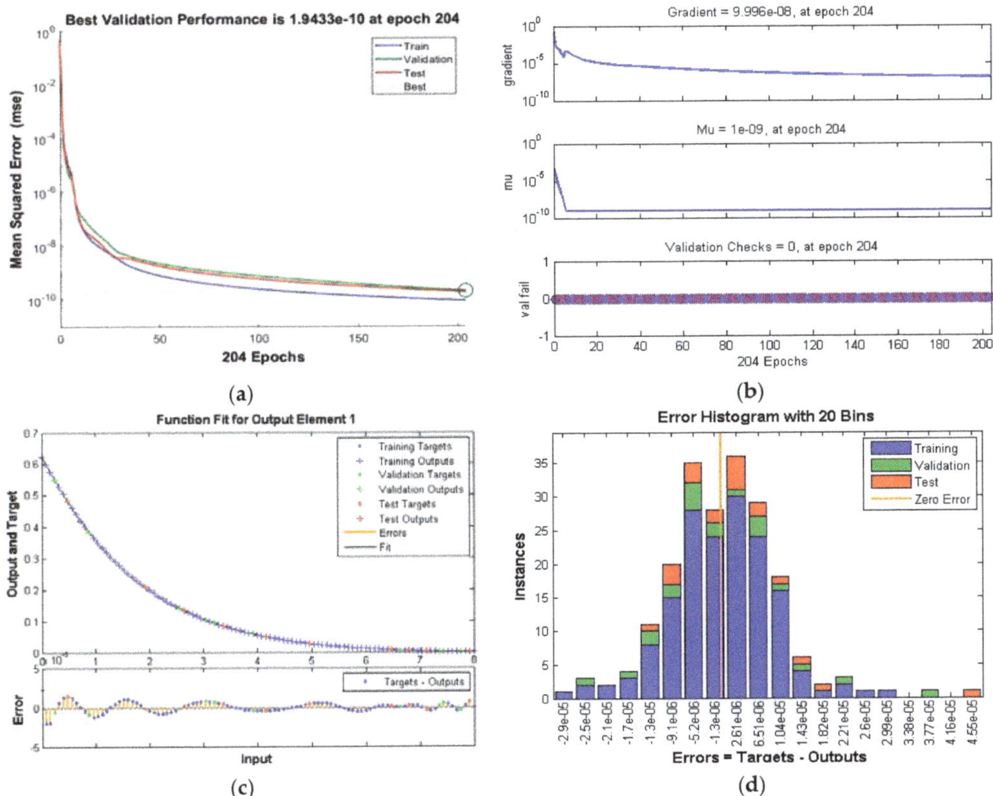

**Figure 4.** *Cont.*

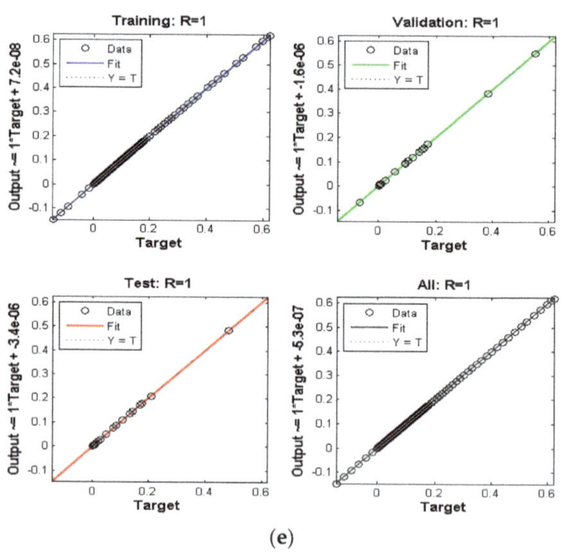



(**e**)

**Figure 4.** Scenario I Case I of TD-PNFM. (**a**) MSE Results: Scenario I Case I; (**b**) Transition state: Scenario I Case I; (**c**) Fitness: Scenario I Case I; (**d**) Error Histogram: Scenario I Case I; (**e**)regression: Scenario I Case I.

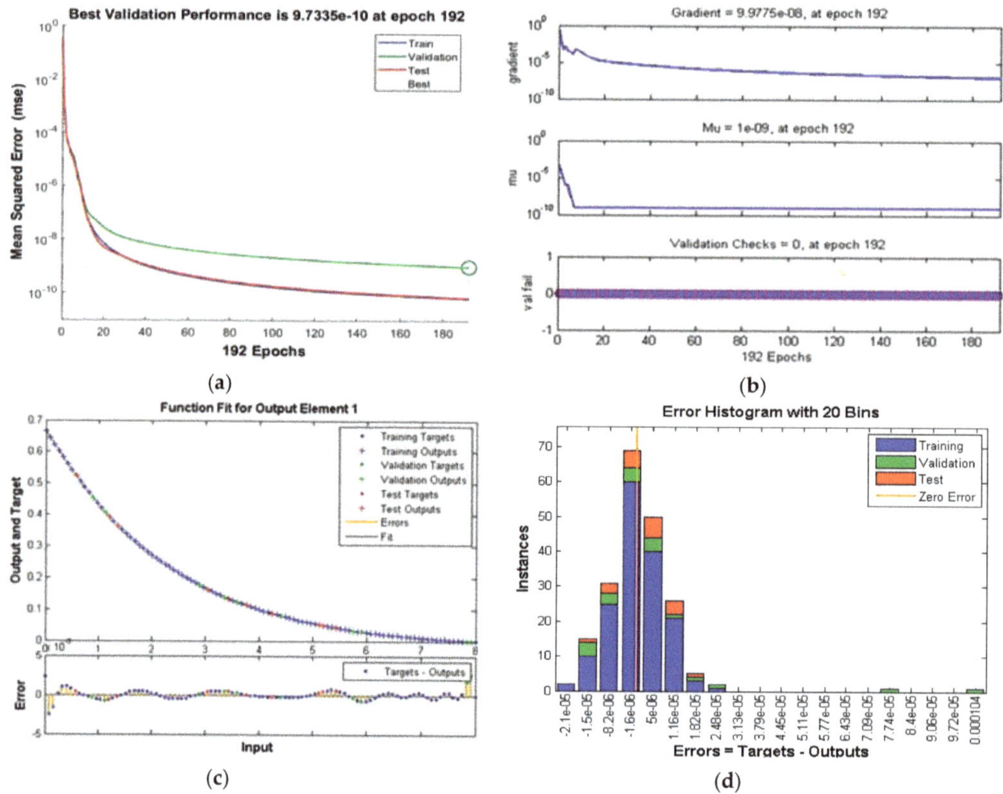



(**a**)  (**b**)  (**c**)  (**d**)

**Figure 5.** *Cont.*

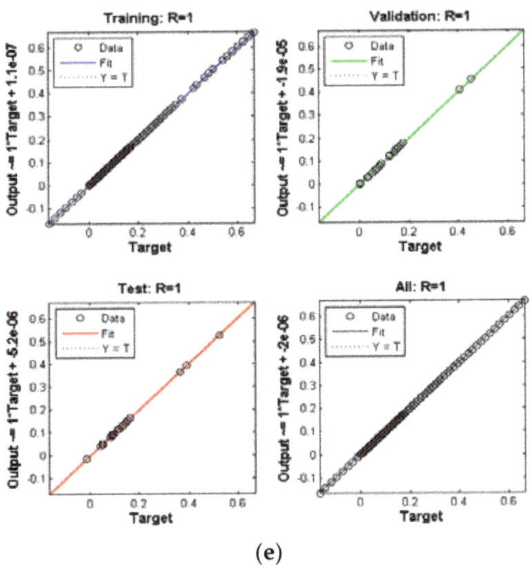

Figure 5. Scenario II Case I of TD-PNFM. (**a**) MSE Results: Scenario II Case I; (**b**) Transition state: Scenario II Case I; (**c**) Fitness: Scenario II Case I; (**d**) Error Histogram: Scenario II Case I; (**e**) regression: Scenario II Case I.

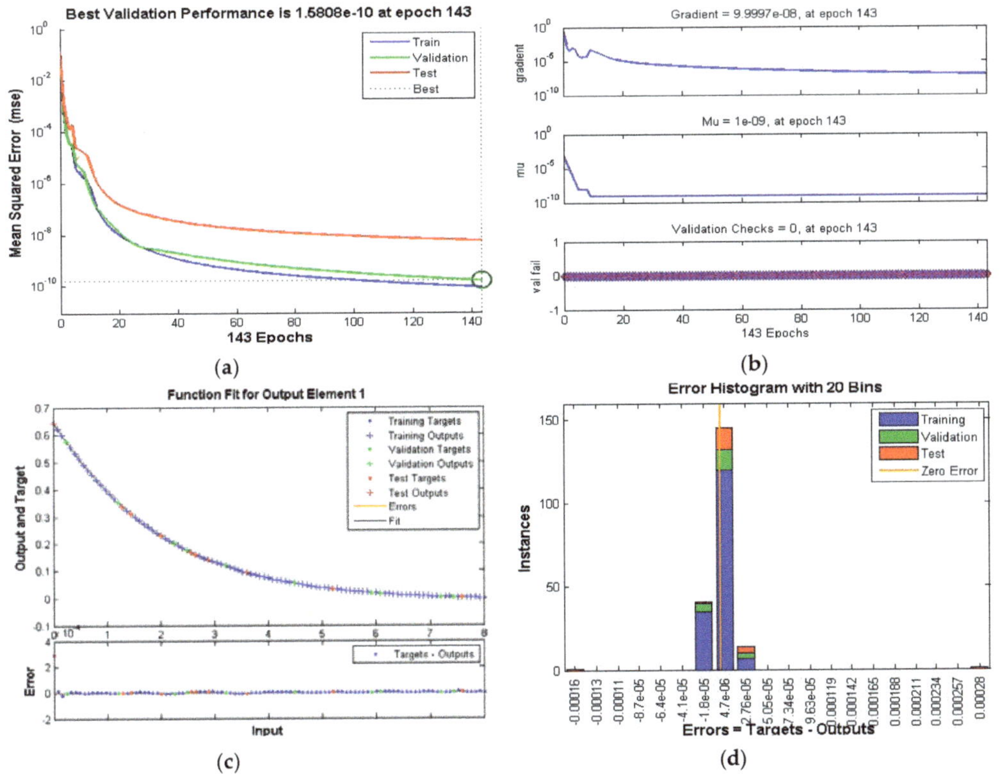

**Figure 6.** *Cont.*

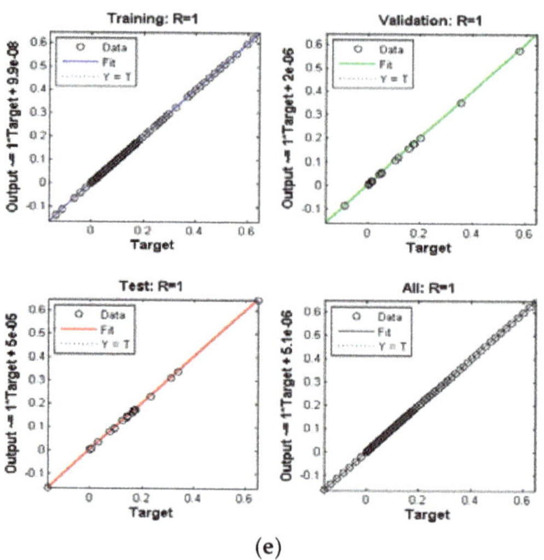



(**e**)

**Figure 6.** Scenario III Case I of TD-PNFM. (**a**) MSE Results: Scenario III Case I; (**b**) Transition state: Scenario III Case I; (**c**) Fitness: Scenario III Case I; (**d**) Error Histogram: Scenario III Case I; (**e**)regression: Scenario III Case I.

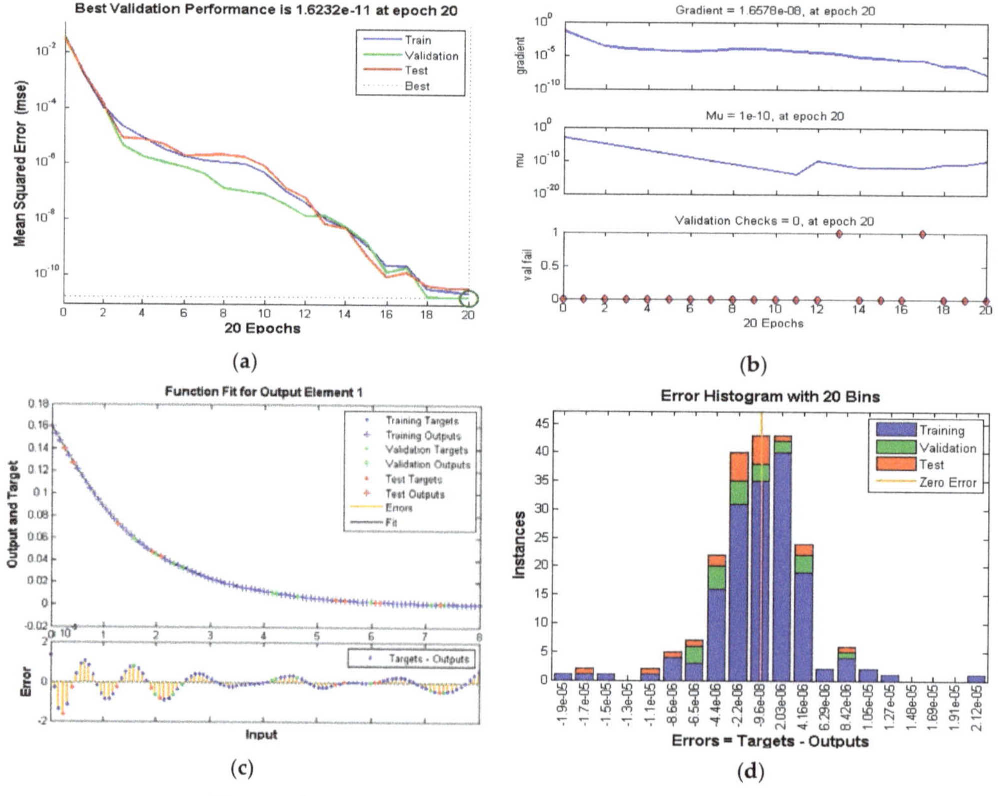



(**a**)    (**b**)

(**c**)    (**d**)

**Figure 7.** *Cont.*

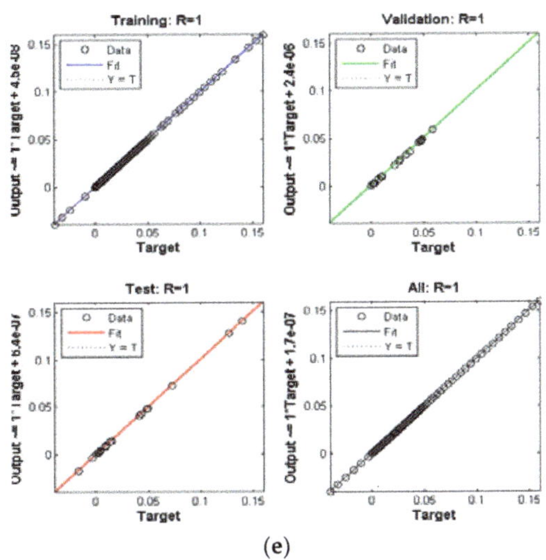

**Figure 7.** Scenario IV Case I of TD-PNFM. (**a**) MSE Results: Scenario IV Case I; (**b**) Transition state: Scenario IV Case I; (**c**) Fitness: Scenario IV Case I; (**d**) Error Histogram: Scenario IV Case I; (**e**)regression: Scenario IV Case I.

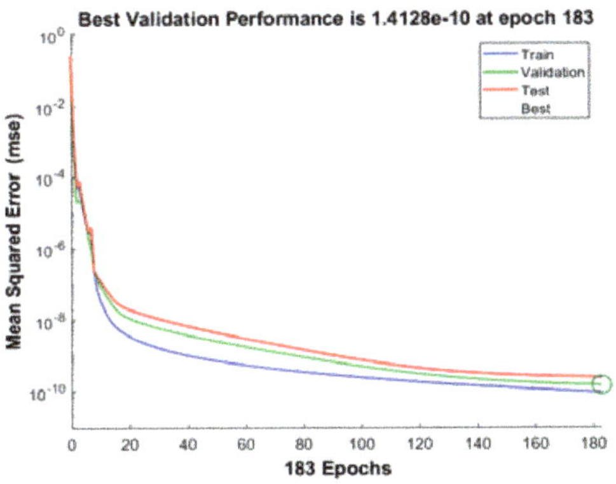

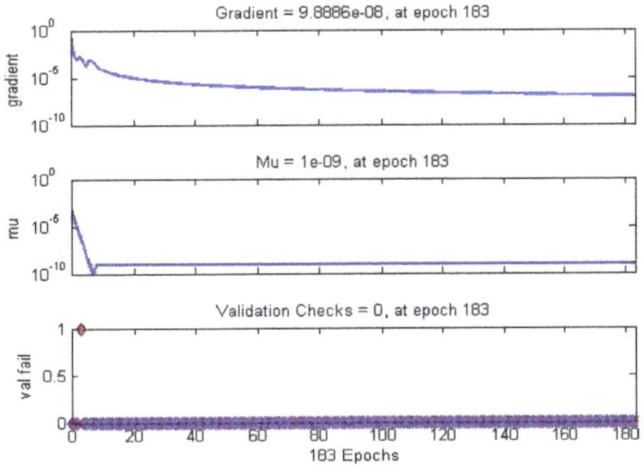

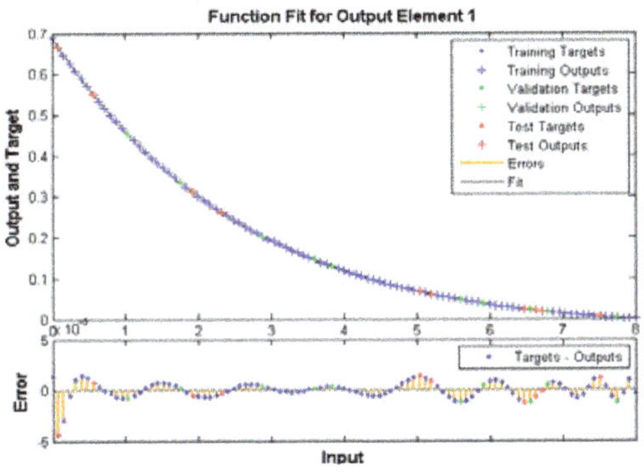

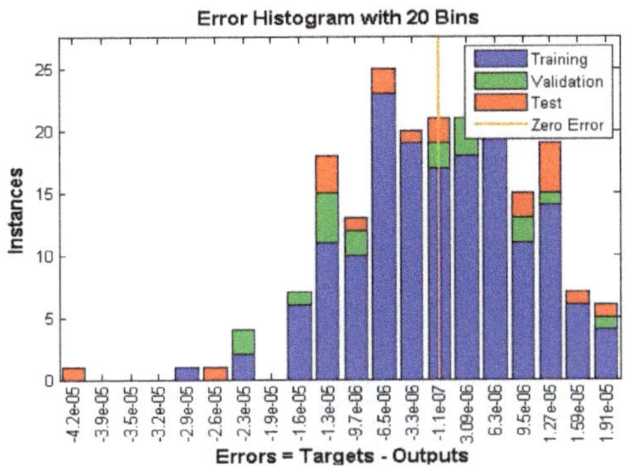

**Figure 8.** *Cont.*

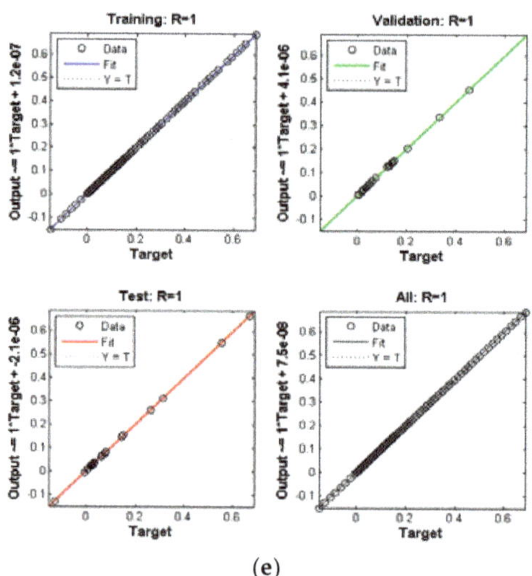

(**e**)

**Figure 8.** Scenario V Case I of TD-PNFM. (**a**) MSE Results: Scenario V Case I; (**b**) Transition state: Scenario V Case I; (**c**) Fitness: Scenario V Case I; (**d**) Error Histogram: Scenario V Case I; (**e**)regression: Scenario V Case I.

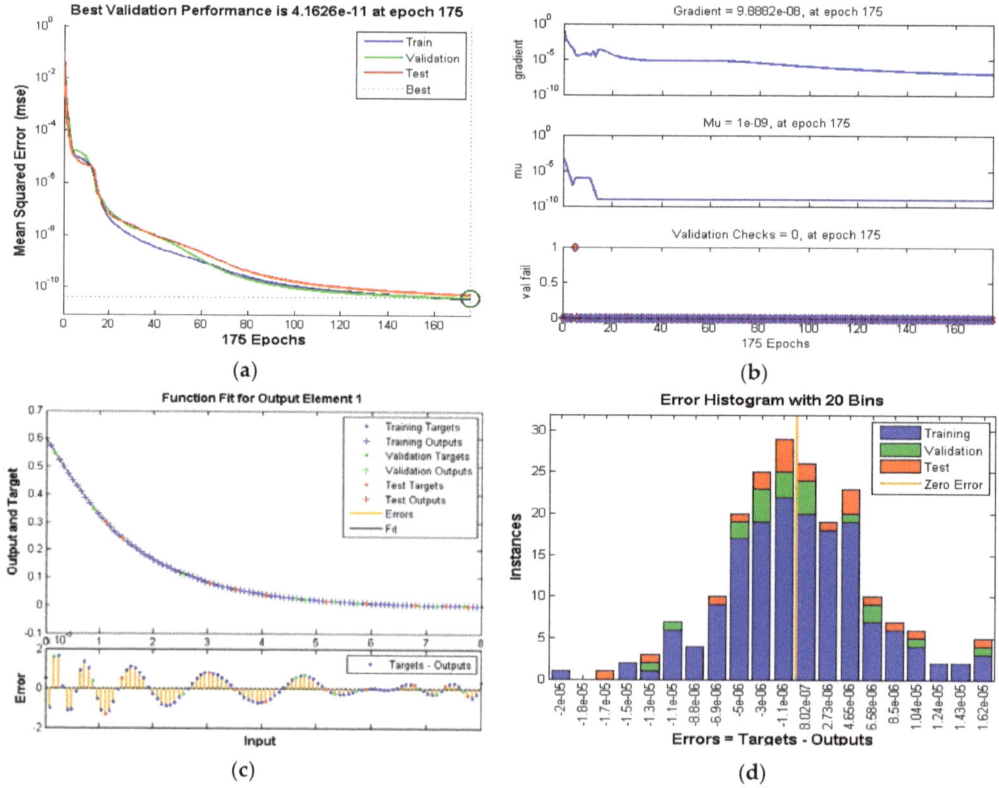

**Figure 9.** *Cont.*

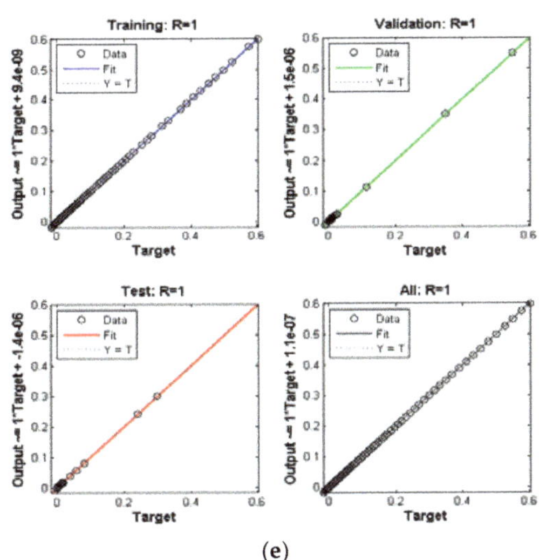

(e)

**Figure 9.** Scenario VI Case I of TD-PNFM. (**a**) MSE Results: Scenario VI Case I; (**b**) Transition state: Scenario VI Case I; (**c**) Fitness: Scenario VI Case I; (**d**) Error Histogram: Scenario VI Case I; (**e**) regression: Scenario VI Case I.

**Table 2.** Outcomes of TLM-BANN of TD-PNFM.

| Scenario | Case | MSE Data | | | Performance | Gradient | Mu | Final Epoch | Time |
|---|---|---|---|---|---|---|---|---|---|
| | | **Training** | **Validation** | **Testing** | | | | | |
| 1 | 1 | $8.42 \times 10^{-11}$ | $1.94 \times 10^{-10}$ | $1.78 \times 10^{-10}$ | $8.42 \times 10^{-11}$ | $1.00 \times 10^{-07}$ | $1.00 \times 10^{-09}$ | 204 | 5 s |
| | 2 | $3.02 \times 10^{-11}$ | $1.01 \times 10^{-10}$ | $5.59 \times 10^{-10}$ | $3.02 \times 10^{-11}$ | $9.85 \times 10^{-08}$ | $1.00 \times 10^{-09}$ | 188 | 2 s |
| | 3 | $2.82 \times 10^{-09}$ | $3.07 \times 10^{-09}$ | $4.17 \times 10^{-09}$ | $1.75 \times 10^{-09}$ | $6.68 \times 10^{-06}$ | $1.00 \times 10^{-09}$ | 43 | <1 s |
| | 4 | $2.26 \times 10^{-11}$ | $2.17 \times 10^{-11}$ | $6.99 \times 10^{-09}$ | $2.26 \times 10^{-11}$ | $9.79 \times 10^{-08}$ | $1.00 \times 10^{-09}$ | 134 | 1 s |
| 2 | 1 | $6.30 \times 10^{-11}$ | $9.73 \times 10^{-10}$ | $6.64 \times 10^{-11}$ | $6.30 \times 10^{-11}$ | $9.98 \times 10^{-08}$ | $1.00 \times 10^{-09}$ | 192 | 2 s |
| | 2 | $4.98 \times 10^{-11}$ | $1.16 \times 10^{-10}$ | $5.26 \times 10^{-11}$ | $4.98 \times 10^{-11}$ | $9.88 \times 10^{-08}$ | $1.00 \times 10^{-09}$ | 27 | 2 s |
| | 3 | $7.24 \times 10^{-12}$ | $1.40 \times 10^{-11}$ | $1.06 \times 10^{-11}$ | $7.24 \times 10^{-12}$ | $9.90 \times 10^{-08}$ | $1.00 \times 10^{-10}$ | 164 | 2 s |
| | 4 | $1.19 \times 10^{-08}$ | $7.42 \times 10^{-09}$ | $1.46 \times 10^{-08}$ | $5.11 \times 10^{-09}$ | $1.45 \times 10^{-05}$ | $1.00 \times 10^{-09}$ | 25 | <1 s |
| 3 | 1 | $8.81 \times 10^{-11}$ | $1.58 \times 10^{-10}$ | $5.75 \times 10^{-09}$ | $8.81 \times 10^{-11}$ | $1.00 \times 10^{-07}$ | $1.00 \times 10^{-09}$ | 143 | 1 s |
| | 2 | $8.70 \times 10^{-10}$ | $3.56 \times 10^{-09}$ | $1.67 \times 10^{-09}$ | $8.71 \times 10^{-10}$ | $1.00 \times 10^{-07}$ | $1.00 \times 10^{-08}$ | 278 | 3 s |
| | 3 | $4.60 \times 10^{-11}$ | $6.79 \times 10^{-11}$ | $1.05 \times 10^{-09}$ | $4.60 \times 10^{-11}$ | $9.96 \times 10^{-08}$ | $1.00 \times 10^{-09}$ | 204 | 2 s |
| | 4 | $4.58 \times 10^{-11}$ | $1.91 \times 10^{-10}$ | $2.47 \times 10^{-10}$ | $4.58 \times 10^{-11}$ | $9.89 \times 10^{-08}$ | $1.00 \times 10^{-09}$ | 139 | 1 s |
| 4 | 1 | $2.25 \times 10^{-11}$ | $1.62 \times 10^{-11}$ | $3.48 \times 10^{-11}$ | $2.25 \times 10^{-11}$ | $1.66 \times 10^{-08}$ | $1.00 \times 10^{-10}$ | 20 | <1 s |
| | 2 | $4.71 \times 10^{-11}$ | $2.35 \times 10^{-11}$ | $9.08 \times 10^{-12}$ | $4.71 \times 10^{-11}$ | $9.92 \times 10^{-08}$ | $1.00 \times 10^{-11}$ | 106 | 1 s |
| | 3 | $5.50 \times 10^{-11}$ | $4.25 \times 10^{-11}$ | $5.92 \times 10^{-11}$ | $5.50 \times 10^{-11}$ | $9.93 \times 10^{-08}$ | $1.00 \times 10^{-09}$ | 212 | 2 s |
| | 4 | $1.50 \times 10^{-08}$ | $2.71 \times 10^{-08}$ | $1.58 \times 10^{-08}$ | $1.12 \times 10^{-08}$ | $2.11 \times 10^{-08}$ | $1.00 \times 10^{-08}$ | 68 | <1 s |
| 5 | 1 | $8.90 \times 10^{-11}$ | $1.41 \times 10^{-10}$ | $2.30 \times 10^{-10}$ | $8.91 \times 10^{-11}$ | $9.89 \times 10^{-08}$ | $1.00 \times 10^{-09}$ | 183 | 2 s |
| | 2 | $7.41 \times 10^{-11}$ | $8.25 \times 10^{-11}$ | $1.50 \times 10^{-10}$ | $7.41 \times 10^{-11}$ | $9.99 \times 10^{-08}$ | $1.00 \times 10^{-09}$ | 191 | 2 s |
| | 3 | $4.80 \times 10^{-11}$ | $7.13 \times 10^{-11}$ | $3.81 \times 10^{-10}$ | $4.81 \times 10^{-11}$ | $9.91 \times 10^{-08}$ | $1.00 \times 10^{-09}$ | 165 | 2 s |
| | 4 | $5.51 \times 10^{-11}$ | $5.72 \times 10^{-11}$ | $1.69 \times 10^{-10}$ | $5.52 \times 10^{-11}$ | $9.90 \times 10^{-08}$ | $1.00 \times 10^{-09}$ | 142 | 1 s |
| 6 | 1 | $3.87 \times 10^{-11}$ | $4.16 \times 10^{-11}$ | $5.47 \times 10^{-11}$ | $3.88 \times 10^{-11}$ | $9.89 \times 10^{-08}$ | $1.00 \times 10^{-09}$ | 175 | 2 s |
| | 2 | $2.27 \times 10^{-08}$ | $9.76 \times 10^{-09}$ | $2.75 \times 10^{-08}$ | $7.32 \times 10^{-09}$ | $9.00 \times 10^{-06}$ | $1.00 \times 10^{-09}$ | 26 | <1 s |
| | 3 | $4.07 \times 10^{-11}$ | $7.95 \times 10^{-10}$ | $7.77 \times 10^{-11}$ | $4.07 \times 10^{-11}$ | $9.84 \times 10^{-08}$ | $1.00 \times 10^{-09}$ | 169 | 2 s |
| | 4 | $2.30 \times 10^{-10}$ | $3.84 \times 10^{-09}$ | $2.97 \times 10^{-10}$ | $2.30 \times 10^{-10}$ | $9.92 \times 10^{-08}$ | $1.00 \times 10^{-08}$ | 308 | 4 s |

The solution for three-dimension Prandtl nanofluid model for the temperature profile is depicted in Figure 10a,c,e,g,i,k, whereas the AE analysis plots for the temperature profile is given in Figure 10b,d,f,h,j,l. Similarly the solution plots for the concentration profile is shown in Figure 11a,c,e,g. The AE analysis plots are given in Figure 11b,d,f,h. The solution plots for the velocity distribution are shown in Figure 12a,b. The outcomes for the skin friction and Nusselt number are depicted in Figures 13a–c and 14a, respectively.

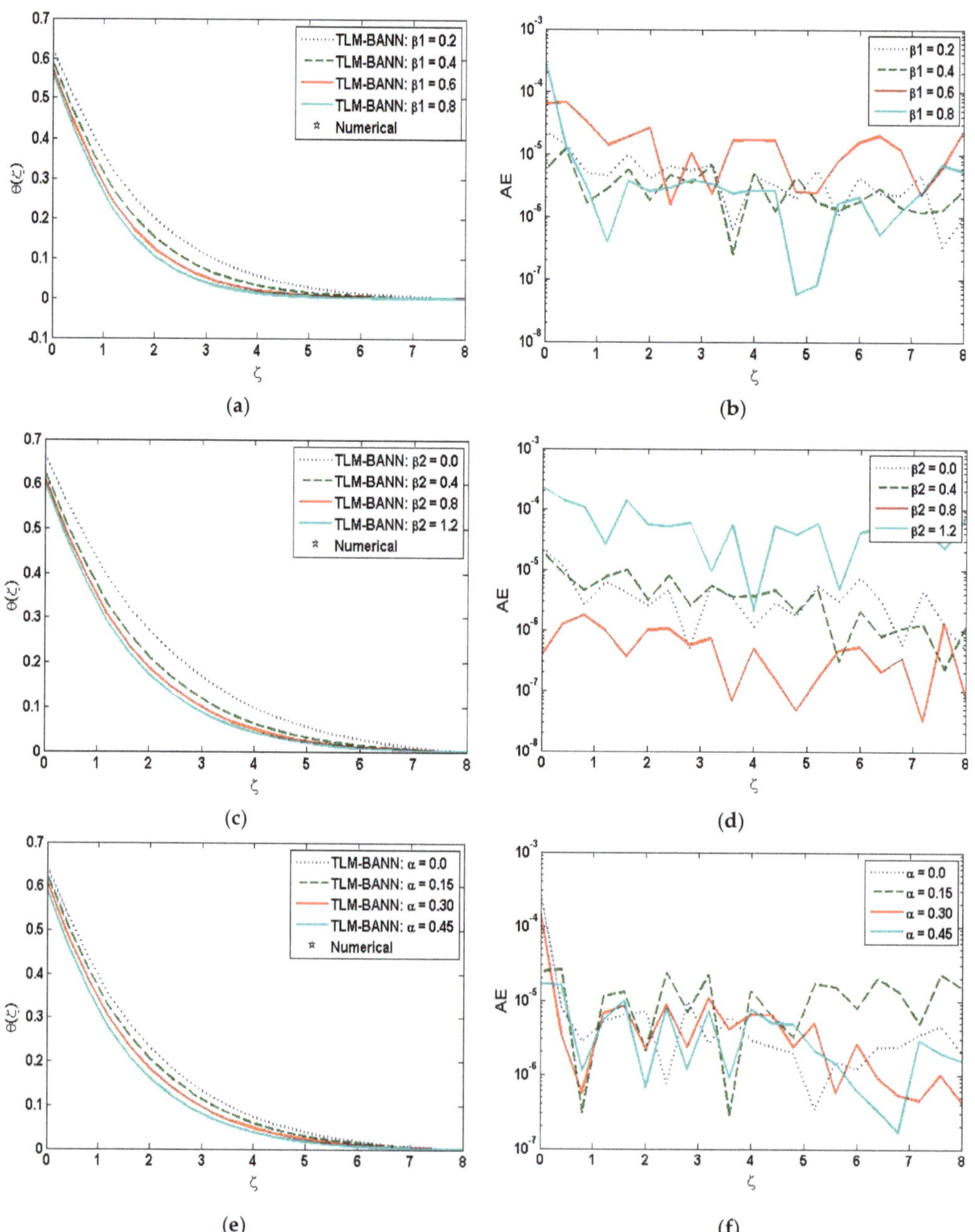

**Figure 10.** *Cont.*

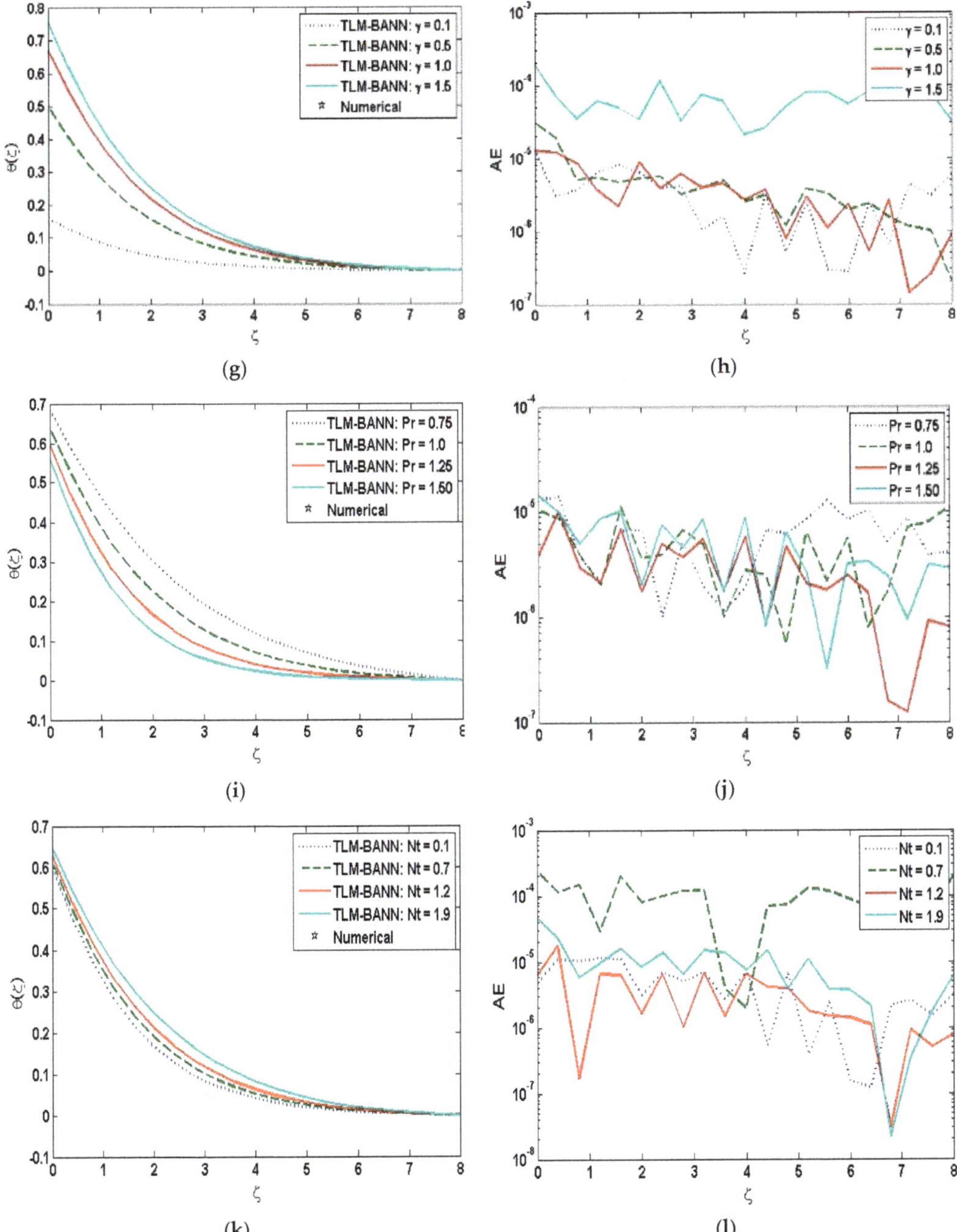

**Figure 10.** Assessment of TLM-BANN for $\theta$ with reference dataset of TD-PNFM (**a**) Variation of $\beta_1$ for $\theta$; (**b**) AE analysis in variation of $\beta_1$ for $\theta$; (**c**) Variation of $\beta_2$ for $\theta$; (**d**) AE analysis in variation of $\beta_2$ for $\theta$; (**e**) Variation of $\alpha$ for $\theta$; (**f**) AE analysis in variation of $\alpha$ for $\theta$; (**g**) Variation of $\gamma$ for $\theta$; (**h**) AE analysis in variation of $\gamma$ for $\theta$; (**i**) Variation of Pr for $\theta$; (**j**) AE analysis in variation of Pr for $\theta$; (**k**) Variation of $N_t$ for $\theta$ (**l**) AE analysis in variation of $N_t$ for $\theta$.

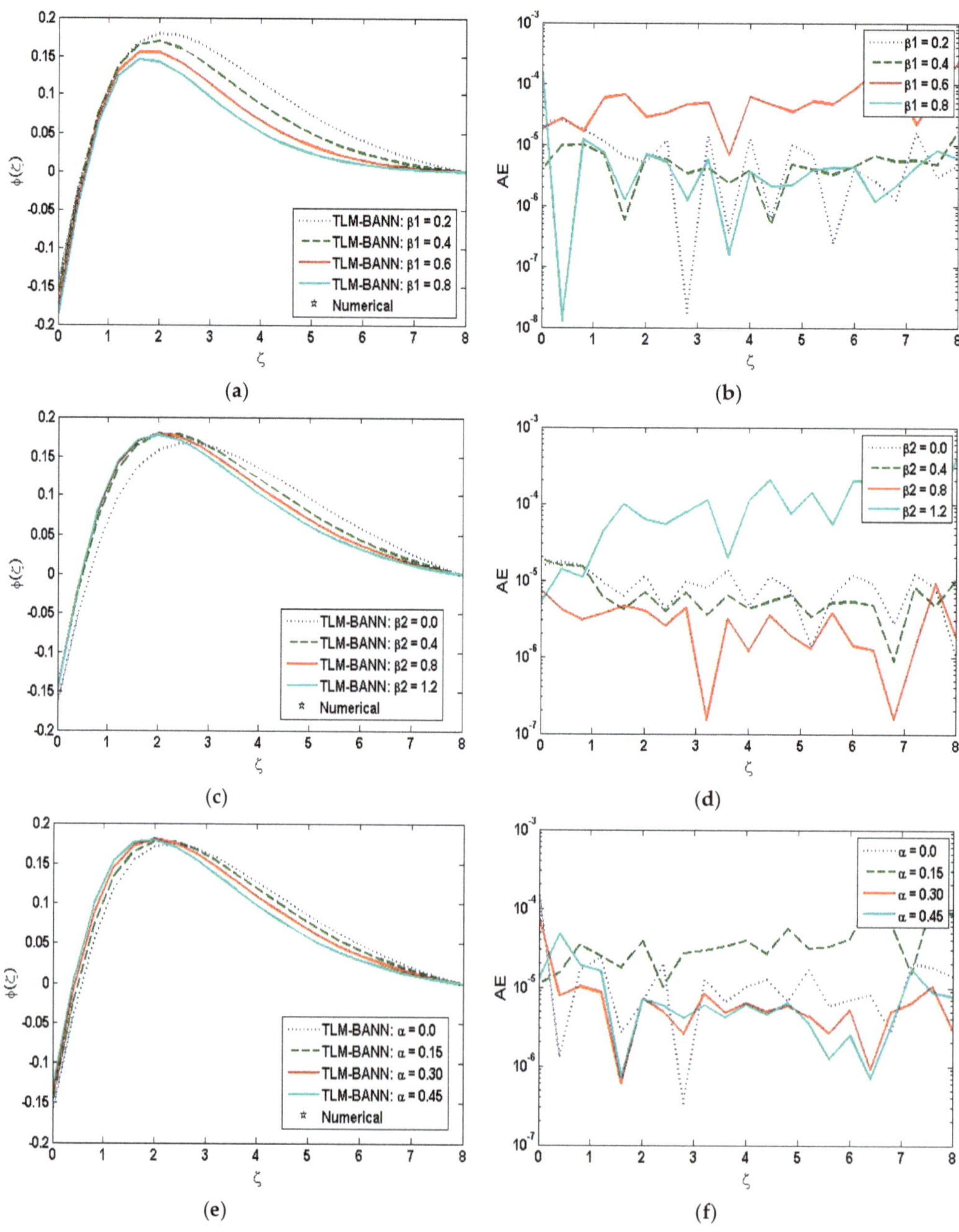

**Figure 11.** *Cont.*

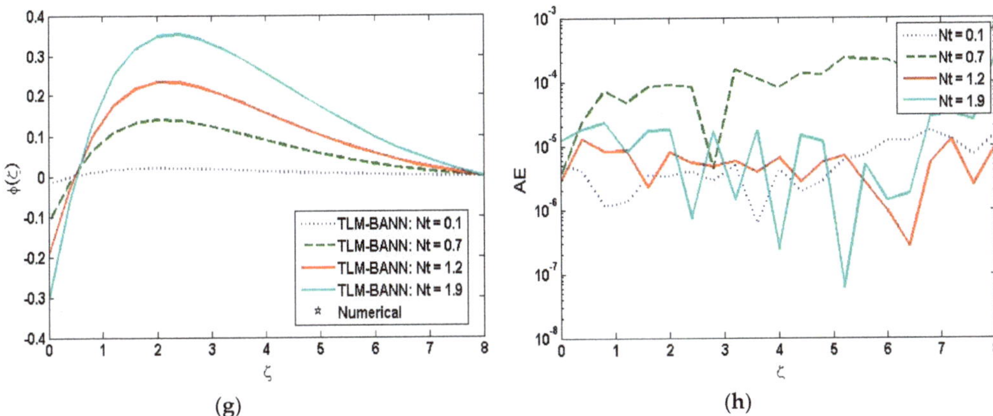



(**g**)        (**h**)

**Figure 11.** Assessment of TLM-BANN for $\varphi$ with reference dataset of TD-PNFM. (**a**) Variation of $\beta_1$ for $\varphi$; (**b**) AE analysis in variation of $\beta_1$ for $\varphi$; (**c**) Variation of $\beta_2$ for $\varphi$; (**d**) AE analysis in variation of $\beta_2$ for $\varphi$; (**e**) Variation of $\alpha$ for $\varphi$; (**f**) AE analysis in variation of $\alpha$ for $\varphi$; (**g**) Variation of $N_t$ for $\varphi$; (**h**) AE analysis in variation of $N_t$ for $\varphi$.

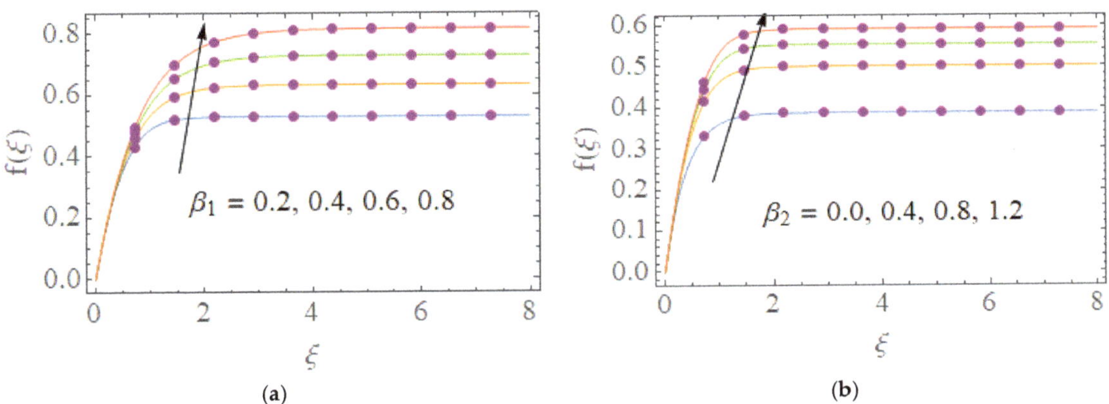



(**a**)        (**b**)

**Figure 12.** Assessment of TLM-BANN for $f$ with reference dataset of TD-PNFM. (**a**) Variation of $\beta_1$ for $f$; (**b**) Variation of $\beta_2$ for $f$.

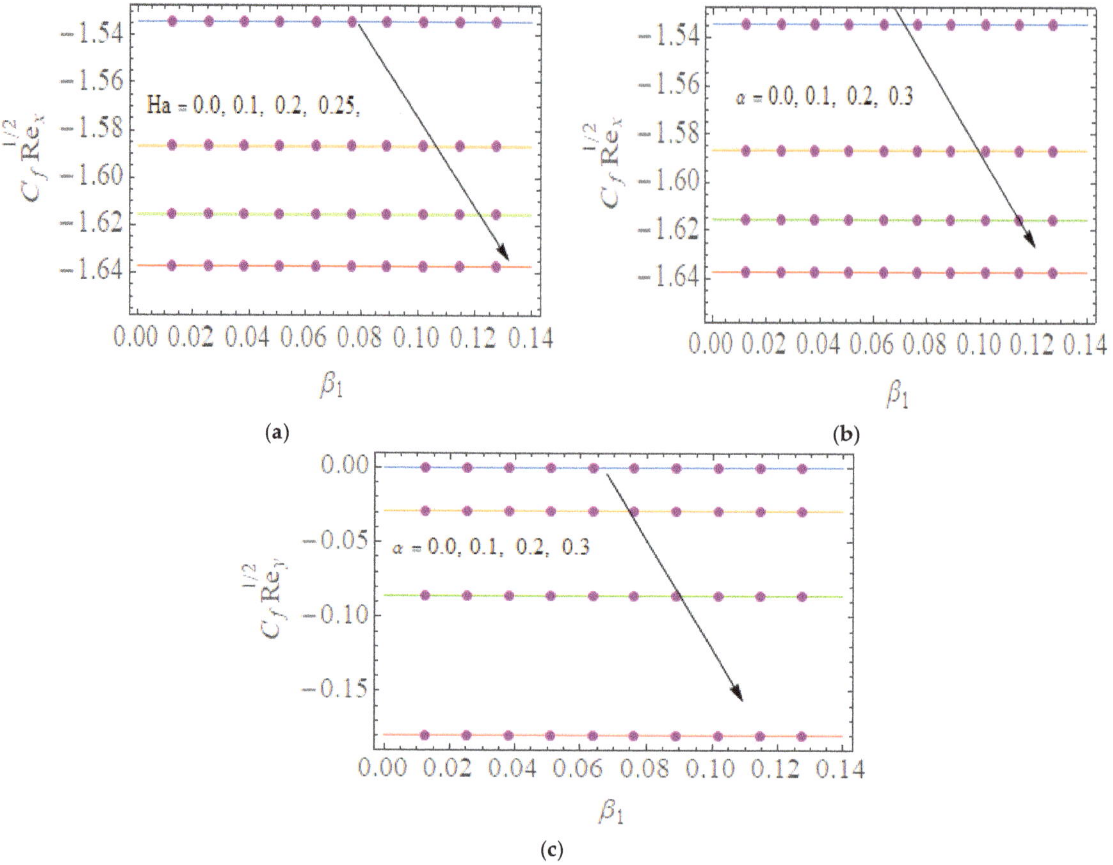

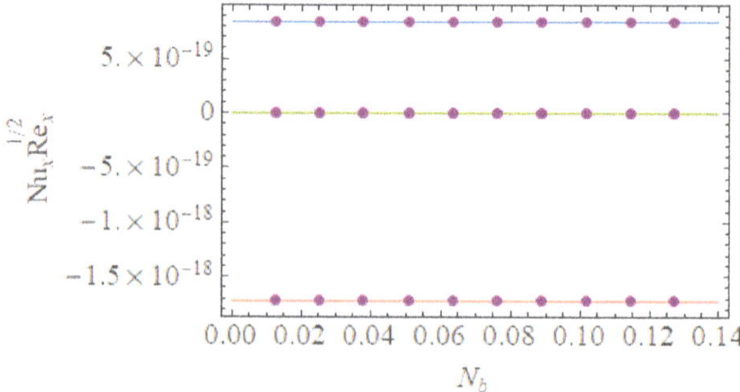

**Figure 13.** Assessment of TLM-BANN for $f$ with reference dataset of TD-PNFM. (**a**) Variation of $Ha$ for $C_f \mathrm{Re}_x^{1/2}$; (**b**) Variation of $\alpha$ for $C_f \mathrm{Re}_x^{1/2}$; (**c**) Variation of $\alpha$ for $C_f \mathrm{Re}_y^{1/2}$.

**Figure 14.** Assessment of TLM-BANN for $f$ with reference dataset of TD-PNFM. Variation of $N_t$ for $Nu_x \mathrm{Re}_x^{-1/2}$.

### 4.1. Impact of Variation on $\theta(\zeta)$

The MATLAB software is utilize to interpret the effects of variation $\beta_1$, $\beta_2$, $\alpha$, Pr, $\gamma$ and $N_t$ on temperature profile. Figure 10 shows the solution plots and AE analysis plots for the temperature profile by varying of Prandtl fluid number, flexible number, ratio parameter,

Prandtl number, Biot number and thermophoresis number. The absolute error values for the following variations are $10^{-8}$ to $10^{-3}$, $10^{-8}$ to $10^{-3}$, $10^{-7}$ to $10^{-3}$, $10^{-7}$ to $10^{-3}$, $10^{-7}$ to $10^{-4}$ and $10^{-8}$ to $10^{-3}$, respectively.

One may witness that:

- The temperature profile increases with the increase in Biot number.
- Temperature distribution shows decreasing behavior with the increase in Prandtl fluid number and flexible number.
- Increase in thermophoresis number leads to an increase in temperature profile.
- Temperature profile decreases with the increase in the values of Prandtl number and ratio parameter.

### 4.2. Impact of Variation on

Software MATLAB is used to interpret the AE analysis plots and the solution plots to examine the variation of $\beta_1$, $\beta_2$, $\alpha$ and $N_t$ on the concentration profile. Figure 11 depicts the solution and AE analysis plots for the concentration profile by the variation of Prandtl fluid number, flexible number, ratio parameter and thermophoresis number. It can be easily seen that Prandtl fluid number, flexible number and ratio parameter have the same impact on the concentration profile, but thermophoresis number has the opposite effect, which is an upsurge in the trend of concentration distribution. The absolute error values for the following variations are $10^{-8}$ to $10^{-3}$, $10^{-7}$ to $10^{-3}$, $10^{-7}$ to $10^{-3}$ and $10^{-8}$ to $10^{-3}$, respectively.

One may witness that:

- The concentration profile shows decreasing behavior when the value of Prandtl fluid parameter increases.
- Increase in flexible number causes a decrease in concentration profile.
- Increasing values of ratio parameter leads to a decrease in the behavior of concentration profile.
- The concentration profile increases with the increase in thermophoresis parameter. The reason is, raising thermophoresis parameter induces a rise in the system's thermal conductivity, which adds to an increase in concentration.

### 4.3. Impact of Variation on $f(\zeta)$

The effects of variation $\beta_1$ and $\beta_2$ on velocity distribution is shown in Figure 12. Figure 12 shows the solution plots for the velocity profile by varying of Prandtl fluid number and flexible number.

One may witness that:

- Velocity distribution shows an increasing trend with the increase in Prandtl fluid number.
- Increase in flexible number leads to an increase in velocity profile.

### 4.4. Skin Friction Coefficient

The effects of variation of $Ha$ and $\alpha$ on skin friction ($C_f \mathrm{Re}_x^{1/2}$ and $C_f \mathrm{Re}_y^{1/2}$) is shown in Figure 13. Figure 13 shows the plots for the skin friction by varying of Hartmann number and ratio parameter.

One may witness that:

- $C_f \mathrm{Re}_x^{1/2}$ shows a declining trend with the increase in Hartmann number and ratio parameter against Prandtl fluid parameter.
- Rise in ratio parameter leads to a decrease in $C_f \mathrm{Re}_y^{1/2}$ against Prandtl fluid parameetr.

### 4.5. Nusselt Number

The effects of variation of thermophoresis parameter on Nusselt number ($Nu_x \mathrm{Re}_x^{-1/2}$) is shown in Figure 14. Figure 14 shows the plots for the Nusselt number by varying of thermophoresis parameter against Brownian motion parameter.

One may witness that:

- $Nu_x\mathrm{Re}_x^{-1/2}$ shows a falling pattern with the increase in thermophoresis parameter against Brownian motion parameter.

## 5. Conclusions

The technique of Levenberg-Marquardt with backpropagated neural network is used to study the three-dimensional Prandtl nanofluid flow model with a convectively heated surface by the variations of Prandtl fluid number, flexible number, ratio parameter, Prandtl number, Biot number and thermophoresis number. The governing PDEs are transformed to system of ordinary differential equations by applying suitable transformation. The Mathematica command 'ND Solve' is utilized to compute the reference dataset for 101 points, in which there are 81 points for training, 10 point for testing and 10 points for validation. The regression analysis, error histogram and MSE data validates the performance of TLM-BANN.

There are few points drawn from the results are written below:

- The temperature profile shows increasing behavior with the increase in Biot number.
- Temperature distribution is decreasing with the increase in Prandtl fluid number and flexible number.
- Increase in thermophoresis number leads to an increase in temperature profile.
- Temperature profile decreases when the values of Prandtl number and ratio parameter increase.
- The concentration profile decreasing when the value of Prandtl fluid parameter increases.
- Increase in flexible number causes a decrease in concentration profile.
- Increasing values of ratio parameter causes a decrease in concentration profile.
- The concentration profile increases with the increasing values of thermophoresis parameter.
- Velocity distribution shows an increasing trend for the upsurge in the values of Prandtl fluid parameter and flexible parameter.
- Skin friction coefficient declines for the increase in Hartmann number and ratio parameter
- Nusselt number falls for the rising values of thermophoresis parameter against $Nb$.

## 6. Future Work

In future, some new soft computing architectures may be implemented for different fluidic models provided in [50–52] successfully.

**Author Contributions:** Conceptualization, M.S., G.Z. and M.A.Z.R.; writing—original draft preparation, M.S., G.Z., M.A.Z.R. and K.S.N.; software, M.S., G.Z. and K.S.N.; methodology, M.S. and M.A.Z.R.; formal analysis, K.S.N., A.-H.A.-A. and I.S.Y.; writing—review and editing, M.S., K.S.N., A.-H.A.-A. and I.S.Y.; funding acquisition, A.-H.A.-A. and I.S.Y.; investigation, M.S. and M.A.Z.R. All authors have read and agreed to the published version of the manuscript.

**Funding:** The authors express their appreciation to the Deanship of Scientific Research at King Khalid University for funding this work through the research groups program under Grant No. R.G.P.2/111/41. The authors extend their appreciation to the Deputyship for Research & Innovation, Ministry of Education, in Saudi Arabia, to fund this research work through the project number (IFP-KKU-2020/9).

**Institutional Review Board Statement:** Not applicable.

**Informed Consent Statement:** Not applicable.

**Data Availability Statement:** Not applicable.

**Acknowledgments:** The authors express their appreciation to the Deanship of Scientific Research at King Khalid University for funding this work through the research groups program under Grant No. R.G.P.2/111/41. The authors extend their appreciation to the Deputyship for Research & Innovation, Ministry of Education, in Saudi Arabia, to fund this research work through the project number (IFP-KKU-2020/9).

**Conflicts of Interest:** The authors declare no conflict of interest.

**Nomenclature**

| | |
|---|---|
| $\beta_1$ | Prandtl fluid number |
| $\alpha$ | Ratio parameter |
| Pr | Prandtl number |
| $\gamma$ | Biot number |
| $N_b$ | Brownian motion parameter |
| $N_t$ | Thermophoresis Parameter |
| TD-PNFM | three dimensional Prandtl nanofluid flow model |
| $Sc$ | Schmidt number |
| $\theta$ | Temperature profile |
| $\varphi$ | Concentration profile |
| $Ha$ | Hartman number |
| $\beta_2$ | Flexible number |
| MSE | Mean square error |
| TLM-BANN | technique of Levenberg Marquardt with backpropagated artificial neural network |

# References


1. Choi, S.U.; Eastman, J.A. *Enhancing Thermal Conductivity of Fluids with Nanoparticles (No. ANL/MSD/CP-84938; CONF-951135-29);* Argonne National Lab: Argonne, IL, USA, 1995.
2. Wang, X.Q.; Mujumdar, A.S. Heat transfer characteristics of nanofluids: A review. *Int. J. Ther. Sci.* **2007**, *46*, 1–19. [CrossRef]
3. Das, S.K.; Choi, S.U.; Patel, H.E. Heat transfer in nanofluids—A review. *Heat Transf. Eng.* **2006**, *27*, 3–19. [CrossRef]
4. Wen, D.; Lin, G.; Vafaei, S.; Zhang, K. Review of nanofluids for heat transfer applications. *Particuology* **2009**, *7*, 141–150. [CrossRef]
5. Duangthongsuk, W.; Wongwises, S. Heat transfer enhancement and pressure drop characteristics of $TiO_2$–water nanofluid in a double-tube counter flow heat exchanger. *Int. J. Heat Mass Transf.* **2009**, *52*, 2059–2067. [CrossRef]
6. Leong, K.Y.; Saidur, R.; Kazi, S.N.; Mamun, A.H. Performance investigation of an automotive car radiator operated with nanofluid-based coolants (nanofluid as a coolant in a radiator). *Appl. Ther. Eng.* **2010**, *30*, 2685–2692. [CrossRef]
7. Saidur, R.; Leong, K.Y.; Mohammed, H.A. A review on applications and challenges of nanofluids. *Renew. Sust. Energ. Rev.* **2011**, *15*, 1646–1668. [CrossRef]
8. Yousefi, T.; Shojaeizadeh, E.; Veysi, F.; Zinadini, S. An experimental investigation on the effect of pH variation of MWCNT–$H_2O$ nanofluid on the efficiency of a flat-plate solar collector. *Sol. Energ.* **2012**, *86*, 771–779. [CrossRef]
9. Kakaç, S.; Pramuanjaroenkij, A. Review of convective heat transfer enhancement with nanofluids. *Int. J. Heat Mass Transf.* **2009**, *52*, 3187–3196. [CrossRef]
10. Turkyilmazoglu, M. Nanofluid flow and heat transfer due to a rotating disk. *Comput. Fluids* **2014**, *94*, 139–146. [CrossRef]
11. Sheikholeslami, M.; Hatami, M.; Ganji, D.D. Nanofluid flow and heat transfer in a rotating system in the presence of a magnetic field. *J. Mol. Liq.* **2014**, *190*, 112–120. [CrossRef]
12. Wang, C.Y. The three-dimensional flow due to a stretching flat surface. *Phys. Fluids* **1984**, *27*, 1915–1917. [CrossRef]
13. Ariel, P.D. The three-dimensional flow past a stretching sheet and the homotopy perturbation method. *Comput. Math Appl.* **2007**, *54*, 920–925. [CrossRef]
14. Xu, H.; Liao, S.J.; Pop, I. Series solutions of unsteady three-dimensional MHD flow and heat transfer in the boundary layer over an impulsively stretching plate. *Eur. J. Mech. B Fluids* **2007**, *26*, 15–27. [CrossRef]
15. Liu, I.C.; Wang, H.H.; Peng, Y.F. Flow and heat transfer for three-dimensional flow over an exponentially stretching surface. *Chem. Eng. Commun.* **2013**, *200*, 253–268. [CrossRef]
16. Hayat, T.; Zahir, H.; Tanveer, A.; Alsaedi, A. Influences of Hall current and chemical reaction in mixed convective peristaltic flow of Prandtl fluid. *J. Magn. Magn. Mater.* **2016**, *407*, 321–327. [CrossRef]
17. Hayat, T.; Asghar, S.; Tanveer, A.; Alsaedi, A. Homogeneous–heterogeneous reactions in peristaltic flow of Prandtl fluid with thermal radiation. *J. Mol. Liq.* **2017**, *240*, 504–513. [CrossRef]
18. Kumar, K.G.; Rudraswamy, N.G.; Gireesha, B.J. Effects of mass transfer on MHD three dimensional flow of a Prandtl liquid over a flat plate in the presence of chemical reaction. *Results Phys.* **2017**, *7*, 3465–3471. [CrossRef]
19. Hayat, T.; Aziz, A.; Muhammad, T.; Alsaedi, A. Three-dimensional flow of Prandtl fluid with Cattaneo-Christov double diffusion. *Results Phys.* **2018**, *9*, 290–296. [CrossRef]
20. Nadeem, S.; Sadaf, H. Exploration of single wall carbon nanotubes for the peristaltic motion in a curved channel with variable viscosity. *J. Braz. Soc. Mech. Sci. Eng.* **2017**, *39*, 117–125. [CrossRef]
21. Akbar, N.S.; Khan, Z.H.; Haq, R.U.; Nadeem, S. Dual solutions in MHD stagnation-point flow of Prandtl fluid impinging on shrinking sheet. *Appl. Math. Mech.* **2014**, *35*, 813–820. [CrossRef]
22. Akbar, N.S. Blood flow analysis of Prandtl fluid model in tapered stenosed arteries. *Ain Shams Eng. J.* **2014**, *5*, 1267–1275. [CrossRef]

23. Sooppy Nisar, K.; Bilal, S.; Shah, I.A.; Awais, M.; Khan, I.; Thonthong, P. Hydromagnetic flow of Prandtl nanofluid past cylindrical surface with chemical reaction and convective heat transfer aspects. *Math. Probl. Eng.* **2021**, *2021*. [CrossRef]
24. Hamid, M.; Zubair, T.; Usman, M.; Khan, Z.H.; Wang, W. Natural convection effects on heat and mass transfer of slip flow of time-dependent Prandtl fluid. *Comput. Des. Eng.* **2019**, *6*, 584–592. [CrossRef]
25. Soomro, F.A.; Haq, R.U.; Khan, Z.H.; Zhang, Q. Passive control of nanoparticle due to convective heat transfer of Prandtl fluid model at the stretching surface. *Chin. J. Phys.* **2017**, *55*, 1561–1568. [CrossRef]
26. Acharya, N. Spectral quasi linearization simulation on the radiative nanofluid spraying over a permeable inclined spinning disk considering the existence of heat source/sink. *Appl. Math. Comput.* **2021**, *411*, 126547. [CrossRef]
27. Acharya, N. Spectral quasi linearization simulation of radiative nanofluidic transport over a bended surface considering the effects of multiple convective conditions. *Eur. J. Mech. B Fluids.* **2020**, *84*, 139–154. [CrossRef]
28. Sabu, A.S.; Wakif, A.; Areekara, S.; Mathew, A.; Shah, N.A. Significance of nanoparticles' shape and thermo-hydrodynamic slip constraints on MHD alumina-water nanoliquid flows over a rotating heated disk: The passive control approach. *Int. Commun. Heat Mass Transf.* **2021**, *129*, 105711. [CrossRef]
29. Virmani, K.; Deepak, C.; Sharma, S.; Chadha, U.; Selvaraj, S.K. Nanomaterials for automotive outer panel components: A review. *Eur. Phys. J. Plus* **2021**, *136*, 1–29. [CrossRef]
30. Uddin, I.; Ullah, I.; Ali, R.; Khan, I.; Nisar, K.S. Numerical analysis of nonlinear mixed convective MHD chemically reacting flow of Prandtl–Eyring nanofluids in the presence of activation energy and Joule heating. *Therm. Anal. Calorim.* **2021**, *145*, 495–505. [CrossRef]
31. Ullah, M.Z.; Alghamdi, M.; Alshomrani, A.S. Significance of heat generation/absorption in three-dimensional flow of Prandtl nanofluid with convectively heated surface. *Phys. Scr.* **2019**, *95*, 015703. [CrossRef]
32. Patil, A.B.; Humane, P.P.; Patil, V.S.; Rajput, G.R. MHD Prandtl nanofluid flow due to convectively heated stretching sheet below the control of chemical reaction with thermal radiation. *Int. J. Ambient Energy.* **2021**, 1–13. [CrossRef]
33. Hosseinzadeh, K.; Gholinia, M.; Jafari, B.; Ghanbarpour, A.; Olfian, H.; Ganji, D.D. Nonlinear thermal radiation and chemical reaction effects on Maxwell fluid flow with convectively heated plate in a porous medium. *Heat Transf. -Asian Res.* **2019**, *48*, 744–759. [CrossRef]
34. Ahmed, N.; Khan, U.; Mohyud-Din, S.T. Unsteady radiative flow of chemically reacting fluid over a convectively heated stretchable surface with cross-diffusion gradients. *Int. J. Therm. Sci.* **2017**, *121*, 182–191. [CrossRef]
35. Alamri, S.Z.; Khan, A.A.; Azeez, M.; Ellahi, R. Effects of mass transfer on MHD second grade fluid towards stretching cylinder: A novel perspective of Cattaneo–Christov heat flux model. *Phys. Lett. A* **2019**, *383*, 276–281. [CrossRef]
36. Yousif, M.A.; Ismael, H.F.; Abbas, T.; Ellahi, R. Numerical study of momentum and heat transfer of MHD Carreau nanofluid over an exponentially stretched plate with internal heat source/sink and radiation. *Heat Transf Res.* **2019**, *50*. [CrossRef]
37. Ellahi, R.; Zeeshan, A.; Hussain, F.; Abbas, T. Thermally charged MHD bi-phase flow coatings with non-Newtonian nanofluid and hafnium particles along slippery walls. *Coatings* **2019**, *9*, 300. [CrossRef]
38. Ellahi, R.; Alamri, S.Z.; Basit, A.; Majeed, A. Effects of MHD and slip on heat transfer boundary layer flow over a moving plate based on specific entropy generation. *J. Taibah Univ. Sci.* **2018**, *12*, 476–482. [CrossRef]
39. Saeed, A.; Shah, Z.; Islam, S.; Jawad, M.; Ullah, A.; Gul, T.; Kumam, P. Three-dimensional Casson nanofluid thin film flow over an inclined rotating disk with the impact of heat generation/consumption and thermal radiation. *Coatings* **2019**, *9*, 248. [CrossRef]
40. Ilyas, H.; Ahmad, I.; Raja, M.A.Z.; Shoaib, M. A novel design of Gaussian WaveNets for rotational hybrid nanofluidic flow over a stretching sheet involving thermal radiation. *Int. Commun. Heat Mass Transf.* **2021**, *123*, 105196. [CrossRef]
41. Ilyas, H.; Ahmad, I.; Raja, M.A.Z.; Tahir, M.B.; Shoaib, M. Intelligent computing for the dynamics of fluidic system of electrically conducting Ag/Cu nanoparticles with mixed convection for hydrogen possessions. *Int. J. Hydrog. Energy* **2021**, *46*, 4947–4980. [CrossRef]
42. Shoaib, M.; Raja, M.A.Z.; Farhat, I.; Shah, Z.; Kumam, P. Intelligent backpropagated neural networks for numerical computations for mhd squeezing fluid suspended by nanoparticles between two parallel plates. *Res. Sq.* **2021**. [CrossRef]
43. Awan, S.E.; Raja, M.A.Z.; Gul, F.; Khan, Z.A.; Mehmood, A.; Shoaib, M. Numerical computing paradigm for investigation of micropolar nanofluid flow between parallel plates system with impact of electrical MHD and Hall current. *Arab. J. Sci. Eng.* **2021**, *46*, 645–662. [CrossRef]
44. Khan, I.; Raja, M.A.Z.; Shoaib, M.; Kumam, P.; Alrabaiah, H.; Shah, Z.; Islam, S. Design of neural network with levenberg-marquardt and bayesian regularization backpropagation for solving pantograph delay differential equations. *IEEE Access* **2020**, *8*, 137918–137933. [CrossRef]
45. Shoaib, M.; Zubair, G.; Nisar, K.S.; Raja, M.A.Z.; Khan, M.I.; Gowda, R.P.; Prasannakumara, B.C. Ohmic heating effects and entropy generation for nanofluidic system of Ree-Eyring fluid: Intelligent computing paradigm. *Int. Commun. Heat Mass Transf.* **2021**, *129*, 105683. [CrossRef]
46. Shoaib, M.; Raja, M.A.Z.; Zubair, G.; Farhat, I.; Nisar, K.S.; Sabir, Z.; Jamshed, W. Intelligent computing with levenberg–marquardt backpropagation neural networks for third-grade nanofluid over a stretched sheet with convective conditions. *Arab. J. Sci. Eng.* **2021**, *269*, 1–19. [CrossRef] [PubMed]
47. Ullah, H.; Khan, I.; Fiza, M.; Hamadneh, N.N.; Fayz-Al-Asad, M.; Islam, S.; Khan, I.; Raja, M.A.Z.; Shoaib, M. MHD boundary layer flow over a stretching sheet: A new stochastic method. *Math. Probl. Eng.* **2021**, *2021*. [CrossRef]

48. Uddin, I.; Ullah, I.; Raja, M.A.Z.; Shoaib, M.; Islam, S.; Muhammad, T. Design of intelligent computing networks for numerical treatment of thin film flow of Maxwell nanofluid over a stretched and rotating surface. *Surf. Interfaces* **2021**, *24*, 101107. [CrossRef]
49. Ullah, M.Z.; Alghamdi, M. An optimal analysis for 3d flow of Prandtl nanofluid with convectively heated surface. *Commun. Theor. Phys.* **2019**, *71*, 1485. [CrossRef]
50. Uddin, I.; Akhtar, R.; Zhiyu, Z.; Islam, S.; Shoaib, M.; Raja, M.A.Z. Numerical treatment for Darcy-Forchheimer flow of Sisko nanomaterial with nonlinear thermal radiation by lobatto IIIA technique. *Math. Probl. Eng.* **2019**, *2019*. [CrossRef]
51. Ahmad, I.; Cheema, T.N.; Raja, M.A.Z.; Awan, S.E.; Alias, N.B.; Iqbal, S.; Shoaib, M. A novel application of Lobatto IIIA solver for numerical treatment of mixed convection nanofluidic model. *Sci. Rep.* **2021**, *11*, 1–16. [CrossRef]
52. Awais, M.; Raja, M.A.Z.; Awan, S.E.S.M.; Ali, H.M. Heat and mass transfer phenomenon for the dynamics of Casson fluid through porous medium over shrinking wall subject to Lorentz force and heat source/sink. *Alex. Eng. J.* **2021**, *60*, 1355–1363. [CrossRef]

*coatings*

MDPI

*Article*

# Annealing Effect on the Contact Angle, Surface Energy, Electric Property, and Nanomechanical Characteristics of $Co_{40}Fe_{40}W_{20}$ Thin Films


Wen-Jen Liu [1], Yung-Huang Chang [2], Chi-Lon Fern [3], Yuan-Tsung Chen [4,*], Tian-Yi Jhou [4], Po-Chun Chiu [4], Shih-Hung Lin [5], Ko-Wei Lin [3] and Te-Ho Wu [4]

[1] Department of Materials Science and Engineering, I-Shou University, Kaohsiung 840, Taiwan; jurgen@isu.edu.tw
[2] Bachelor Program in Interdisciplinary Studies, National Yunlin University of Science and Technology, 123 University Road, Section 3, Douliou 64002, Taiwan; changyhu@yuntech.edu.tw
[3] Department of Materials Science and Engineering, National Chung Hsing University, Taichung City 402, Taiwan; fengcl@yuntech.edu.tw (C.-L.F.); kwlin@dragon.nchu.edu.tw (K.-W.L.)
[4] Graduate School of Materials Science, National Yunlin University of Science and Technology, 123 University Road, Section 3, Douliou 64002, Taiwan; M10847001@yuntech.edu.tw (T.-Y.J.); M10947005@yuntech.edu.tw (P.-C.C.); wuth@yuntech.edu.tw (T.-H.W.)
[5] Department of Electronic Engineering, National Yunlin University of Science and Technology, 123 University Road, Section 3, Douliou 64002, Taiwan; isshokenmei@yuntech.edu.tw
* Correspondence: ytchen@yuntech.edu.tw; Tel.: +886-5-534-2601



**Citation:** Liu, W.-J.; Chang, Y.-H.; Fern, C.-L.; Chen, Y.-T.; Jhou, T.-Y.; Chiu, P.-C.; Lin, S.-H.; Lin, K.-W.; Wu, T.-H. Annealing Effect on the Contact Angle, Surface Energy, Electric Property, and Nanomechanical Characteristics of $Co_{40}Fe_{40}W_{20}$ Thin Films. *Coatings* **2021**, *11*, 1268. https://doi.org/10.3390/coatings11111268

Academic Editor: Angela De Bonis

Received: 7 October 2021
Accepted: 18 October 2021
Published: 20 October 2021

**Publisher's Note:** MDPI stays neutral with regard to jurisdictional claims in published maps and institutional affiliations.


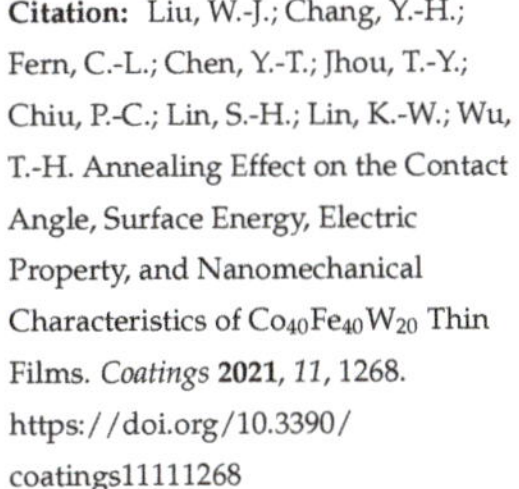




**Abstract:** This study investigated $Co_{40}Fe_{40}W_{20}$ single-layer thin films according to their corresponding structure, grain size, contact angle, and surface energy characteristics. $Co_{40}Fe_{40}W_{20}$ alloy thin films of different thicknesses, ranging from 10 to 50 nm, were sputtered on Si(100) substrates by DC magnetron sputtering. The thin films were annealed under three conditions: as-deposited, 250 °C, and 350 °C temperatures, respectively. The Scherrer equation was applied to calculate the grain size of $Co_{40}Fe_{40}W_{20}$ thin films. The results show that the grain size of CoFe(110) increased simultaneously with the increase of post-annealing temperature, suggesting that the crystallinity of $Co_{40}Fe_{40}W_{20}$ thin films increased with the post-annealing temperature. Moreover, the contact angles of all $Co_{40}Fe_{40}W_{20}$ thin films were all less than 90°, suggesting that $Co_{40}Fe_{40}W_{20}$ thin films show changes in the direction of higher hydrophilicity. However, we found that their contact angles decreased as the grain size of CoFe increased. Finally, the Young equation was applied to calculate the surface energy of $Co_{40}Fe_{40}W_{20}$ thin films. After post-annealing, the surface energy of $Co_{40}Fe_{40}W_{20}$ thin films increased with the rising post-annealing temperature. This is the highest value of surface energy observed for 350 °C. In addition, the surface energy increased as the contact angle of $Co_{40}Fe_{40}W_{20}$ thin films decreased. The high surface energy means stronger adhesion, allowing the formation of multilayer thin films with magnetic tunneling junctions (MTJs). The sheet resistance of the as-deposited and thinner CoFeW films is larger than annealed and thicker CoFeW films. When the thickness is from 10 nm to 50 nm, the hardness and Young's modulus of the CoFeW film also show a saturation trend.

**Keywords:** annealed $Co_{40}Fe_{40}W_{20}$ thin films; magnetic tunnel junctions (MTJs); X-ray diffraction (XRD); contact angle; surface energy; nanomechanical properties


## 1. Introduction

Magnetic nanofilms have attracted increasing attention in recent years, especially in the CoFeB system. CoFeB is a soft ferromagnetic material, which has been extensively applied in spintronic devices. When a CoFeB film is thinner than 1.7 nm, it shows a perpendicular magnetic anisotropy (PMA), implying its easy axis is obviously out-of-plane [1,2]. The PMA in Ta/CoFeB/MgO stack, first identified by Ikeda et al., has promising applications in spin-transfer torque devices [3]. Moreover, magnetic tunnel junctions

(MTJs) have been widely investigated because of their wide applications in spintronic magnetic devices, such as reading heads of hard disks and magnetoresistance random access memories (MRAM) [4–7]. The CoFeB film was combined with the MgO layer to form MTJs. A PMA in magnetic films is an essential characteristic for MTJs [8,9]. On the other hand, CoFeB magnetic film has been extensively used as a free or pinned layer in MTJs. CoFeB and MgO layers via post-annealing result in a high spin polarization and high tunneling magnetoresistance (TMR) [10–15]. Meng et al. reported that post-annealing at 250 °C could effectively increase TMR about nine times [16]. The CoFeB-MgO-based MTJs are considered as a potential candidate for MRAM with high PMA, lower critical switching current density ($J_c$), and high TMR [17,18]. Researchers have paid great attention to increasing the PMA and the thermal stability of MTJs structure. Inserting other metal spacer layers in the MTJs system was common and effective in improving its various properties [19–21]. Tungsten (W) has a high melting point, high mechanical strength, and high-temperature thermal conductivity. In recent years, people have paid more and more attention to adding new elements to material in magnetic fields. Especially, the study of adding other elements to the CoFe matrix has also increased. However, few researchers have paid attention to adding W to CoFe alloy. The research on the efficacy of CoFeW is noteworthy because it is a promising material. It is usually inserted into MTJ as a free layer or pinned layer or is combined with other layers in a multilayer structure. It can be widely used in magnetic and semiconductor applications. In our previous studies, the relationship between magnetism and structure of CoFeW films, as deposited and annealed, has been studied [22]. This study is an extension of our previous research to investigate contact angle and surface energy. However, few research studies have focused on the adhesive properties of CoFeW films. In addition, the effectiveness of CoFeW-based multilayer structure is sensitive to the temperature environment. It is of great significance to study the characteristics of CoFeW films deposited by direct-current (DC) magnetron sputtering in as-deposited and annealing conditions. The strong adhesion is a critical property for MTJ. Strong adhesion is a critical property for MTJs. Therefore, this study measured the contact angle of $Co_{40}Fe_{40}W_{20}$ thin films with different thicknesses and heat treatment conditions by a measuring analyzer via deionized (DI) water and glycerol. The Young equation was used to calculate the surface energy. Finally, this study discussed the correlation between grain size, contact angle, surface energy, and adhesion. The findings can provide a reference to future research on MTJs.

## 2. Materials and Methods

A CoFeW film with a thickness of 10–50 nm was sputtered onto a Si(100) substrate at room temperature by magnetron direct current (DC) sputtering with a power of 50 W power under the following three conditions: (a) the deposited films were kept at room temperature, (b) annealed at a treatment temperature ($T_A$) at 250 °C for 1 h, and (c) annealed at 350 °C for 1 h. The chamber base pressure was $2 \times 10^{-7}$ Torr, and the Ar working pressure was $3 \times 10^{-3}$ Torr. The pressure under the ex-situ annealed condition is $3 \times 10^{-3}$ Torr, and the selected Ar gas is used. The composition of the CoFeW alloy target was 40 at% Co, 40 at% Fe, 20 at% W. Furthermore, the structure of the CoFeW films was determined by using the grazing incidence X-ray Diffraction (GIXRD) patterns obtained by CuK$\alpha$1 (PAN analytical X'pert PRO MRD, Malvern Panalytical Ltd, Cambridge, United Kingdom) and a low angle diffraction incidence at a two-degree angle. To determine the thickness accurately, a high-resolution cross-sectional field emission scanning electron microscopy (SEM, Hitachi SU 8200, Tokyo, Japan) was used to study the calibration thickness of the corresponding sputtering time. Before the measurement, the contact angle was properly air-cleaned on the surface. Then, the contact angles of CoFeW film were measured with deionized (DI) water and glycerol. The contact angles were then measured when the samples were removed from the chamber. Lastly, the surface energy was calculated based on the contact angle. The sheet resistance (Rs) was measured using a conventional four-point technique. The hardness and Young's modulus of the CoFeW film were measured using the MTS Nano Indenter XP with a Berkovich tip. An indentation load of 1 mN was used to limit the penetration depth of the indenter to less than 10% of the film thickness. The indenter repeated the measurement at 10 positions for each sample. The indentation load was increased in 40 steps, and the indentation depth was measured at each step. Six indentations were studied in each sample, and the standard deviation was averaged to obtain more accurate results.

## 3. Results

### 3.1. Full-Width at Half Maximum (FWHM) and Grain Size Distribution

The corresponding X-ray diffraction patterns (XRD) of the CoFeW films have been proved in our previous literature [22]. From a previous study, it was indicated that the CoFe(110) peak and specific $Fe_2O_3(320)$, $WO_3(002)$, $Co_2O_3(422)$, and $Co_2O_3(511)$ oxide peaks are displayed in XRD. The corresponding joint committee on powder diffraction standards (JCPDS) cards are #JCPDS 49-1567, #JCPDS 24-0081, #JCPDS 05-0363, and #JCPDS 02-0770. The reason for the formation of the oxide peak can be reasonably inferred to be that there may still be oxygen in the chamber system of the sputtering system. In addition, both the adventitious oxide on the Si(100) substrate and the oxygen contamination on the sputtering target contribute to the formation of oxidation peaks. According to the above reason, the main peak of CoFeW film is CoFe(110), and the FWHM of CoFe(110) were analyzed. Moreover, it can be calculated grain size and FWHM of CoFe(110) by XRD data. Figure 1a shows the FWHM of the as-deposited and post-annealed CoFeW thin films. The results indicate that the FWHM decreased as CoFeW thickness increased. In addition, Figure 1a shows that the FWHM decreased as the post-annealing temperature raised. Finally, we calculated the grain size of CoFe(110) by using the FWHM determined by XRD and the Scherrer equation.

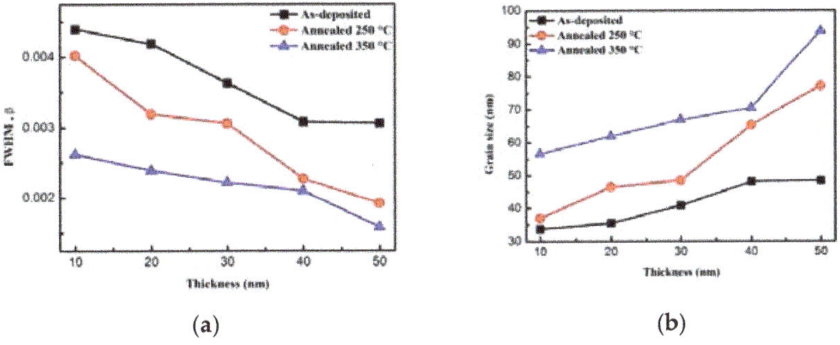


**Figure 1.** (**a**) FWHM of CoFeW thin films. (**b**) Grain size of CoFeW thin films.

The Scherrer equation is [23,24]:

$$D = K\lambda/\beta\cos\theta \tag{1}$$

where the quantity D is grain size, K (0.89) is the Scherrer's constant, $\lambda$ is the X-ray wavelength of the Cu Kα1 line, $\beta$ is the FWHM diffraction of CoFe(110) peak, and $\theta$ is the half-angle of the diffraction peak. The FWHM of the CoFe(110) peak was used to estimate the grain sizes at the as-deposited and the two annealing temperatures. Figure 1b shows the grain size of CoFeW films after deposition and annealing. The as-deposited CoFeW thin films possessed the smallest grain size. However, as post-annealing temperature increased, we found that the grain size increased, and was affected by the thicknesses of CoFeW thin films, as shown in Figure 1b. Moreover, the post-annealing of CoFeW thin films at 350 °C when the thickness was 50 nm possessed the highest observed grain size value—this is because post-annealing provided energy to CoFeW thin films, resulting in an increased grain size of CoFeW thin films.

### 3.2. SEM Image

High-resolution cross-sectional field emission scanning electron microscope (SEM) images of the as-deposited and annealed 350 °C samples are shown in Figure 2a,b. It can be seen from the results of the SEM that the annealed CoFeW film is denser than the as-deposited state. A high-density CoFeW film is obtained by using annealing treatment. The plan views of the as-deposited and annealed 350 °C samples are shown in Figure 2c,d. From the SEM results, the as-deposited CoFeW film showed a loose surface morphology.

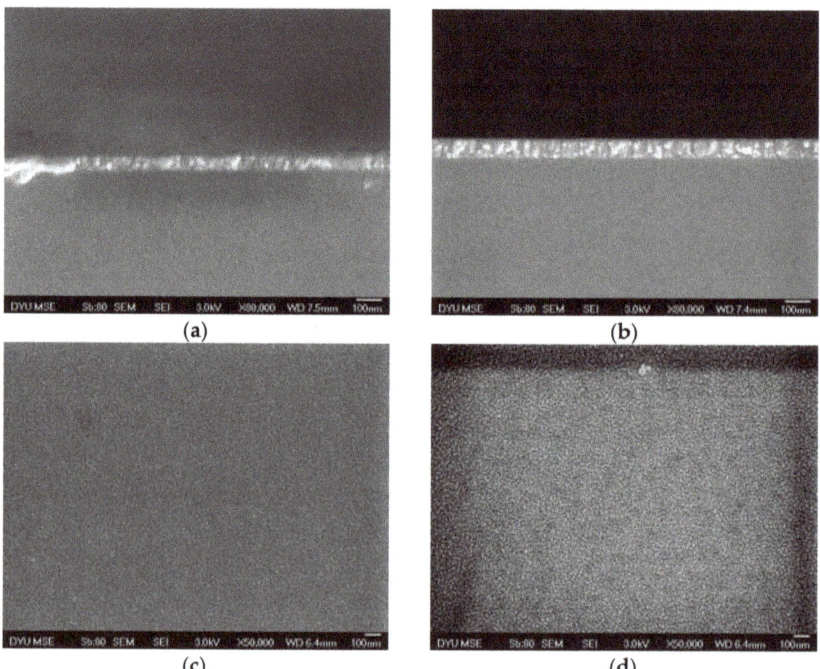

**Figure 2.** Cross-sectional SEM images of CoFeW 50 nm. (**a**) RT and (**b**) annealed at 350 °C. SEM micrographs of CoFeW 50 nm. (**c**) RT and (**d**) annealed at 350 °C.

### 3.3. Contact Angle

Figure 3A–C depicts the contact angle ($\theta$) of CoFeW films under three conditions, namely the as-deposited state, the temperature of 250 °C and 350 °C. The contact angles were measured with DI water and glycerol. The data in Table 1 show that the contact angles of the as-deposited CoFeW using DI Water were 81.2°, 80.7°, 81.5°, 82.9°, and 82.0°. On the other hand, the contact angles with glycerol were 76.7°, 78.1°, 79.6°, 75.5°, and 75.9°. Table 1 shows the contact angles of the $Co_{40}Fe_{40}W_{20}$ films after annealing at 350 °C, measured with DI water and glycerol. As seen, the contact angles of the as-deposited CoFeW with DI water are 66.9°, 69.8°, 66.6°, 69.1°, and 66.8°, and those with glycerol are 62.3°, 65.0°, 60.7°, 67.6°, and 65.3°. Based on the above results, the contact angles of all $Co_{40}Fe_{40}W_{20}$ thin films are less than 90°, suggesting that $Co_{40}Fe_{40}W_{20}$ thin films show changes in the direction of higher hydrophilicity. Table 1 shows that the contact angle decreased with the increase of post-annealing temperature. It shows the same behavior as $C_{60}$ and ZnO [25–28].

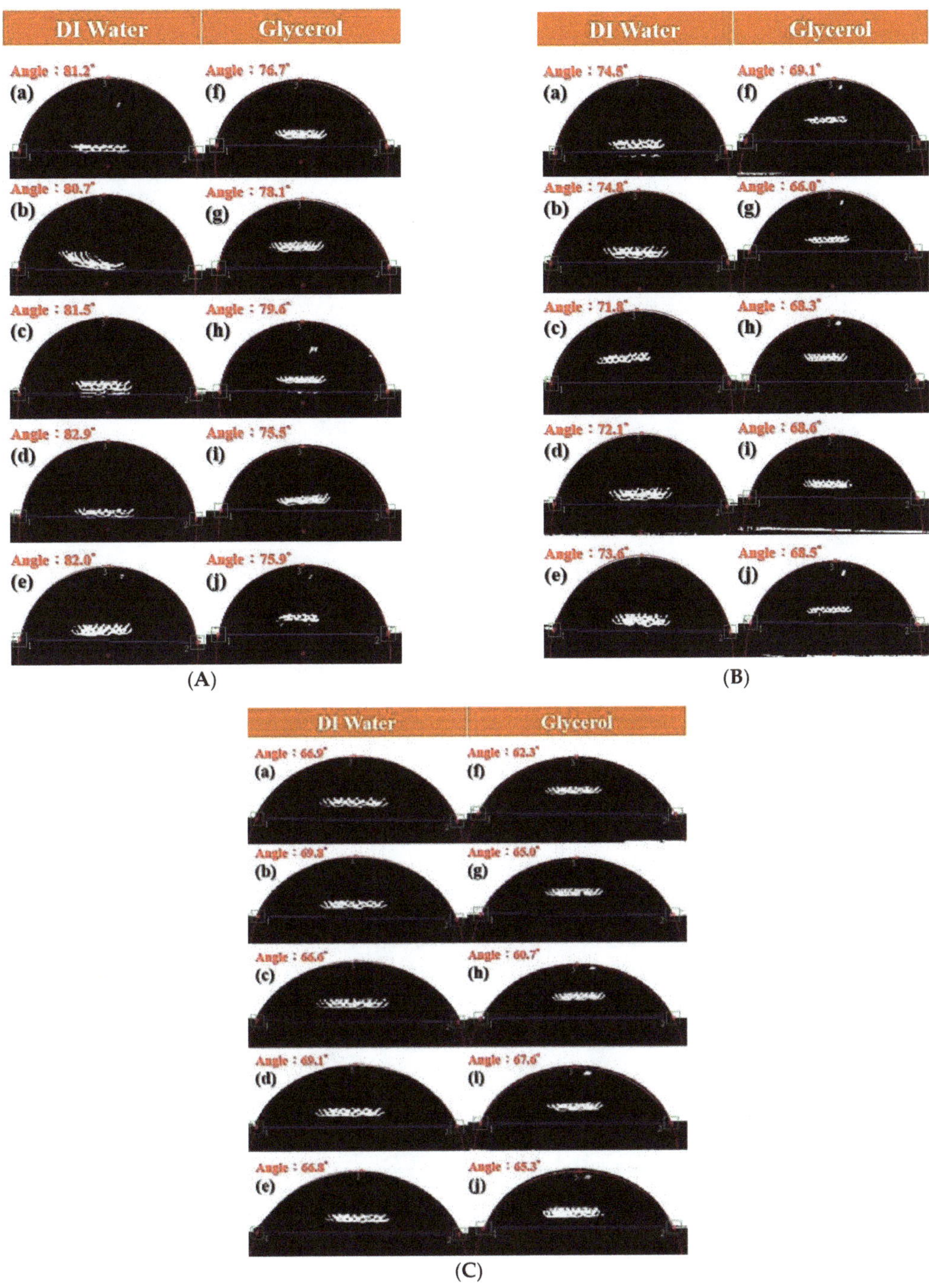

**Figure 3.** Contact angles of the $Co_{40}Fe_{40}W_{20}$ thin films under three conditions: (**A**) as-deposited, (**B**) after post-annealing at 250 °C, (**C**) after post-annealing at 350 °C with DI water: (a) 10 nm, (b) 20 nm, (c) 30 nm, (d) 40 nm, and (e) 50 nm. Contact angles of the $Co_{40}Fe_{40}W_{20}$ thin films with glycerol: (f) 10 nm, (g) 20 nm, (h) 30 nm, (i) 40 nm, and (j) 50 nm.

**Table 1.** Comparing contact angle, surface energy, and grain size for $Co_{40}Fe_{40}W_{20}$ thin films from different fabrication processes.

| Process | Thickness | Contact Angle with DI Water ($\theta$) | Contact Angle with Glycerol ($\theta$) | Surface Energy ($mJ/mm^2$) | Grain Size (nm) |
|---|---|---|---|---|---|
| As-deposited | 10 nm | 81.2° | 76.7° | 24.55 | 33.6 |
| | 20 nm | 80.7° | 78.1° | 24.56 | 35.4 |
| | 30 nm | 81.5° | 79.6° | 24.00 | 40.8 |
| | 40 nm | 82.9° | 75.5° | 25.10 | 48.1 |
| | 50 nm | 82.0° | 75.9° | 24.76 | 48.4 |
| Post-annealing 250 °C | 10 nm | 74.5° | 69.1° | 29.77 | 36.8 |
| | 20 nm | 74.8° | 66.0° | 31.79 | 46.5 |
| | 30 nm | 71.8° | 68.3° | 31.37 | 48.4 |
| | 40 nm | 72.1° | 68.6° | 31.13 | 65.2 |
| | 50 nm | 73.6° | 68.5° | 30.33 | 77.1 |
| Post-annealing 350 °C | 10 nm | 66.9° | 62.3° | 35.42 | 56.5 |
| | 20 nm | 69.8° | 65.0° | 33.17 | 61.9 |
| | 30 nm | 66.6° | 60.7° | 36.02 | 66.9 |
| | 40 nm | 69.1° | 67.6° | 33.69 | 70.4 |
| | 50 nm | 66.8° | 65.3° | 35.58 | 93.7 |

*3.4. Surface Energy*

The surface energy of films is significant as it relates to the adhesion of thin films. When CoFeW thin films are used as a seed layer or buffer layer, strong adhesion of thin films is essential. The data of contact angles are used to calculate the surface energy using the Young equation [29,30].

The Young Equation (2) is

$$\sigma_{sg} = \sigma_{sl} + \sigma_{lg} \cos\theta \tag{2}$$

where $\sigma_{sg}$ is the surface free energy of the solid, and $\sigma_{sl}$ denotes the interfacial tension between liquid and solid. $\sigma_{lg}$ is the surface tension of the liquid, and $\theta$ is contact angle. Figure 4 exhibits the surface energy of the $Co_{40}Fe_{40}W_{20}$ thin films. These data are further shown in Table 1. It can be observed that the surface energy of annealed CoFeW films was higher than that of the deposited films. As the post-annealing temperature increased, the surface energy also increased, and the surface energy of as-deposited CoFeW thin films reached 25.1 $mJ/mm^2$ at 40 nm, which was the highest value. When the post-annealing temperature was at 250 °C, the highest surface energy at 20 nm was 31.79 $mJ/mm^2$. When the post-annealing temperature reached 350 °C, the surface energy was 36.02 $mJ/mm^2$ at 30 nm. According to the XRD results [22], the formation of oxide layers on the thin films' surface resulted in decreased contact angles and increased surface energy [31]. These CoFeW thin film data are shown in Table 1. According to the grain size result of Figure 1b and Table 1, the annealed treatment produces larger grain size distribution than as-deposited treatment. From the result of Figure 4, it suggests that the surface energy of annealed 350 °C is larger about 1.5 times than as-deposited condition. It can be reasonably concluded that when the grains are arranged in the material, the large grains of annealed material show large gaps between adjacent grains. The gaps become larger, and the support between the crystal grains will be reduced, so the crystal grain size increases and leads to a contact angle reduction trend. When water drops on the surface, it is easy for water drops to flow into the gap, resulting in a small contact angle, strong adhesion, and high surface energy. It shows the same behavior as PTFE films [32]. In contrast, the small grains of as-deposited material show small gaps and the grains are tightly arranged. When the water droplets fall on the surface, the water droplets have difficulty flowing to the gaps, resulting in a large contact angle, weak adhesion, and a low surface energy.

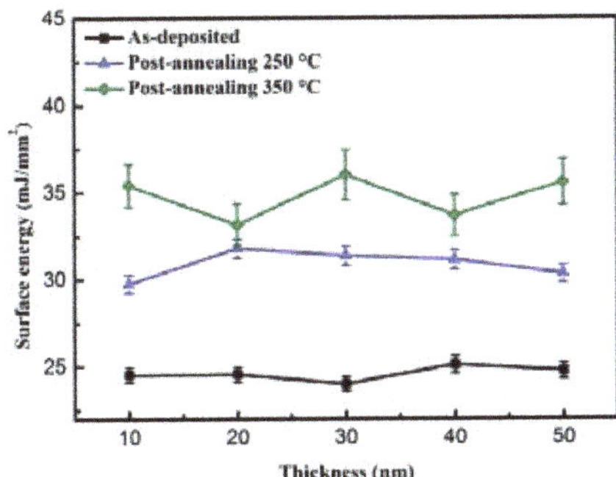

**Figure 4.** The surface energy of CoFeW thin films.

Figure 5a–e shows the relationship between contact angle, surface energy, and post-annealing temperature. It can be found that the surface energy of the CoFeW films increased as their contact angles decreased. It results from the Young equation [29,30]. After the post-annealing, the surface energy of CoFeW films tends to increase, suggesting that the post-annealing is significant for the surface energy of the films. When CoFeW thin films have high surface energy, the adhesion is the strongest. The adhesion depends on surface energy, but about adhesion force decides the so-called weak boundary layer. It means there can be strong adhesion from the top layer with high surface energy what decreases cohesion to the layer below, and total adhesion between layers is very low due to the presence of created weak boundary layers. The outcomes represent that a thin film is easier to combine with other layers in forming MTJs. The CoFeW film can be a free or pinned layer in MTJ. To detect the surface energy and adhesion of CoFeW performance, Table 2 is compared with other specific CoFeBY and CoFeW materials under various Si and Glass substrates. Table 2 demonstrates that the surface energy of current research is larger than other CoFeBY and CoFeW materials, which indicates that the CoFeW film is more compatible with other layers for MTJ.

**Table 2.** Comparing surface energy for specific CoFeW and CoFeBY thin films from different substrates.

| Material | Surface Energy (mJ/mm$^2$) | Factor |
| --- | --- | --- |
| Si(100)/Co$_{40}$Fe$_{40}$W$_{20}$ (✱ Current research) | 24.00–36.02 | Because thin films have largest grain size, there is highest surface energy. |
| Glass/CoFeBY [10] | 23.89–31.07 | The crystallinity of samples was weak. |
| Si(100)/CoFeBY [31] | 24.55–31.85 | The crystallinity of samples was weak. |
| Si(100)/Co$_{40}$Fe$_{40}$W$_{20}$ [33] | 23.61–30.12 | At 42 nm, because the crystal of thin films have highest surface energy in paper. |
| Glass/Co$_{32}$Fe$_{30}$W$_{38}$ [34] | 22.3–28.6 | Crystallinity: Si(100) > glass therefore, the surface energy of Glass/Co$_{32}$Fe$_{30}$W$_{38}$ is lower. |

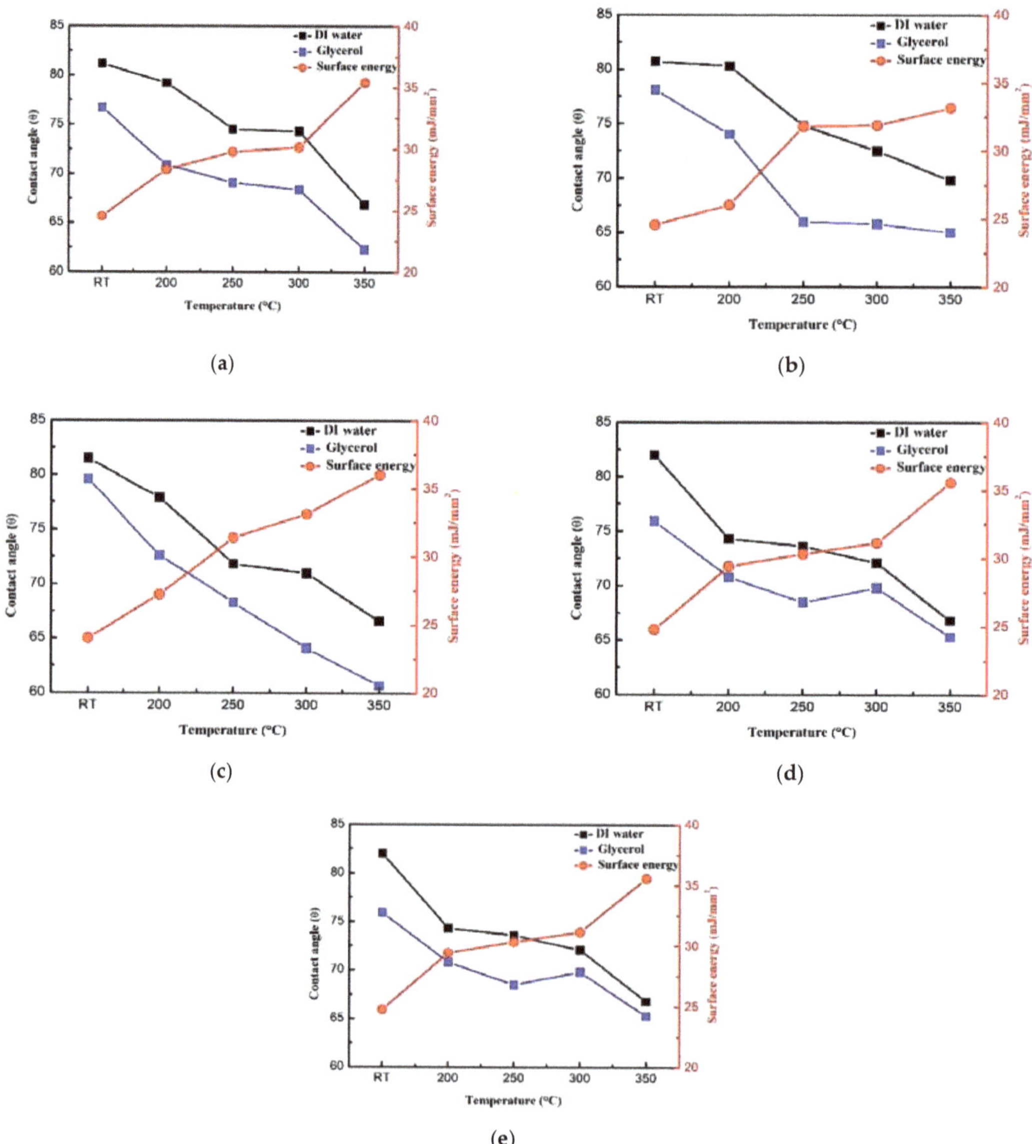

**Figure 5.** The relationship between contact angle, surface energy, and post-annealing temperature. (**a**) 10 nm, (**b**) 20 nm, (**c**) 30 nm, (**d**) 40 nm, (**e**) 50 nm.

### 3.5. Electric Property

The sheet resistance (Rs) is shown in Figure 6 with various thicknesses and temperature conditions. The result of the present study suggests that the sheet resistance is decreased with the increase of thickness and annealed temperatures. The sheet resistance of the as-deposited and thinner CoFeW films is larger than that of the annealed and thicker CoFeW films because the as-deposited and thinner CoFeW films reduce the electron mobility through the films and enhance the scattering of electrons at grain boundaries and impurities, which shows the same behavior as CdS and CoFe films [35,36].

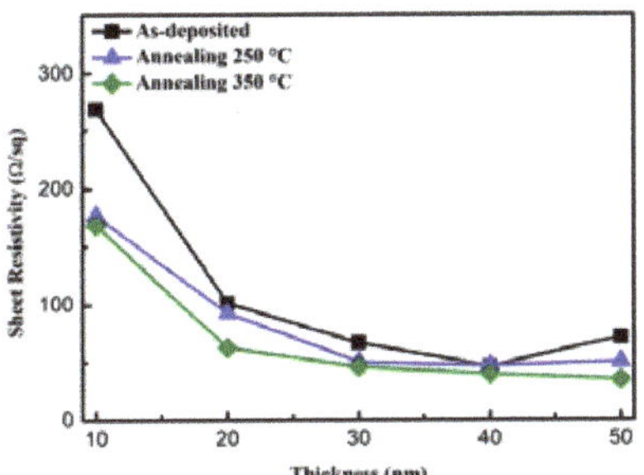

**Figure 6.** The sheet resistance of CoFeW films.

*3.6. Nano-Indentation*

Figure 7a,b show that the hardness and Young's modulus increase as the thickness of the CoFeW film increases. Mostly, the nanoindentation hardness is determined from the loading and unloading curve by the Pharr–Oliver method [37], which indicates the mixed hardness of the silicon substrate and the CoFeW film. Since the thickness of the CoFeW film is too thin, it can be reasonably concluded that there must be a substrate effect in the nanoindentation measurement. In the nanoindentation measurement, the corresponding hardness and Young's modulus values of the substrate are 4.1 and 133.3 Gpa, when the Si(100) substrate is measured. As the thickness increased from 10 nm to 50 nm under as-deposited condition, the hardness and Young's modulus of the CoFeW films increased from 12.9 to 13.4 GPa and 146.8 to 187.2 GPa, respectively. At annealed 250 °C, as the thickness increases from 10 nm to 50 nm, the hardness and Young's modulus of the CoFeW films increased from 10.9 to 13.5 GPa and 169.4 to 188.5 GPa, respectively. At annealed 350 °C, as the thickness increases from 10 to 50 nm, the hardness and Young's modulus of the CoFeW films increased from 12.0 to 13.4 GPa and 163.5 to 186.4 GPa, respectively. When the thickness is from 10 to 50 nm, the hardness and Young's modulus of CoFeW films display a tendency to saturate. According to the result, the Young's modulus of adding the W effect to CoFe films in thicker films is larger than CoFe film. The thinner CoFeW film has a more significant impact on the substrate. It shows a similar behavior as diamond films on crystalline silicon [38].

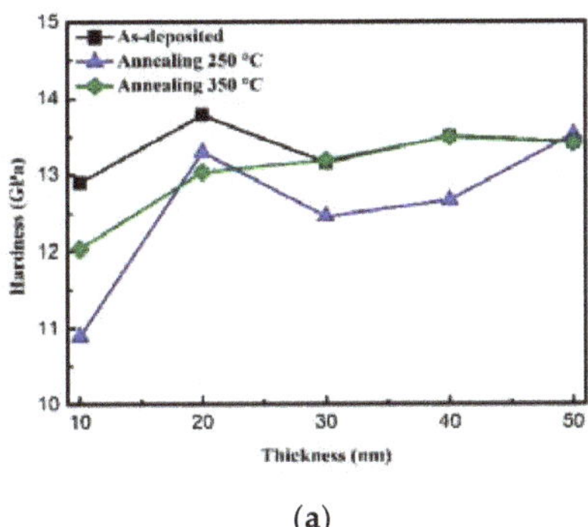

(a)

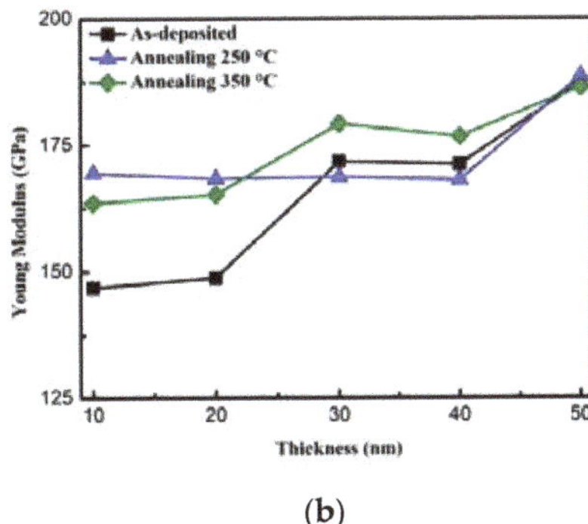

(b)

**Figure 7.** Nano-indentation of CoFeW films. (**a**) Hardness and (**b**) Young's modulus.

## 4. Conclusions

This study investigated the structure and surface property of CoFeW thin films. The CoFeW magnetic thin films were deposited on a Si(100) substrate by sputtering, while the Scherrer equation was estimated to calculate the grain size of CoFe(110). The results indicate that the grain size increased along with the thickness of the CoFeW films, and the grain size was affected by the temperature of post-annealing. When the post-annealing temperature increased, the grain size also increased, suggesting that the post-annealing process provided energy to CoFeW thin films and caused the grain size to grow. The contact angles of all CoFeW samples were less than $90°$, indicating that CoFeW thin films show changes in the direction of higher hydrophilicity. The contact angle decreased along with the increase of post-annealing temperature, while the grain size affected the contact angle of CoFeW films. The contact angle of CoFeW thin films reduced when the grain size of CoFe(110) increased. Moreover, the formation of the oxide layers on the thin films' surface resulted in a decrease in the contact angle and an increase in the surface energy. The results suggest that the surface energy of the annealed CoFeW films was higher than the as-deposited CoFeW samples. When the post-annealing temperature increased, the surface energy increased, thus making the adhesion stronger. In the future, we hope to combine the layered MTJs with free and pinning layers. Due to the scattering of electrons at the grain boundaries or the presence of impurities, the sheet resistance decreases as the film thickness and annealing temperature increase. In addition, as the thickness increases, the hardness and Young's modulus of the CoFeW film shows a tendency to saturate.

**Author Contributions:** Conceptualization, W.-J.L., Y.-H.C., S.-H.L. and Y.-T.C.; methodology, Y.-T.C. and T.-Y.J.; validation, formal analysis, Y.-T.C.; investigation, Y.-T.C. and W.-J.L.; resources, C.-L.F., K.-W.L. and T.-H.W.; writing—original draft preparation, Y.-T.C.; writing—review and editing, Y.-T.C. and W.-J.L.; supervision, Y.-T.C. and P.-C.C.; project administration, Y.-T.C. and T.-H.W.; funding acquisition, W.-J.L. and Y.-H.C. All authors have read and agreed to the published version of the manuscript.

**Funding:** This work was supported by the Ministry of Science and Technology (grant No. MOST108-2221-E-224-015-MY3 and MOST105-2112-M-224-001) and the National Yunlin University of Science and Technology (grant No. 110T06).

**Institutional Review Board Statement:** Not applicable.

**Informed Consent Statement:** Not applicable.

**Data Availability Statement:** The data presented in this study are available on reasonable request from the corresponding author.

**Acknowledgments:** The author thanks Sin-Liang Ou for his help with the SEM experiment and images.

**Conflicts of Interest:** The authors declare that there is no conflict of interest regarding the publication of this paper. The funders had no role in the design of the study; in the collection, analyses, or interpretation of data; in the writing of the manuscript, or in the decision to publish the results.

# References


1. Li, M.; Wang, S.; Zhang, S.; Fang, S.; Feng, G.; Cao, X.; Zhang, P.; Wang, B.; Yu, G. The effect of interfacial oxygen migration on the PMA and thermal stability in MTJ with double MgO layers. *Appl. Surf. Sci.* **2019**, *488*, 30–35. [CrossRef]
2. Iihama, S.; Mizukami, S.; Naganuma, H.; Oogane, M.; Ando, Y.; Miyazaki, T. Gilbert damping constants of Ta/CoFeB/MgO(Ta) thin films measured by optical detection of precessional magnetization dynamics. *Phys. Rev. B* **2014**, *89*, 174416. [CrossRef]
3. Ikeda, S.; Miura, K.; Yamamoto, H.; Mizunuma, K.; Gan, H.D.; Endo, M.; Kanai, S.; Hayakawa, J.; Matsukura, F.; Ohno, H. A perpendicular-anisotropy CoFeB-MgO magnetic tunnel junction. *Nat. Mater.* **2010**, *9*, 721–724. [CrossRef]
4. Wang, S.; Li, M.; Zhang, S.; Fang, S.; Wang, D.; Yu, G. High annealing tolerance and the microstructure study in perpendicular magnetized MgO/CoFeB/MgO structures with thin W spacer layer. *J. Magn. Magn. Mater.* **2019**, *479*, 121–125. [CrossRef]
5. Wolf, S.A.; Awschalom, D.D.; Buhrman, R.A.; Daughton, J.M.; Molnar, S.V.; Roukes, M.L.; Chtchelkanova, A.Y.; Treger, D.M. Spintronics: A spin-based electronics vision for the future. *Science* **2001**, *294*, 1488–1495. [CrossRef]
6. Ikeda, S.; Hayakawa, J.; Ashizawa, Y.; Lee, Y.M.; Miura, K.; Hasegawa, H.; Tsunoda, M.; Matsukura, F.; Ohno, H. Tunnel magnetoresistance of 604% at 300K by suppression of Ta diffusion in CoFeB/MgO/CoFeB pseudo-spin-valves annealed at high temperature. *Appl. Phys. Lett.* **2008**, *93*, 082508. [CrossRef]
7. Parkin, S.S.P.; Kaiser, C.; Panchula, A.; Rice, P.M.; Hughes, B.; Samant, M.; Yang, S.H. Giant tunnelling magnetoresistance at room temperature with MgO (100) tunnel barriers. *Nat. Mater.* **2004**, *3*, 862–867. [CrossRef] [PubMed]

8.   Kiselev, S.I.; Sankey, J.C.; Krivorotov, I.N.; Emley, N.C.; Schoelkopf, R.J.; Buhrman, R.A.; Ralph, D.C. Microwave oscillations of a nanomagnet driven by a spin-polarized current. *Nature* **2003**, *425*, 380–383. [CrossRef]
9.   Parkin, S.S.P.; Hayashi, M.; Thomas, L. Magnetic domain-wall racetrack memory. *Science* **2008**, *320*, 190–194. [CrossRef] [PubMed]
10.  Liu, W.J.; Chang, Y.H.; Chen, Y.T.; Chiang, Y.C.; Tsai, D.Y.; Wu, T.H.; Chi, P.W. Effect of annealing on the characteristics of CoFeBY thin films. *Coatings* **2021**, *11*, 250. [CrossRef]
11.  Li, M.; Shi, H.; Yu, G.; Lu, J.; Chen, X.; Han, G.; Yu, G.; Amiri, P.K.; Wang, K.L. Effects of annealing on the magnetic properties and microstructures of Ta/Mo/CoFeB/MgO/Ta films. *J. Alloys Compd.* **2017**, *692*, 243–248. [CrossRef]
12.  Almasi, H.; Hickey, D.R.; Newhouse-Illige, T.; Xu, M.; Rosales, M.R.; Nahar, S.; Held, J.T.; Mkhoyan, K.A.; Wang, W.G. Enhanced tunneling magnetoresistance and perpendicular magnetic anisotropy in Mo/CoFeB/MgO magnetic tunnel junctions. *Appl. Phys. Lett.* **2015**, *106*, 182406. [CrossRef]
13.  Huang, S.X.; Chen, T.Y.; Chien, C.L. Spin polarization of amorphous CoFeB determined by point-contact Andreev reflection. *Appl. Phys. Lett.* **2008**, *92*, 242509. [CrossRef]
14.  Paluskar, P.V.; Attema, J.J.; de Wijs, G.A.; Fiddy, S.; Snoeck, E.; Kohlhepp, J.T.; Swagten, H.J.M.; de Groot, R.A.; Koopmans, B. Spin tunneling in junctions with disordered ferromagnets. *Phys. Rev. Lett.* **2008**, *100*, 057205. [CrossRef]
15.  Li, M.; Wang, S.; Zhang, S.; Fang, S.; Yu, G. The perpendicular magnetic anisotropies of CoFeB/MgO films with Nb buffer layers. *J. Magn. Magn. Mater.* **2019**, *485*, 187–192. [CrossRef]
16.  Meng, H.; Lum, W.H.; Sbiaa, R.; Lua, S.Y.H.; Tan, H.K. Annealing effects on CoFeB-MgO magnetic tunnel junctions with perpendicular anisotropy. *J. Appl. Phys.* **2011**, *110*, 033904. [CrossRef]
17.  Liu, T.; Cai, J.W.; Sun, L. Large enhanced perpendicular magnetic anisotropy in CoFeB/MgO system with the typical Ta buffer replaced by an Hf layer. *AIP Adv.* **2012**, *2*, 032151. [CrossRef]
18.  Gottwald, M.; Kan, J.J.; Lee, K.; Zhu, X.; Park, C.; Kang, S.H. Scalable and thermally robust perpendicular magnetic tunnel junctions for STT-MRAM. *Appl. Phys. Lett.* **2015**, *106*, 032413. [CrossRef]
19.  Cuchet, L.; Rodmacq, B.; Auffret, S.; Sousa, R.C.; Prejbeanu, I.L.; Dieny, B. Perpendicular magnetic tunnel junctions with a synthetic storage or reference layer: A new route towards Pt- and Pd-free junctions. *Sci. Rep.* **2016**, *6*, 21246. [CrossRef]
20.  Kaidatzis, A.; Bran, C.; Psycharis, V.; Vázquez, M.; Martín, J.M.G.; Niarchos, D. Tailoring the magnetic anisotropy of CoFeB/MgO stacks onto W with a Ta buffer layer. *Appl. Phys. Lett.* **2015**, *106*, 262401. [CrossRef]
21.  Ghaferi, Z.; Sharafi, S.; Bahrololoom, M.E. The role of electrolyte pH on phase evolution and magnetic properties of CoFeW codeposited films. *Appl. Surf. Sci.* **2016**, *375*, 35–41. [CrossRef]
22.  Liu, W.J.; Chang, Y.H.; Chen, Y.T.; Jhou, T.Y.; Chen, Y.H.; Wu, T.H.; Chi, P.W. Impact of annealing on the magnetic and structural of $Co_{40}Fe_{40}W_{20}$ thin films on Si(100) substrate. *Materials* **2021**, *14*, 3017. [CrossRef] [PubMed]
23.  Cullity, B.D. *Elements of X-ray Diffraction*, 2nd ed.; Addison-Wesley: Reading, MA, USA, 1978.
24.  Patterson, A.L. The Scherrer formula for X-Ray particle size determination. *Phys. Rev.* **1939**, *56*, 978–982. [CrossRef]
25.  Kitamura, M.; Kuzumoto, Y.; Kamura, M.; Aomori, S.; Na, J.H.; Arakawa, Y. Low-voltage-operating fullerene C60 thin-film transistors with various surface treatments. *Phys. Stat. Solidi C* **2008**, *5*, 3181–3183. [CrossRef]
26.  Beaini, S.S.; Kronawitter, C.X.; Carey, V.P.; Mao, S.S. ZnO deposition on metal substrates: Relating fabrication, morphology and wettability. *J. Appl. Phys.* **2013**, *113*, 184905. [CrossRef]
27.  Yao, G.; Zhang, M.; Lv, J.; Xu, K.; Shi, S.; Gong, Z.; Tao, J.; Jiang, X.; Yang, L.; Cheng, Y.; et al. Effects of electrodeposition electrolyte concentration on microstructure, optical properties and wettability of ZnO nanorods. *J. Electrochem. Soc.* **2015**, *162*, D300–D304. [CrossRef]
28.  Dhaygude, H.D.; Shinde, S.K.; Velhal, N.B.; Takale, M.V.; Fulari, V.J. Doping effect on SILAR synthesized crystalline nanostructured Cudoped ZnO thin films grown on indium tin oxide (ITO) coated glass substrates and its characterization. *Mater. Res. Express* **2016**, *3*, 086402. [CrossRef]
29.  Owens, D.K.; Wendt, R.C. Estimation of the surface free energy of polymers. *J. Appl. Polym. Sci.* **1969**, *13*, 1741–1747. [CrossRef]
30.  Kaelble, D.H.; Uy, K.C. A Reinterpretation of organic liquid-polytetrafluoroethylene surface interactions. *J. Adhens.* **1970**, *2*, 50–60. [CrossRef]
31.  Liu, W.J.; Chang, Y.H.; Chen, Y.T.; Chiang, Y.C.; Liu, Y.C.; Wu, T.H.; Chi, P.W. Effect of annealing on the structural, magnetic and surface energy of CoFeBY films on Si(100) substrate. *Materials* **2021**, *14*, 987. [CrossRef]
32.  Pachchigar, V.; Ranjan, M.; Mukherjee, S. Role of hierarchical protrusions in water repellent superhydrophobic PTFE surface produced by low energy ion beam irradiation. *Sci. Rep.* **2019**, *9*, 8675. [CrossRef]
33.  Liu, W.J.; Chang, Y.H.; Ou, S.L.; Chen, Y.T.; Liang, Y.C.; Huang, B.J.; Chu, C.L.; Chiang, C.C.; Wu, T.H. Magnetic properties, adhesion, and nanomechanical property of $Co_{40}Fe_{40}W_{20}$ films on Si (100) substrate. *J. Nanomater.* **2020**, *2020*, 1584310. [CrossRef]
34.  Liu, W.J.; Chang, Y.H.; Ou, S.L.; Chen, Y.T.; Li, W.H.; Jhou, T.Y.; Chu, C.L.; Wu, T.H.; Tseng, S.W. Effect of annealing on the structural, magnetic, surface energy and optical properties of $Co_{32}Fe_{30}W_{38}$ films deposited by direct-current magnetron sputtering. *Coatings* **2020**, *10*, 1028. [CrossRef]
35.  Jassim, S.A.J.; Zumaila, A.A.R.A.; Waly, G.A.A.A. Influence of substrate temperature on the structural, optical and electrical properties of CdS thin films deposited by thermal evaporation. *Res. Phys.* **2013**, *3*, 173–178. [CrossRef]
36.  Redjdal, N.; Salah, H.; Hauet, T.; Menari, H.; Chérif, S.M.; Gabouze, N.; Azzaz, M. Microstructural electrical and magnetic properties of $Fe_{35}Co_{65}$ thin films grown by thermal evaporation from mechanically alloyed powders. *Thin Solid Films* **2014**, *552*, 164–169. [CrossRef]

37. Oliver, W.C.; Pharr, G.M. An improved technique for determining hardness and elastic modulus using load and displacement sensing indentation experiments. *J. Mater. Res.* **1992**, *7*, 1564–1583. [CrossRef]
38. Zhu, J.; Han, J.; Liu, A.; Meng, S.; Jiang, C. Mechanical properties and Raman characterization of amorphous diamond films as a function of film thickness. *Surf. Coat. Technol.* **2007**, *201*, 6667–6669. [CrossRef]

*Article*

# Evaluation of the Durability of Slippery, Liquid-Infused Porous Surfaces in Different Aggressive Environments: Influence of the Chemical-Physical Properties of Lubricants

**Federico Veronesi *, Guia Guarini, Alessandro Corozzi and Mariarosa Raimondo**

National Research Council—Institute of Science and Technology for Ceramics CNR-ISTEC, 48022 Faenza, Italy; istec@istec.cnr.it (G.G.); alessandro.corozzi@istec.cnr.it (A.C.); mariarosa.raimondo@istec.cnr.it (M.R.)

* Correspondence: federico.veronesi@istec.cnr.it; Tel.: +39-054-669-9727

**Abstract:** Liquid-repellent surfaces have been extensively investigated due to their potential application in several fields. Superhydrophobic surfaces achieve outstanding water repellence, however their limited durability in severe operational conditions hinders their large-scale application. The Slippery, Liquid-Infused Porous Surface (SLIPS) approach solves many of the durability problems shown by superhydrophobic surfaces due to the presence of an infused liquid layer. Moreover, SLIPS show enhanced repellence towards low surface tension liquids that superhydrophobic surfaces cannot repel. In this perspective, SLIPS assume significant potential for application in harsh environments; however, a systematic evaluation of their durability in different conditions is still lacking in the literature. In this work, we report the fabrication of SLIPS based on a ceramic porous layer infused with different lubricants, namely perfluoropolyethers with variable viscosity and n-hexadecane; we investigate the durability of these surfaces by monitoring the evolution of their wetting behavior after exposure to severe environmental conditions like UV irradiation, chemically aggressive solutions (acidic, alkaline, and saline), and abrasion. Chemical composition and viscosity of the infused liquids prove decisive in determining SLIPS durability; especially highly viscous infused liquids deliver enhanced resistance to abrasion stress and chemical attack, making them candidates for applicable, long-lasting liquid-repellent surfaces.

**Keywords:** liquid-infused surfaces; durability; lubricants; wetting; liquid-repellent coatings

**Citation:** Veronesi, F.; Guarini, G.; Corozzi, A.; Raimondo, M. Evaluation of the Durability of Slippery, Liquid-Infused Porous Surfaces in Different Aggressive Environments: Influence of the Chemical-Physical Properties of Lubricants. *Coatings* **2021**, *11*, 1170. https://doi.org/10.3390/coatings11101170

Academic Editor: Rubén González

Received: 6 September 2021
Accepted: 23 September 2021
Published: 27 September 2021

**Publisher's Note:** MDPI stays neutral with regard to jurisdictional claims in published maps and institutional affiliations.



## 1. Introduction

Liquid-repellent surfaces have drawn huge interest in the last years due to their inherent self-cleaning properties [1–3] and other potential related advantages e.g., anti- or de-icing properties [4–6], drag and friction reduction [7–9], and anti-fouling behavior [10,11]. All these properties might have positive fallout on a wide range of industrial fields and applications like aircraft, building materials, ships, machinery, and countless others. Therefore, the fabrication of liquid-repellent surfaces has become one of the hottest topics in the material science community. The first and most well-known approach to the fabrication of liquid-repellent surfaces starts from the mimicry of the dual-scale (i.e., hierarchical) surface structure observed on the lotus leaf [12]. By coupling the creation of micro/nanoscale surface features with the tailoring of surface chemical composition to achieve low surface energy, superhydrophobic surfaces can be obtained [13,14]. These materials are characterized by high water contact angles, i.e., larger than 150°, and high mobility of water drops that are placed on their surface. Such extreme water repellence is due to the trapping of air pockets between surface morphological features, leading to minimal adhesion between water drops and the solid surface. However, the lotus leaf-like surfaces show some limitations: for instance, they are not able to repel liquids with a surface tension lower than that of water (72 mN·m$^{-1}$ at 25 °C) like oils and alkanes. For that reason, the scientific community started to design surfaces with simultaneous repellence to water (i.e., hydrophobicity) and other liquids, mainly non-polar oils (i.e., oleophobicity). Such combination is commonly termed as amphiphobicity [15,16] or, in case the repellence is extended also to complex fluids like blood or milk, omniphobicity [17,18].

Another limitation of the lotus-mimicking approach is that the trapped air pockets can be displaced by fast impinging drops or when the liquid pressure is too high. Perhaps the most promising fabrication approach to overcome this limit has been proposed by Wong et al. [19] in 2011 and takes inspiration from the surface of Nepenthes plant [20], which is covered by a layer of liquid lubricant that makes it extremely slippery for the insects that land on it. The proposed Slippery, Liquid-Infused Porous Surfaces (SLIPS) display much lower contact angles compared to superhydrophobic surfaces but drops of many liquids can still move very easily on them. Moreover, these materials possess much improved stability when submerged, along with other enhanced properties [21,22].

Due to such exciting potential, SLIPS have drawn large interest and many papers have focused on the design criteria to achieve the best omniphobicity [23]. In principle, SLIPS can be fabricated by using any liquid as infused lubricant, but three fundamental criteria for SLIPS design always hold: the fluid to be repelled must be immiscible with the infused lubricant; the lubricant must wet the solid structures completely, even when no outer liquid is present; the outer liquid must form discrete droplets on the SLIPS. Notwithstanding these design guidelines for SLIPS, the literature still lacks an assessment of their durability in different environments that simulate their performance in real operational conditions. Moreover, a systematic investigation of the relationship between infused lubricant and durability is missing.

In this paper, we report the fabrication of SLIPS infused with different lubricants and study the evolution of their wetting properties after testing in several conditions, namely UV irradiation, immersion in chemically aggressive solutions, and abrasion. Both fluorinated and fluorine-free lubricants were tested, as the former are known for inducing excellent oleophobicity but are also deemed to be environmentally harmful [24], while the latter can only generate hydrophobicity but are much more eco-friendly. The results provide relevant insight on the behavior of SLIPS and draw guidelines for the choice of the lubricants to be infused for long-lasting repellence.

## 2. Materials and Methods

### 2.1. SLIPS Fabrication

The nanostructured pseudoboehmite AlOOH coating was obtained by deposition of an alumina suspension synthesized via a previously reported sol-gel route [25]. Aluminum samples (Al 1050 alloy) were used as substrates; prior to deposition, they were cleaned with an ultrasonic bath in ethanol for 5 min and dried in air. The synthesized alumina nanoparticles were deposited on the substrates via dip-coating in controlled conditions, followed by treatment with boiling water. The complete procedure for the coating fabrication is reported elsewhere [26]. Pseudoboehmite-coated samples were observed by Scanning Electron Microscopy (FESEM Gemini Columns SIGMA, Carl Zeiss Microscopy GmbH, Oberkochen, Germany).

In order to increase the affinity between the nanostructured, porous coating and the infused lubricant, the former was chemically modified depending on the nature of the latter. When n-hexadecane was used as lubricant, alkyl chains were grafted to pseudoboehmite by immersion in a hexadecyltrimethoxysilane ($\geq$85%, Merck, Darmstadt, Germany) solution in ethanol (6 wt.%) for 2 min, followed by drying at 80 °C for 1 h and annealing at 170 °C for 5 min. Otherwise, when Perfluoropolyethers (PFPE) were used as lubricants, fluorinated alkyl chains were grafted to the nanostructured AlOOH by immersion in a commercial solution of fluoroalkylsilane in isopropyl alcohol (Protectosil SC200, Evonik, Essen, Germany) for 2 min, followed by drying in air and annealing at 150 °C for 30 min.

Finally, lubricants were infused in the nanostructured layer by brushing until a continuous, shiny layer covered the sample surface. In order to remove excess oil, the samples were held vertical overnight. Five different lubricants were used: four commercial perfluoropolyethers PFPE with increasing viscosity (Krytox GPL 100, 103, 105, and 107, Chemours, Geneva, Switzerland), or n-hexadecane ($\geq$99%, Merck). The resulting SLIPS were labeled as K100, K103, K105, K107, and HEX, respectively.

### 2.2. SLIPS Characterization

The wetting behavior of SLIPS was determined by measuring the Advancing (ACA) and Receding Contact Angles (RCA) with water and n-hexadecane drops to calculate Contact Angle Hysteresis (CAH) as the difference between the two. Advancing and receding contact angles and contact angle hysteresis with water and n-hexadecane are labeled as $ACA_W$, $RCA_W$, $CAH_W$, $ACA_H$, $RCA_H$, and $CAH_H$ in the text. We chose not to measure "static" contact angles, which are commonly reported in most papers about liquid-repellent surfaces, as they do not adequately describe the behavior of liquid drops on SLIPS. Indeed, static contact angles rarely exceed 120° on SLIPS, while the low CAH well shows the little adhesion of drops on the surface. Contact angle measurements were performed with an optical contact angle system (DSA 30S, Krüss GmbH, Hamburg, Germany). For all contact angle measurements, 5 μL drops were first dispensed with a software-controlled syringe and gently deposited on the surface. Then, 5 μL were added to the drop and the ACA measured during drop expansion. Finally, 5 μL were removed from the drop and the RCA measured to calculate CAH as the difference between ACA and RCA. For each surface, at least five different points were characterized to calculate average ACA, RCA, and CAH with related standard deviations.

The evolution of chemical composition of coated surfaces was monitored via Fourier Transform Infrared spectroscopy (FTIR) using a Nicolet IS5 spectrophotometer (ThermoFisher Scientific, Waltham, MA, USA) in the Attenuated Total Reflection mode. A diamond crystal was pressed against the samples; each spectrum was collected in the 4000–550 $cm^{-1}$ range with a resolution of 4 $cm^{-1}$ and averaged over 16 scans. For all spectra, the non-infused face of the pristine sample was used as background to obtain the spectrum of the infused liquid phase.

### 2.3. Durability Tests

SLIPS were subjected to three types of tests to assess their durability in different aggressive environments.

UV irradiation tests were performed exposing the samples to a lamp (UVGL-58, UVP International, Jena, Germany) with a radiation intensity of 2.0 mW $cm^{-2}$ measured at $\lambda = 354$ nm. The samples were collected, and their wetting properties assessed after 2, 4, 6, and 8 h of irradiation.

Chemical ageing tests were performed by immersing SLIPS in three different aqueous solutions: acidic (pH = 3, obtained by dilution of 3.3 g·$L^{-1}$ of acetic acid, $\geq$99%, Merck), alkaline (pH = 11, obtained by dissolution of 0.04 g·$L^{-1}$ of sodium hydroxide, 99%, Merck), and saline (obtained by dissolution of 100 g·$L^{-1}$ of sodium chloride $\geq$99%, Merck). After fixed amounts of time, samples were withdrawn from the solutions, rinsed with deionized water, dried in air, and characterized in terms of wetting properties, then re-immersed in the solutions.

Abrasion tests were performed as per UNI EN 1096-2 standard, which is used to assess the mechanical properties of coated glasses in the building industry. Samples were fixed to a technical balance with double-sided adhesive tape and a rotating felt disk (diameter 60 mm, rotational speed 60 rpm) was pushed against the sample until a weight of about 400 g, equivalent to a force of about 4 N, was measured by the balance. The abrasion continued for 30 s, then the sample was removed, and its wetting properties re-evaluated.

## 3. Results and Discussion

### 3.1. SLIPS Characterization

For all investigated SLIPS, the same nanostructured pseudoboehmite AlOOH coating was used to retain the infused oil [27]. Such coating is made of randomly oriented lamellae (about 200 nm long and few nm thick) separated by 30–50 nm voids which serve as pockets for lubricant retention. The morphology of the, as-prepared, pseudoboehmite coating on aluminum is reported in Figure 1.

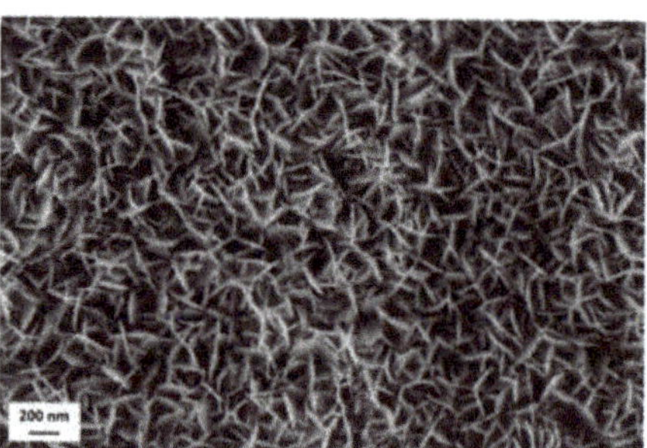

**Figure 1.** Scanning electron micrograph of the nanostructured pseudoboehmite layer. Scale bar is reported.

AlOOH coatings were infused with either PFPE or n-hexadecane. Table 1 reports all measured contact angle values for as-fabricated SLIPS.

**Table 1.** Kinematic viscosity of the infused oils at $T$ = 40 °C and wetting data of related infused surfaces (ACA: Advancing Contact Angle, RCA: Receding Contact Angle, and CAH: Contact Angle Hysteresis; subscript W: Water and subscript H: Hexadecane). Standard deviations are reported as errors. Viscosity values were obtained from the suppliers.

| Infused Liquid | Viscosity [mm²/s] | $ACA_W$ [°] | $RCA_W$ [°] | $CAH_W$ [°] | $ACA_H$ [°] | $RCA_H$ [°] | $CAH_H$ [°] |
|---|---|---|---|---|---|---|---|
| Krytox 100 | 7.8 | 124.2 ± 0.7 | 122.2 ± 0.4 | 1.9 ± 0.9 | 71.4 ± 0.4 | 69.5 ± 0.4 | 1.9 ± 0.5 |
| Krytox 103 | 30 | 121.8 ± 0.6 | 117.7 ± 1.3 | 4.2 ± 1.9 | 70.7 ± 0.4 | 68.1 ± 0.9 | 2.6 ± 0.6 |
| Krytox 105 | 160 | 121.5 ± 1.5 | 115.6 ± 0.9 | 5.9 ± 0.7 | 71.8 ± 1.1 | 65.5 ± 1.0 | 6.3 ± 0.8 |
| Krytox 107 | 450 | 120.9 ± 0.4 | 107.8 ± 1.0 | 13.1 ± 1.3 | 73.5 ± 0.9 | 61.1 ± 1.8 | 12.4 ± 1.3 |
| Hexadecane | 2.8 | 106.2 ± 2.5 | 103.2 ± 1.8 | 3.0 ± 0.8 | ≈0 | ≈0 | N/A |

All coatings infused with PFPE displayed excellent amphiphobicity, as both water and n-hexadecane drops did not stick to these surfaces leading to low contact angle hysteresis values. The increase in CAH observed for PFPE-infused SLIPS from Krytox 100 to 107 has already been reported in the literature [28] and is due to the increasing chain length of PFPE molecules, which leads to higher oil viscosity: the recession of water drops on these surfaces is hindered by pinning of their contact line, thus diminishing drop mobility and receding contact angle [29].

Hexadecane-infused SLIPS also showed excellent mobility of water drops, with a $CAH_W$ of 3.0°. Obviously, these surfaces were completely oleophilic as hexadecane drops quickly wetted the surface, therefore it was not possible to measure contact angles with hexadecane.

*3.2. Durability Tests*

3.2.1. Response to UV Irradiation

All fabricated SLIPS were exposed to UV radiation and their wetting behavior was evaluated after 2, 4, 6, and 8 h of irradiation. However, as the changes in contact angles were limited, only the values measured after 8 h are reported for brevity. Figure 2 compares the CAH values of the fabricated SLIPS before and after UV irradiation.

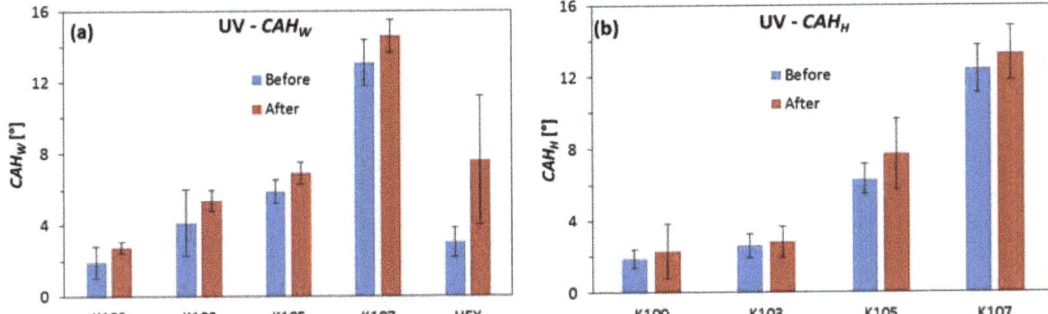

**Figure 2.** Contact angle hysteresis with (**a**) water (CAH$_W$) and (**b**) n-hexadecane drops (CAH$_H$) for SLIPS samples before (blue) and after UV irradiation (red) for 8 h. Standard deviations are reported as error bars.

Krytox-infused SLIPS had almost constant CAH values with both liquids, thus displaying excellent resistance to prolonged UV exposure. These liquids have limited UV absorption [30], therefore they do not undergo radiolysis. Moreover, even though the actual values of vapor pressure for commercial Krytox lubricants are unknown, PFPE usually possesses extremely low vapor pressure, in the order of $10^{-8}$ Torr at 20 °C [31].

On the other hand, hexadecane-infused SLIPS behaved differently. After 6 h, contact angle hysteresis with water (CAH$_W$) increased from 2.6° to 3.8°; then, after 8 h, it further increased to 7.6° along with standard error (from 0.2° to 3.6°). Both phenomena suggest the formation of defects on the SLIPS, probably due to evaporation of the infused oil which has significant vapor pressure at room temperature ($1.4 \times 10^{-4}$ Torr) [32]. The underlying surface might have been exposed, acting as a surface defect point that caused the increase in CAH$_W$. These results suggest that hexadecane-infused SLIPS are not ideal candidates for applications in which the surface is exposed to air for prolonged time, due to the high evaporation rate of the lubricant. Notably, hexadecane is the least volatile among alkanes that are liquid at room temperature; therefore, the same considerations apply to the lower homologues of hexadecane (i.e., tetradecane, dodecane, decane, etc.). On the contrary, PFPE-based SLIPS showed remarkable stability when exposed to UV and represent a good choice for application on surfaces that remain often dry.

### 3.2.2. Response to Chemical Ageing

In order to evaluate their resistance to chemically aggressive environments, SLIPS samples were immersed in either acidic, alkaline, or saline solutions and their wetting properties were evaluated after different immersion times. Figure 3 recaps the behavior of SLIPS samples in terms of CAH with water (left-hand graph) and n-hexadecane drops (right-hand graph).

In both cases, K100 samples proved remarkably less durable than other Krytox-infused samples. Despite their low initial CAH$_W$ and CAH$_H$ values, after only 3 days of immersion in the acidic solution their amphiphobicity was lost, with CAH$_W$ = 89° and CAH$_H$ = 41°. After 14 days of immersion, RCA values with both liquids were rapidly decreasing, therefore data acquisition was interrupted. This behavior indicates that the infused Krytox 100 oil was removed by the acidic solution, thus exposing the underlying coating and degrading it as suggested by the low receding contact angle values with water (RCA$_W$ = 39°) and hexadecane (RCA$_H$ = 19°). On the contrary, the other Krytox-infused SLIPS showed significant durability in the acidic environment. Notably, increasing the viscosity of the infused liquid led to improved retention of the amphiphobic behavior, with K107 sample being the most stable SLIPS. It is worth to highlight the different trends of CAH with water and n-hexadecane. The former increases with time, probably due to the formation of polar -OH hydroxyl groups in PFPE chains which increase the interaction with water molecules via hydrogen bonding. To confirm this hypothesis, we performed FTIR analysis on the fabricated surfaces. First, the spectrum of each surface was obtained prior to immersion in the acidic solution, then it was acquired after 7 and 60 days of immersion. Figure 4 reports the spectra obtained for K107 as an example.

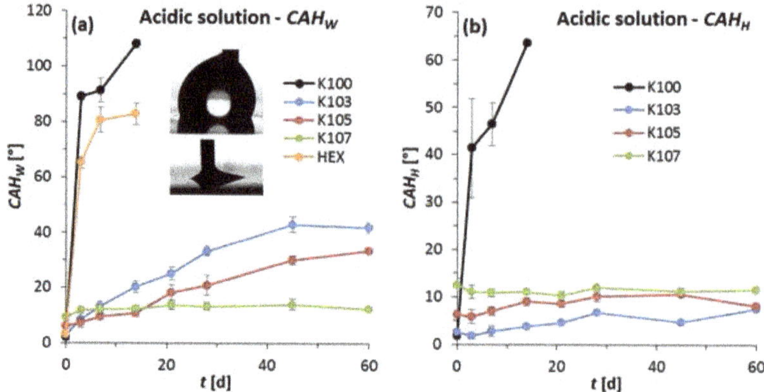

**Figure 3.** Contact angle hysteresis with (**a**) water (CAH$_W$) and (**b**) n-hexadecane drops (CAH$_H$) for SLIPS samples immersed in an acidic solution (pH = 3) as a function of immersion time t: K100 (black), K103 (blue), K105 (red), K107 (green), and HEX (orange). Standard deviations are reported as error bars. Inset: examples of frames for the measurement of RCA$_W$ on K107 (top) and HEX samples (bottom) after 7 days of immersion.

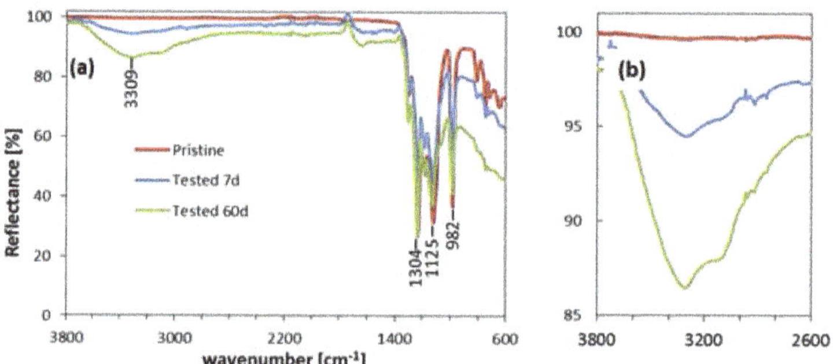

**Figure 4.** (**a**) Fourier Transform Infrared spectra (FTIR) of K107 sample before (red) and after immersion in acidic solution for 7 (blue) and 60 days (green). (**b**) Detail of the spectra in the 3800–2600 cm$^{-1}$ range. The positions of the most relevant peaks are reported.

After 7 days of immersion, a broad band centered at 3309 cm$^{-1}$ appeared; it can be assigned to the stretching vibration of -OH groups formed on polyether chains [33,34]. Notably, after 60 days the intensity of the band increased, as highlighted in the inset spectra, suggesting that further hydroxylation of the chain had occurred. Fourier Transform IR (FTIR) spectra for all Krytox-infused surfaces after 7 days are compared in Figure 5.

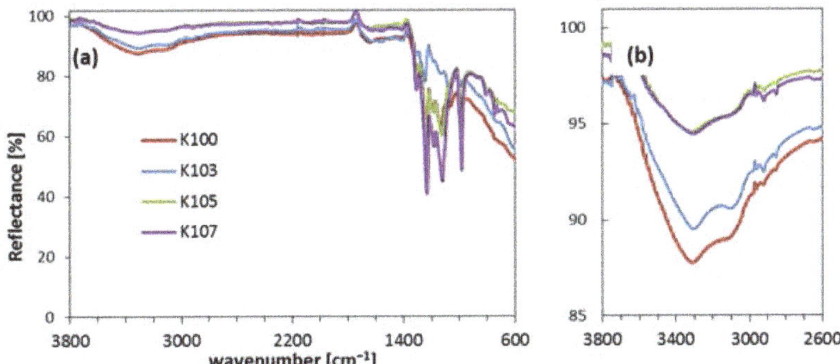

**Figure 5.** (**a**) Fourier Transform Infrared spectra (FTIR) of K100 (red), K103 (blue), K105 (green), and K107 samples (purple) after immersion in acidic solution for 7 days. (**b**) Detail of the spectra in the 3800–2600 cm$^{-1}$ range.

Focusing on the -OH stretching band (Figure 5b), it is clear that K100 suffered the highest degree of hydroxylation, followed by K103, while K105 and K107 showed similar signal. It was not possible to compare FTIR spectra after more prolonged immersion time due to the severe degradation of K100 samples.

On the other hand, $CAH_H$ did not change significantly, probably because the aforementioned polar -OH groups do not interact with non-polar hexadecane molecules, thus they do not affect contact angle values. Moreover, the different size of water and n-hexadecane molecules can contribute to explain this phenomenon. According to Wang et al. [35], small water molecules can penetrate the damaged PFPE network, resulting into pinning phenomena and increased $CAH_W$; on the other hand, large n-hexadecane molecules cannot do the same and $CAH_H$ remains substantially unaltered.

Regarding HEX samples, they showed a quick increase in $CAH_W$, probably due to n-hexadecane displacement and degradation of the underlying coating as observed for K100. In fact, n-hexadecane is even less viscous than Krytox 100 and can be displaced by water shortly after immersion.

Similar CAH trends were observed for SLIPS immersed in alkaline solution, as displayed in Figure 6.

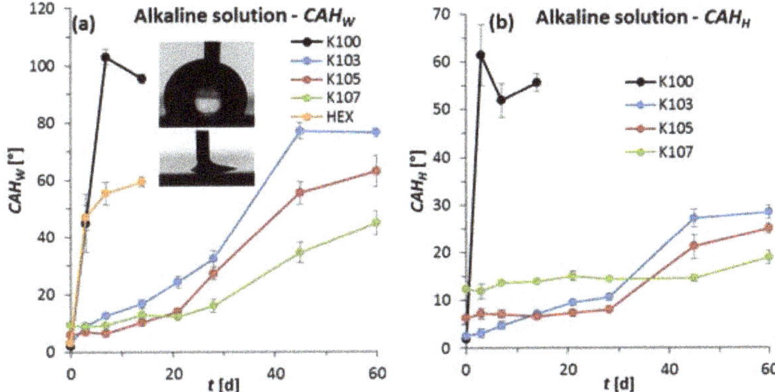

**Figure 6.** Contact angle hysteresis with (**a**) water ($CAH_W$) and (**b**) n-hexadecane drops ($CAH_H$) for SLIPS samples immersed in a alkaline solution (pH = 11) as a function of immersion time t: K100 (black), K103 (blue), K105 (red), K107 (green), and HEX (orange). Standard deviations are reported as error bars. Inset: examples of frames for the measurement of $RCA_W$ on K107 (top) and HEX samples (bottom) after 7 days of immersion.

K100 and HEX samples showed quick degradation of their liquid-repellent behavior, with both CAH values rapidly increasing in few days. The unexpected drop in CAH for K100 after 7–14 days is due to the fact that ACA decreases more than RCA in that period. For other PFPE-infused SLIPS, the increase in $CAH_W$ was more evident than in the acidic environment: $CAH_W$ value for the K103 sample reached 77° after 45 days. Increasing lubricant viscosity led to smaller increase in $CAH_W$, with K107 displaying the best value at 45° after 60 days. As described before for the immersion in acidic solution, $CAH_H$ increased less than $CAH_W$, also, in alkaline conditions, and the same trend with viscosity was observed, as $CAH_H$ for K107 remained constant after 60 days. FTIR spectra were collected after 7 days (Figure 7).

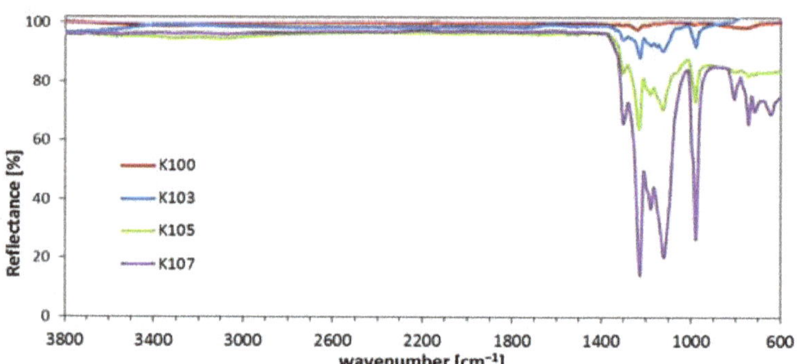


**Figure 7.** Fourier Transform Infrared spectra (FTIR) of K100 (red), K103 (blue), K105 (green), and K107 samples (purple) after immersion in alkaline solution for 7 days.

FTIR spectra confirm that less viscous PFPE oils are more prone to degradation than the more viscous ones: K100 gave no signal, suggesting a complete loss of PFPE; on the other hand, increasing oil viscosity led to more intense signal below 1400 cm$^{-1}$, with K107 being unaltered (see Figure 4 for comparison). From these results, it seems clear that PFPE is more susceptible to alkaline environments than to acidic ones, as already reported in the literature [36]. Indeed, polyether chains are intrinsically hydrophilic and prone to hydrolysis in alkaline conditions [37]; probably, substitution of the polymer backbone with fluorine atoms might only temporarily delay hydrolysis. Judging from FTIR spectra, low-viscosity PFPE oils are totally depleted in alkaline conditions, while in acidic solution they, rather, seem to be hydroxylated but not removed.

The immersion tests in saline solution showed similar trends to those observed for the alkaline conditions (Figure 8).

Once again, K100 and HEX quickly lost their amphiphobicity, with steep increase in CAH after only 3 days; the decrease in $CAH_H$ for K100 was due to the drop in $ACA_W$. Meanwhile, K103 lost its amphiphobicity more gradually, stabilizing its CAH values after 28–45 days; on the other hand, K105 and K107 retained their wetting properties for the entire testing period. Especially K107 had its CAH values almost completely unaltered after 60 days of immersion in the saline solution. FTIR spectra after 7 days (not reported for brevity) were compared also for these samples; as observed in alkaline solution, lubricant depletion increased with decreasing PFPE viscosity. Such result is remarkable in perspective of an application of these coatings in marine environment and in corrosive conditions in general; the retention of the amphiphobic behavior indirectly suggests anti-corrosion properties for K107 coatings. The anti-corrosion properties of SLIPS have already been reported [38,39], but never in such harsh conditions as usually mild NaCl solutions or seawater are used; in our tests, NaCl concentration was almost three times larger than in seawater, therefore corrosion is expected to be more severe.

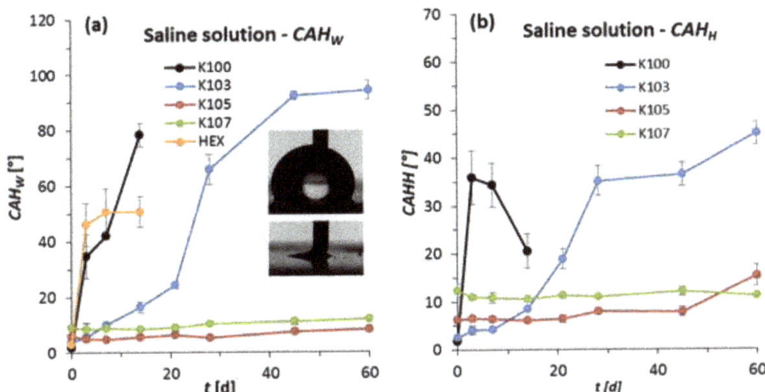

**Figure 8.** Contact angle hysteresis with (**a**) water (CAH$_W$) and (**b**) n-hexadecane drops (CAH$_H$) for SLIPS samples immersed in a saline solution (NaCl 100 g L$^{-1}$) as a function of immersion time t: K100 (black), K103 (blue), K105 (red), K107 (green), and HEX (orange). Standard deviations are reported as error bars. Inset: examples of frames for the measurement of RCA$_W$ on K107 (top) and HEX samples (bottom) after 7 days of immersion.

### 3.2.3. Response to Abrasion

Mechanical stresses are the most common cause of performance loss in liquid-repellent coatings; therefore, it is necessary to address their response to such stresses in perspective of future applications. We chose to perform abrasion tests as per the UNI EN 1096-2 standard because it is widely applied on coatings for the building industry; moreover, this test applies compression and shear stress on the surface contemporarily, thus effectively simulating complex operational conditions. CAH values for the tested SLIPS before and after abrasion tests are reported in Figure 9.

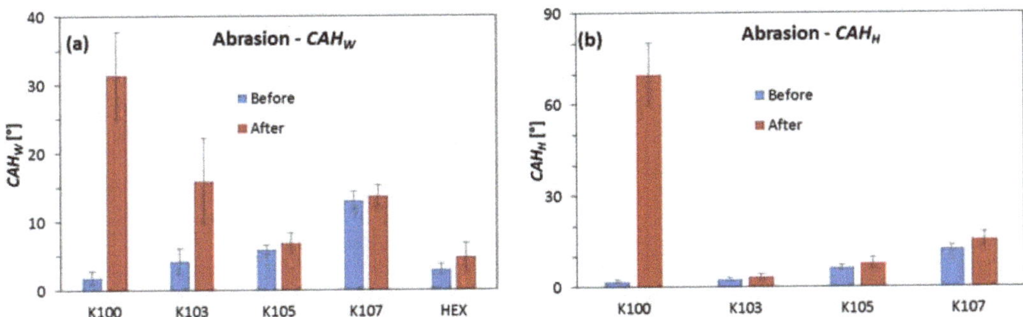

**Figure 9.** Contact angle hysteresis with (**a**) water (CAH$_W$) and (**b**) n-hexadecane drops (CAH$_H$) for SLIPS samples before (blue) and after abrasion (red) as per the UNI EN 1096-2 standard. Standard deviations are reported as error bars.

Among PFPE-infused SLIPS, a relationship between lubricant viscosity and increase in CAH$_W$ was observed (Figure 10): the sample infused with the least viscous oil (K100) showed the most significant increase in CAH values, eventually losing its amphiphobicity.

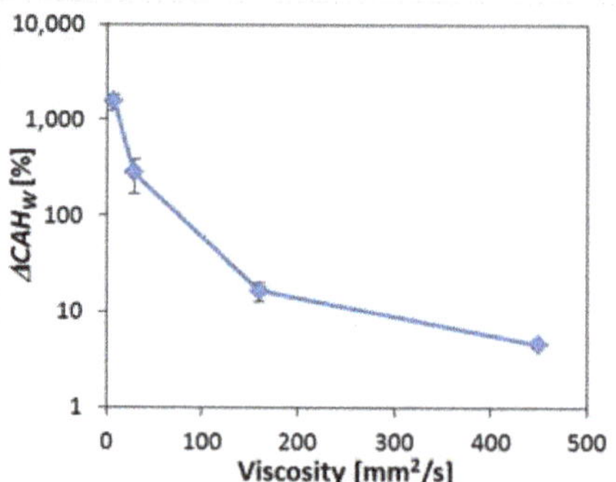

**Figure 10.** Increase in contact angle hysteresis with water ($\Delta$CAH$_W$) as a function of oil viscosity for PFPE-infused SLIPS. Standard deviations are reported as error bars.

With increasing lubricant viscosity, the increase in CAH (especially with water) became less evident, with K105 and K107 samples retaining their amphiphobicity after the abrasion tests. In the past years, several papers [40,41] investigated the response of PFPE oils under abrasion because of their application in hard disk drives; it was demonstrated that, in such conditions, these materials can be involved in tribochemical degradation reactions, which can be significantly catalyzed by Lewis acids like $Al_2O_3$ [42]. The most important degradation mechanisms include triboelectrical reactions (with creation of radical species) and mechanical cleavage, due to the friction between PTFE and solid surface asperities [43]. The degradation rate depends on the length of polymer chains: short macromolecules like those in K100 have higher mobility (i.e., lower viscosity) which lead to higher degradation rates. Increasing the chain length can slow down degradation reactions (especially mechanical cleavage) [44], thus explaining the retention of amphiphobic properties of K105 and K107. It is also necessary to consider that abrasion tests cause friction and related increase in surface temperature; Krytox 100 is more susceptible to temperature increase than higher Krytox oils, with obvious negative effect on the amphiphobicity of K100.

On the other hand, HEX samples showed limited increase in $CAH_W$ after the test, although n-hexadecane has lower viscosity than Krytox 100. These results can be explained considering that the C-C bonds in the n-hexadecane molecule are less prone to mechanical cleavage than the C-O bonds in PFPE polymer chain [45]; therefore, HEX SLIPS are less prone than K100 ones to mechanical degradation caused by abrasion.

### 4. Conclusions

We reported the fabrication of SLIPS based on a nanostructured, porous alumina coating infused with low surface tension lubricants with different chemical composition (PFPE, n-hexadecane) and viscosity (from 2.8 to 450 mm$^2$/s). Wetting characterization of as-produced SLIPS showed that contact angle hysteresis with water and hexadecane drops increased with the viscosity of the infused liquid. However, higher viscosity led to enhanced resistance to almost all of the tested severe environmental conditions: n-hexadecane and the least viscous PFPE (Krytox 100) proved unstable in acidic, alkaline, and saline environments, leading to complete loss of repellence in few days. On the other hand, SLIPS infused with the most viscous PFPE oil (Krytox 107) retained their amphiphobic behavior for up to 60 days, especially in acidic and highly saline solutions. These results lead to consider these surfaces as potential candidates for application in the naval and maritime industries. The same trend was observed in abrasion tests: the high mobility of short polymer chains in the low-viscosity Krytox 100 made them susceptible to tribochemical degradation, which in turn did not affect the repellence of highly viscous Krytox 107. Even the chemical composition of the infused liquid affected SLIPS durability under abrasion: despite its very low viscosity, hexadecane proved stable in such conditions due to the non-cleavable C-C bonds forming its chain. These considerations are extremely relevant for the application of SLIPS in devices like hard drives or lubricated joints for mechanics. On the other hand, hexadecane-infused SLIPS proved the least stable in UV irradiation tests due to the high volatility of the infused lubricant, which makes these coatings unsuitable for applications in which the surface is expected to be exposed to air (i.e., not being wetted by another liquid or in an open ambient). These results once again highlight the need for careful consideration of the application conditions when designing a liquid-infused coating.

**Author Contributions:** Conceptualization, F.V. and M.R.; methodology, F.V. and M.R.; software, F.V.; validation, F.V. and A.C.; formal analysis, F.V.; investigation, G.G.; resources, F.V. and M.R.; data curation, F.V. and G.G.; writing—original draft preparation, F.V.; writing—review and editing, F.V., A.C., and M.R.; visualization, F.V. and A.C.; supervision, F.V. and M.R.; project administration, F.V. and M.R.; funding acquisition, M.R. All authors have read and agreed to the published version of the manuscript.

**Funding:** This research received no external funding.

**Institutional Review Board Statement:** Not applicable.

**Informed Consent Statement:** Not applicable.

**Data Availability Statement:** Data available in a publicly accessible repository. The data presented in this study are openly available in https://doi.org/10.6084/m9.figshare.16566960.v1, accessed on 6 September 2021.

**Acknowledgments:** The authors want to acknowledge the entire Smart Surfaces Group at CNR-ISTEC for their precious work.

**Conflicts of Interest:** The authors declare no conflict of interest.

## References

1. Bhushan, B.; Jung, Y.C. Natural and Biomimetic Artificial Surfaces for Superhydrophobicity, Self-Cleaning, Low Adhesion, and Drag Reduction. *Prog. Mater. Sci.* **2011**, *56*, 1–108. [CrossRef]
2. Yang, C.Y.; Chuang, S.I.; Lo, Y.H.; Cheng, H.M.; Duh, J.G.; Chen, P.Y. Stalagmite-like Self-Cleaning Surfaces Prepared by Silanization of Plasma-Assisted Metal-Oxide Nanostructures. *J. Mater. Chem. A* **2016**, *4*, 3406–3414. [CrossRef]
3. Liu, H.; Zhang, P.; Liu, M.; Wang, S.; Jiang, L. Organogel-Based Thin Films for Self-Cleaning on Various Surfaces. *Adv. Mater.* **2013**, *25*, 4477–4481. [CrossRef]
4. del Cerro, D.A.; Römer, G.R.B.E.; Huis In't Veld, A.J. Erosion Resistant Anti-Ice Surfaces Generated by Ultra Short Laser Pulses. *Phys. Procedia* **2010**, *5*, 231–235. [CrossRef]
5. Farhadi, S.; Farzaneh, M.; Kulinich, S.A. Anti-Icing Performance of Superhydrophobic Surfaces. *App. Surf. Sci.* **2011**, *257*, 6264–6269. [CrossRef]
6. Ge, L.; Ding, G.; Wang, H.; Yao, J.; Cheng, P.; Wang, Y. Anti-Icing Property of Superhydrophobic Octadecyltrichlorosilane Film and Its Ice Adhesion Strength. *J. Nanomater.* **2013**, *2013*, 278936. [CrossRef]
7. Golovin, K.B.; Gose, J.; Perlin, M.; Ceccio, S.L.; Tuteja, A. Bioinspired Surfaces for Turbulent Drag Reduction. *Philos. Trans. R. Soc. A Math. Phys. Eng. Sci.* **2016**, *374*, 20160189. [CrossRef] [PubMed]
8. Rosenberg, B.J.; van Buren, T.; Fu, M.K.; Smits, A.J. Turbulent Drag Reduction over Air- and Liquid-Impregnated Surfaces. *Phys. Fluids* **2016**, *28*, 015103. [CrossRef]
9. Rizzo, G.; Massarotti, G.P.; Bonanno, A.; Paoluzzi, R.; Raimondom, M.; Blosi, M.; Veronesi, F.; Caldarelli, A.; Guarini, G. Axial Piston Pumps Slippers with Nanocoated Surfaces to Reduce Friction. *Int. J. Fluid Power* **2015**, *16*, 1–10. [CrossRef]
10. Gittens, J.E.; Smith, T.J.; Suleiman, R.; Akid, R. Current and Emerging Environmentally-Friendly Systems for Fouling Control in the Marine Environment. *Biotechnol. Adv.* **2013**, *31*, 1738–1753. [CrossRef]
11. Oldani, V.; del Negro, R.; Bianchi, C.L.; Suriano, R.; Turri, S.; Pirola, C.; Sacchi, B. Surface Properties and Anti-Fouling Assessment of Coatings Obtained from Perfluoropolyethers and Ceramic Oxides Nanopowders Deposited on Stainless Steel. *J. Fluor. Chem.* **2015**, *180*, 7–14. [CrossRef]
12. Barthlott, W.; Neinhuis, C. Purity of the Sacred Lotus, or Escape from Contamination in Biological Surfaces. *Planta* **1997**, *202*, 1. [CrossRef]
13. Zorba, V.; Stratakis, E.; Barberoglou, M.; Spanakis, E.; Tzanetakis, P.; Anastasiadis, S.H.; Fotakis, C. Biomimetic Artificial Surfaces Quantitatively Reproduce the Water Repellency of a Lotus Leaf. *Adv. Mater.* **2008**, *20*, 4049–4054. [CrossRef]
14. Ebert, D.; Bhushan, B. Durable Lotus-Effect Surfaces with Hierarchical Structure Using Micro- and Nanosized Hydrophobic Silica Particles. *J. Colloid Interface Sci.* **2012**, *368*, 584–591. [CrossRef]

15. Guo, M.; Ding, B.; Li, X.; Wang, X.; Yu, J.; Wang, M. Amphiphobic Nanofibrous Silica Mats with Flexible and High-Heat-Resistant Properties. *J. Phys. Chem. C* **2010**, *114*, 916–921. [CrossRef]

16. Zhou, H.; Wang, H.; Niu, H.; Gestos, A.; Lin, T. Robust, Self-Healing Superamphiphobic Fabrics Prepared by Two-Step Coating of Fluoro-Containing Polymer, Fluoroalkyl Silane, and Modified Silica Nanoparticles. *Adv. Funct. Mater.* **2013**, *23*, 1664–1670. [CrossRef]

17. Ma, W.; Higaki, Y.; Otsuka, H.; Takahara, A. Perfluoropolyether-Infused Nano-Texture: A Versatile Approach to Omniphobic Coatings with Low Hysteresis and High Transparency. *Chem. Commun.* **2013**, *49*, 597–599. [CrossRef] [PubMed]

18. Wang, J.; Kato, K.; Blois, A.P.; Wong, T.S. Bioinspired Omniphobic Coatings with a Thermal Self-Repair Function on Industrial Materials. *ACS Appl. Mater. Interfaces* **2016**, *8*, 8265–8271. [CrossRef]

19. Wong, T.-S.; Kang, S.H.; Tang, S.K.Y.; Smythe, E.J.; Hatton, B.D.; Grinthal, A.; Aizenberg, J. Bioinspired Self-Repairing Slippery Surfaces with Pressure-Stable Omniphobicity. *Nature* **2011**, *477*, 443–447. [CrossRef]

20. Riedel, M.; Eichner, A.; Jetter, R. Slippery Surfaces of Carnivorous Plants: Composition of Epicuticular Wax Crystals in Nepenthes Alata Blanco Pitchers. *Planta* **2003**, *218*, 87–97. [CrossRef] [PubMed]

21. Niemelä-Anttonen, H.; Koivuluoto, H.; Tuominen, M.; Teisala, H.; Juuti, P.; Haapanen, J.; Harra, J.; Stenroos, C.; Lahti, J.; Kuusipalo, J.; et al. Icephobicity of Slippery Liquid Infused Porous Surfaces under Multiple Freeze–Thaw and Ice Accretion–Detachment Cycles. *Adv. Mater. Interfaces* **2018**, *5*, 1–8. [CrossRef]

22. Yang, S.; Qiu, R.; Song, H.; Wang, P.; Shi, Z.; Wang, Y. Slippery Liquid-Infused Porous Surface Based on Perfluorinated Lubricant/Iron Tetradecanoate: Preparation and Corrosion Protection Application. *Appl. Surf. Sci.* **2015**, *328*, 491–500. [CrossRef]

23. Preston, D.J.; Song, Y.; Lu, Z.; Antao, D.S.; Wang, E.N. Design of Lubricant Infused Surfaces. *ACS Appl. Mater. Interfaces* **2017**, *9*, 42383–42392. [CrossRef] [PubMed]

24. Togasawa, R.; Tenjimbayashi, M.; Matsubayashi, T.; Moriya, T.; Manabe, K.; Shiratori, S. A Fluorine-Free Slippery Surface with Hot Water Repellency and Improved Stability against Boiling. *ACS Appl. Mater. Interfaces* **2018**, *10*, 4198–4205. [CrossRef]

25. Yamaguchi, N.; Tadanaga, K.; Matsuda, A.; Minami, T.; Tatsumisago, M. Formation of Anti-Reflective Alumina Films on Polymer Substrates by the Sol-Gel Process with Hot Water Treatment. *Surf. Coat. Technol.* **2006**, *201*, 3653–3657. [CrossRef]

26. Raimondo, M.; Blosi, M.; Caldarelli, A.; Guarini, G.; Veronesi, F. Wetting Behavior and Remarkable Durability of Amphiphobic Aluminum Alloys Surfaces in a Wide Range of Environmental Conditions. *Chem. Eng. J.* **2014**, *258*, 101–109. [CrossRef]

27. Kim, P.; Kreder, M.J.; Alvarenga, J.; Aizenberg, J. Hierarchical or Not? Effect of the Length Scale and Hierarchy of the Surface Roughness on Omniphobicity of Lubricant-Infused Substrates. *Nano Lett.* **2013**, *13*, 1793–1799. [CrossRef] [PubMed]

28. Kim, D.; Lee, M.; Kim, J.H.; Lee, J. Dynamic Contact Angle Measurements on Lubricant Infused Surfaces. *J. Colloid Interface Sci.* **2021**, *586*, 647–654. [CrossRef]

29. Daniel, D.; Mankin, M.N.; Belisle, R.A.; Wong, T.-S.; Aizenberg, J. Lubricant-Infused Micro/Nano-Structured Surfaces with Tunable Dynamic Omniphobicity at High Temperatures. *Appl. Phys. Lett.* **2013**, *102*, 2–6. [CrossRef]

30. Faucitano, A.; Buttafava, A.; Karolczak, S.; Guarda, P.A.; Marchionni, G. The Chemical Effects of Ionizing Radiations of Fluorinated Ethers. *J. Fluor. Chem.* **2004**, *125*, 221–241. [CrossRef]

31. Bassi, M. Estimation of the Vapor Pressure of PFPEs by TGA. *Thermochim. Acta* **2011**, *521*, 197–201. [CrossRef]

32. Parks, G.S.; Moore, G.E. Vapor Pressure and Other Thermodynamic Data for N-Hexadecane and n-Dodecylcyclohexane near Room Temperature. *J. Chem. Phys.* **1949**, *17*, 1151–1153. [CrossRef]

33. Sawatari, C.; Kondo, T. Interchain Hydrogen Bonds in Blend Films of Poly(Vinyl Alcohol) and Its Derivatives with Poly(Ethylene Oxide). *Macromolecules* **1999**, *32*, 1949–1955. [CrossRef]

34. Zheng, S.; Al, S.; Guo, Q. Poly(Hydroxyether of Phenolphthalein) and Its Blends with Poly(Ethylene Oxide). *J. Polym. Sci. Part B Polym. Phys.* **2003**, *41*, 466–475. [CrossRef]

35. Wang, Y.; Dugan, M.; Urbaniak, B.; Li, L. Fabricating Nanometer-Thick Simultaneously Oleophobic/Hydrophilic Polymer Coatings via a Photochemical Approach. *Langmuir* **2016**, *32*, 6723–6729. [CrossRef]

36. Yang, Y.-W.; Hentschel, J.; Chen, Y.-C.; Lazari, M.; Zeng, H.; Michael Van Dam, R.; Guan, Z. "Clicked" Fluoropolymer Elastomers as Robust Materials for Potential Microfluidic Device Applications. *J. Mater. Chem.* **2012**, *22*, 1100. [CrossRef]

37. Li, S.M.; Chen, X.H.; Gross, R.A.; Mccarthy, S.P. Hydrolytic Degradation of PCL/PEO Copolymers in Alkaline Media. *J. Mater. Sci. Mater. Med.* **2000**, *11*, 227–233. [CrossRef] [PubMed]

38. Lee, J.; Shin, S.; Jiang, Y.; Jeong, C.; Stone, H.A.; Choi, C.H. Oil-Impregnated Nanoporous Oxide Layer for Corrosion Protection with Self-Healing. *Adv. Funct. Mater.* **2017**, *27*, 1606040. [CrossRef]

39. Ma, Q.; Wang, W.; Dong, G. Facile Fabrication of Biomimetic Liquid-Infused Slippery Surface on Carbon Steel and Its Self-Cleaning, Anti-Corrosion, Anti-Frosting and Tribological Properties. *Colloids Surf. A Physicochem. Eng. Asp.* **2019**, *577*, 17–26. [CrossRef]

40. Homola, A.M. Lubrication Issues in Magnetic Disk Storage Devices. *IEEE Trans. Magn.* **1996**, *32*, 1812–1818. [CrossRef]

41. Tagawa, N.; Tani, H. Tribological Characteristics of Novel Branched Perfluoropolyether Lubricant Films in Hard Disk Drives. *Microsyst. Technol.* **2013**, *19*, 1513–1518. [CrossRef]

42. Zhang, J.; Cheng, T.; Cheng, P.; Chao, J. Relationship between the Molecular Structures of Lubricants and Their Performance at the Head-Disk Interface of Hard Disk Drives. *Wear* **2003**, *254*, 321–331. [CrossRef]

43. Tao, Z.; Bhushan, B. Bonding, Degradation, and Environmental Effects on Novel Perfluoropolyether Lubricants. *Wear* **2005**, *259*, 1352–1361. [CrossRef]

44. Min, K.; Han, J.; Park, B.; Cho, E. Characterization of Mechanical Degradation in Perfluoropolyether Film for Its Application to Antifingerprint Coatings. *ACS Appl. Mater. Interfaces* **2018**, *10*, 37498–37506. [CrossRef] [PubMed]
45. Chen, X.; Kawai, K.; Zhang, H.; Fukuzawa, K.; Koga, N.; Itoh, S.; Azuma, N. ReaxFF Reactive Molecular Dynamics Simulations of Mechano-Chemical Decomposition of Perfluoropolyether Lubricants in Heat-Assisted Magnetic Recording. *J. Phys. Chem. C* **2020**, *124*, 22496–22505. [CrossRef]

*Article*

# Nanomechanical Concepts in Magnetically Guided Systems to Investigate the Magnetic Dipole Effect on Ferromagnetic Flow Past a Vertical Cone Surface

Auwalu Hamisu Usman [1,2], Zahir Shah [3], Poom Kumam [4,5], Waris Khan [6] and Usa Wannasingha Humphries [1,*]

1 Department of Mathematics, Faculty of Science, King Mongkut's University of Technology Thonburi, 126 Pracha Uthit Road, Bang Mod, Thung Khru, Bangkok 10140, Thailand; auwalu.mth@mail.kmutt.ac.th
2 Department of Mathematical Sciences, Faculty of Physical Sciences, Bayero University, Kano 700241, Nigeria
3 Department of Mathematical Sciences, University of Lakki Marwat, Lakki Marwat 28420, Pakistan; zahir1987@yahoo.com
4 KMUTTFixed Point Research Laboratory, Room SCL 802 Fixed Point Laboratory, Department of Mathematics, Faculty of Science, King Mongkut's University of Technology Thonburi (KMUTT), Bangkok 10140, Thailand; poom.kum@kmutt.ac.th
5 Center of Excellence in Theoretical and Computational Science (TaCS-CoE), Faculty of Science, King Mongkut's University of Technology Thonburi (KMUTT), 126 Pracha-Uthit Road, Bang Mod, Thung Khru, Bangkok 10140, Thailand
6 Department of Mathematics and Statistics, Hazara University Mansehra, Khyber Pakhtunkhwa 21120, Pakistan; wariskhan758@yahoo.com
* Correspondence: usa.wan@kmutt.ac.th; Tel.: +66-024708831

Citation: Usman, A.H.; Shah, Z.; Kumam, P.; Khan, W.; Humphries, U.W. Nanomechanical Concepts in Magnetically Guided Systems to Investigate the Magnetic Dipole Effect on Ferromagnetic Flow Past a Vertical Cone Surface. *Coatings* **2021**, *11*, 1129. https://doi.org/10.3390/coatings11091129

Academic Editor: Eduardo Guzmán

Received: 16 August 2021
Accepted: 12 September 2021
Published: 16 September 2021

**Publisher's Note:** MDPI stays neutral with regard to jurisdictional claims in published maps and institutional affiliations.

**Abstract:** Because of the floating magnetic nanomaterial, ferrofluids have magneto-viscous properties, enabling controllable temperature changes as well as nano-structured fluid characteristics. The study's purpose is to evolve and solve a theoretical model of bioconvection nanofluid flow with a magnetic dipole effect in the presence of Curie temperature and using the Forchheimer-extended Darcy law subjected to a vertical cone surface. The model also includes the nonlinear thermal radiation, heat suction/injection, viscous dissipation, and chemical reaction effects. The developed model problem is transformed into nonlinear ordinary differentials, which have been solved using the homotopy analysis technique. In this problem, the behavior of function profiles are graphically depicted and explained for a variety of key parameters. For a given set of parameters, tables representthe expected numerical values and behaviors of physical quantities. The nanofluid velocity decreases as the ferrohydrodynamic, local inertia, and porosity parameters increase and decrease when the bioconvection Rayleigh number increases. Many key parameters improved the thermal boundary layer and temperature. The concentration is low when the chemical reaction parameter and Schmidt number rises. Furthermore, as the bioconvection constant, Peclet and Lewis numbers rise, so does the density of motile microorganisms.

**Keywords:** ferromagnetic; nanofluid; bioconvection; porous medium; heat suction/injection; magnetic dipole

## 1. Introduction

Fluids that are often magnetized by the existence of an exterior magnetic field are known as ferrofluids, which is an abbreviation for fluid and ferromagnetic particles. These fluids are made up of colloidal fluids formed of nanosized ferromagnetic or ferrimagnetic particles that have been stopped inside the fluid transporter. Brownian motion causes particle suspension and must not start moving under normal conditions. Besides that, to avoid clogging, each ferromagnetic particle is encased in a solvent, and the nano-scaled ferromagnetic particles have a weak magnetic attraction whenever the surfactant's Van der Waals force adequately stopped aggregation or clustering. Numerous applications of ferromagnetic fluids have emerged in a variety of fields. Heat transfer agents, angular momentum changers, friction reducers, and so on are used in electronic equipment, analytical techniques, and medical science; some examples can be found in the references [1–3]. Because of these numerous applications, many researchers and scientists have been focused on this subject. Andersson and Vanes [4] first investigated the influence caused by magnetic dipoles on ferrofluids. Zeeshan et al. [5] investigated the convective heat transfer flow of ferromagnetic fluids with partial slip effects using a stretching sheet. Hayat et al. [6] reported on radiation and magnetic dipole effects of Williamson ferromagnetic fluid flow across a stretched surface.

A nanofluid is a nanometer-sized particle suspended in a fluid. Choi [7] established the basic extension of "nanofluid", and scientific results verified that heat transfer can be significantly enhanced through the mixture of tiny metallic nanomaterials with the working fluids. A few studies in particular on nanofluids have been conducted. Ellahi [8] performed an analytical study and concluded that the temperature variable and viscosity affects MHD non-Newtonian nanofluid flow in a pipe. Ellahi et al. [9] presented peristaltic nanofluid flow with entropy generation via a medium of porosity. Hayat et al. [10] investigated the flow of third-grade nanofluids caused by a rotating stretchable disk containing a heat source and a chemical reaction. Awais et al. [11] explored the effects of magnetohydrodynamics on peristaltic ciliary-induced flow coatings to rheological hybrid nanofluids. Reddy et al. [12] studied the boundary layer naturally convective MHD nanofluid flow along a vertical cone under the influence of chemical reaction and heat suction/injection.

Due to disorganized frameworks and destabilization, low-density microorganisms remain on the surface of a fluid, causing bioconvection. Because nanoparticles move differently than motile microorganisms, the cumulative importance of nanomaterials and bioconvection is such that they play a vital role in microfluidic devices. Bioconvection is a novel manufacturing and fluid mechanic with a biological phenomenon involving gyrotactic microorganisms. As a result, it becomes an interesting field of research to which many researchers continue to pay attention. Alsaedi et al. [13] investigated stratified magnetohydrodynamic nanofluid flow, causing bioconvection in gyrotactic microorganisms. Hayat et al. [14] researched the magnetohydrodynamic (MHD) nonlinear radiative nanofluid flow with gyrotactic microorganisms. Nadeem et al. [15] reported on the Rosseland assessment for ferromagnetic fluid with involvement of magnetic dipoles and gyrotactic microorganisms. Bhatti and Michaelides [16] researched thermo-bioconvection nanofluid flows across a Riga plate as a function of Arrhenius activation energy. Waqas et al. [17] have also numerically simulated the magnetized non-Newtonian bioconvection nanofluid flow along stretching cylinders/plates.

Combining mass and heat fluxes in liquid saturated porous media is crucial among a wide range of engineering procedures such as heating systems, oil and gas reservoirs, and chemical catalytic reactor designs [18,19]. The dragging force, the Darcy–Forchheimer technique, is a widely popular method for simulating fluid passed through a porous medium with high velocity. In the literature, flow through a cone in Darcy–Forchheimer porous media has already been analyzed by many researchers. Kumar et al. [20] investigated the non-Darcy MHD viscoelastic fluid flow through a flat plate and a vertical cone. Chamkha et al. [21] explored the non-Newtonian natural convective nanofluid flow over a saturated cone in a non-Darcy porous medium with uniform volume fraction and heat fluxes. Mallikarjuna et al. [22] researched the impacts of radiation, thermophoresis and transpiration on convective non-Darcy flow via a rotating cone. Durairaj et al. [23] investigated the chemically reacting Casson fluid of a non-Darcy porous medium flow through a flat plate and a vertical cone saturated with heat generating/absorbing. Patrulescu et al. [24] investigated a convection flow due to a vertical plate embedded in a bi-disperse non-Darcy porous medium.

According to a recent literature review, despite important applications in extrusion systems, geothermics, organic compounds, geophysics, improved manufacturing techniques, material processing, and improved energy generation, research on viscous ferrofluid flows via a linear vertical cone with consideration of Darcy–Forchheimer porous media has been studied by very few researchers in the past. The aim of this research is to use the Forchheimer-extended Darcy law to explore the effect of magnetic dipole and porosity relations in the boundary layer of a ferromagnetic nanofluid flow via a vertical cone surface. The study of the effect of the magnetic dipole on ferromagnetic nanofluid flow via the vertical cone surface makes this work different from the existing literature. A nonlinear ordinary differential equation replaces the governing equations and is solved using HAM techniques. Initially, Liao [25–27] presented a homotopy analysis method with HAM. The method has fast convergent solutions with many advantages over some existing methods. Various researchers have been drawn to it as a result of its rapid convergence [28–31]. The results collected for all associated parameters on all profiles are shown graphically. The validation of the results by comparing them to previously published material in the literature is an important feature of the presented model. In this regard, illustrious coherence has been attained.

## 2. Materials and Methods

An incompressible electrically conducting viscous nanofluid flow via a vertical cone with bioconvection is explored in two dimensions as an axisymmetric, steady, natural convective ferrofluid flow. Furthermore, it is postulated that temperature and concentration are non-uniform at the surface due to the influence of heat generation/absorption, chemical reaction and viscous dissipation. The flow is electrically magnetized by a magnetic dipole, and a Darcy–Forchheimer porous medium model is also used. Thermal radiation exists as a unidirectional flux in the transverse to the cone surface ($s$-direction). In comparison to the $s$-direction, the radiation heat flux in the $x$-direction is considered neglected. The $x$-axis of the chosen coordinate system corresponds to the direction of flow over the cone surface. $T_w$ is taken to be the temperature at the cone's surface ($s = 0$), and the concentration is governed by the condition $D_B \frac{\partial C}{\partial s} + \frac{D_T}{T_\infty} \frac{\partial T}{\partial s} = 0$ at the cone's surface, where $T_\infty$ is the temperature and $C_\infty$ is the concentration and $N_\infty$ density of microorganisms in the ambient nanofluid.

The boundary layer equation [12,14,21] based on the assumptions stated above are the equations of continuity and momentum as well as energy, concentration, and microorganisms:

$$\frac{\partial (ru)}{\partial x} + \frac{\partial \{(rw)\}}{\partial s} = 0, \tag{1}$$

$$w\frac{\partial u}{\partial s} + u\frac{\partial u}{\partial x} = \frac{\mu_f}{\rho_f}\frac{\partial^2 u}{\partial s^2} - \frac{\mu_f}{k_o^*}u - \frac{\rho_f C_b}{\sqrt{k_o^*}}u^2 + \lambda_o M\frac{\partial H}{\partial x}$$
$$+ g\left[\beta_T(T - T_\infty) + \beta_C(C - C_\infty) + \beta_N(N - N_\infty)\right]cos(\alpha), \tag{2}$$

$$w\frac{\partial T}{\partial s} + u\frac{\partial T}{\partial x} = \alpha_f \frac{\partial^2 T}{\partial s^2} + \frac{\mu_f}{(\rho c_p)_f}\left(\frac{\partial u}{\partial s}\right)^2 + \left(u\frac{\partial H}{\partial x} + w\frac{\partial H}{\partial s}\right)\frac{\lambda_0}{(\rho c_p)_f}T\frac{\partial M}{\partial T}$$
$$\tau\left[D_B\frac{\partial C}{\partial s}\frac{\partial T}{\partial s} + \frac{D_T}{T_\infty}\left(\frac{\partial T}{\partial s}\right)^2\right] - \frac{1}{(\rho c_p)_f}\frac{\partial q_r}{\partial s}, \tag{3}$$

$$w\frac{\partial C}{\partial s} + u\frac{\partial C}{\partial x} = D_B\frac{\partial^2 C}{\partial s^2} + \frac{D_T}{T_\infty}\frac{\partial^2 T}{\partial s^2} - K_r(C - C_\infty), \tag{4}$$

$$w\frac{\partial N}{\partial s} + u\frac{\partial N}{\partial x} + \frac{bW_c}{C_w - C_\infty}\left(\frac{\partial C}{\partial s}\frac{\partial N}{\partial s} + N\frac{\partial^2 C}{\partial s^2}\right) = D_n\frac{\partial^2 N}{\partial s^2}, \tag{5}$$

with initial boundary conditions

$$u = 0, \quad w = W_w, \quad T = T_w, \quad -D_B \frac{\partial C}{\partial s} = \frac{D_T}{T_\infty} \frac{\partial T}{\partial s}, \quad N = N_w \ at \quad s = 0, \tag{6}$$

$$u \to 0, \quad w \to 0, \quad T \to T_\infty, \quad C \to C_\infty, \quad N \to N_\infty, \quad as \quad s \to \infty, \tag{7}$$

where $(u,w)$ are the velocity components in the $x$-direction (radial) and $s$-direction (transverse), respectively; $T$, $C$, $N$ are the temperature, concentration, and gyrotactic microorganism, respectively; the diffusion coefficients named Brownian, thermophoresis, and microorganism correspond to $D_B$, $D_T$, and $D_n$, respectively; while $\tau$ is the ratio of heat capacitance, fluid density is $\rho_f$, thermal conductivity of fluid is $k_f$, electrical conductivity of fluid is $\sigma$, the dynamic viscosity is $\mu_f$, thermal diffusivity of base fluid is $\alpha_f$, magnetic permeability is $\lambda_o$, heat capacitance of fluid is $(\rho c_p)_f$, first order chemical reaction parameter is $K_r$, speed of gyrotactic cell is $W_c$, and $b$ is chemotaxis.

*Magnetic Dipole*

The magnetic field features impacted the ferrofluid flow with magnetic dipole effects detected mostly by magnetic scalar potential $\Phi_1$, as given in Equation (8):

$$\Phi_1 = \frac{\gamma}{2\pi} \frac{x}{x^2 + (s+c)^2}, \tag{8}$$

Considering $H_x$ and $H_s$ to be the components of magnetic field, with $\gamma$ as the magnetic field strength at the source, see Equations (9) and (10):

$$H_x = -\frac{\partial \Phi_1}{\partial x} = \frac{\gamma}{2\pi} \frac{x^2 - (s+d)^2}{[x^2 + (s+d)^2]^2} \tag{9}$$

$$H_s = -\frac{\partial \Phi_1}{\partial s} = \frac{\gamma}{2\pi} \frac{2x(s+d)}{[x^2 + (s+d)^2]^2}. \tag{10}$$

As the strength of a magnetic body is normally approximately equal to the $H_x$ and $H_s$ gradients, it is therefore given as in (11):

$$H = \sqrt{H_x^2 + H_s^2}. \tag{11}$$

Equation (12) displays the approximate linearized relation of the magnetization $M$ as function of temperature $T$,

$$M = -K_1(T_\infty - T), \tag{12}$$

with $K_1$ identified as the ferromagnetic coefficient. Figure 1 depicts the physical configuration of the heated ferrofluid.

Considering the following transformations, given the stream function as $\Phi(x,s)$, such that

$$u = \frac{1}{r} \frac{\partial \Phi}{\partial s} \qquad\qquad w = -\frac{1}{r} \frac{\partial \Phi}{\partial x} \tag{13}$$

with $\Phi(x,s) = \nu_f r Ra_x f(\zeta)$ and $Ra_x$ is the Rayleigh number given by $Ra_x = \frac{\rho_f \beta_T g(T_w - T_\infty)x^3 cos(\alpha)}{\nu_f^2}$, therefore, the following are given:

$$u = \frac{\nu_f Ra_x^{\frac{1}{2}}}{x} f'(\zeta), \quad w = -\frac{\nu_f Ra_x^{\frac{1}{4}}}{x}\left(\zeta f'(\zeta) - f(\zeta)\right), \quad \zeta = \frac{s}{x} Ra_x^{\frac{1}{4}},$$

$$(T_w - T_\infty)\theta(\zeta) = (T - T_\infty), \quad (C_w - C_\infty)\phi(\zeta) = (C - C_\infty), \quad (N_w - N_\infty)\chi(\zeta) = (N - N_\infty). \tag{14}$$

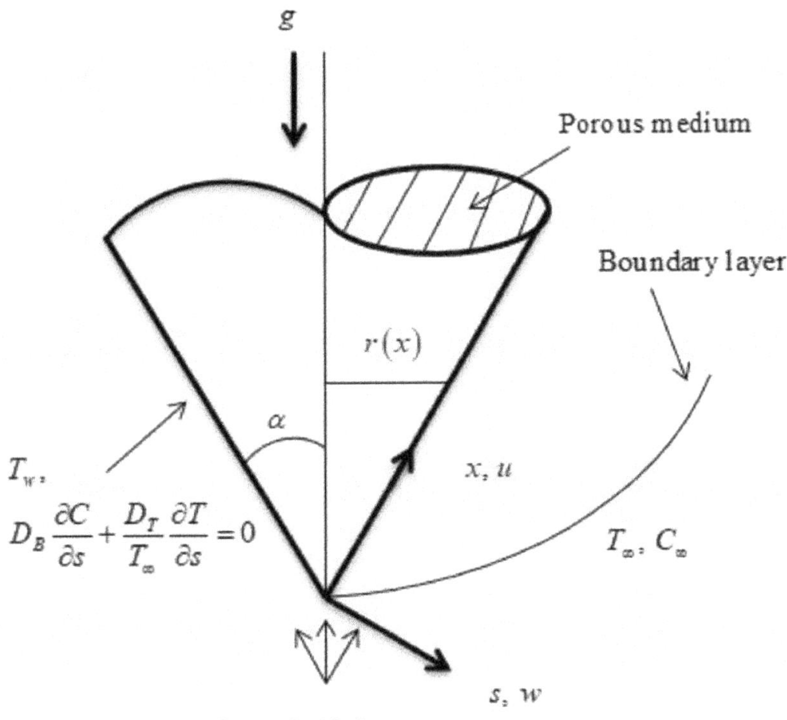



**Figure 1.** A picture scheme of the problem.

Taking $r$ to be approximately the cone local radius, for the thermal boundary layer becoming thin, it will be along the $x$ coordinate with $r = x\sin(\alpha)$.

By using the above transformations, Equation (1) will be satisfactory, and Equations (2)–(5) will be

$$f''' - P_1 f' + f f'' - F_r f'^2 + \frac{2\beta}{(\zeta + \alpha_1)^4}(1+\theta) + N_c\phi + Ra_b\chi = 0, \tag{15}$$

$$(1 + Rd)\theta'' + Prf\theta' + Nb\phi'\theta' + Nt(\theta')^2 + \frac{2Pr\beta\lambda(\theta - \epsilon)(f - \zeta f')}{(\zeta + d\alpha_1)^3}$$

$$+ Pr\beta\lambda(\theta - \epsilon)\left[\frac{2f'}{(\zeta + \alpha_1)^4} + \frac{4(\zeta f' - f)}{(\zeta + \alpha_1)^5}\right] + PrEc(f'')^2 = 0, \tag{16}$$

$$\phi'' + \frac{Nt}{Nb}\theta'' + Scf\phi' - \delta Sc\phi = 0, \tag{17}$$

$$\chi'' - Pe[\phi'\chi' + \phi''\chi + \delta_n\phi''] + Lbf\chi' = 0. \tag{18}$$

Moreover, with the new boundary conditions:

$$f' = 1, f = S, \theta = 1, Nb\phi' + Nt\theta = 0, \chi = 1 \, at \, \zeta = 0,$$
$$f' \to 0, \theta \to 0, \phi \to 0, \chi \to 0, \, as \, \zeta \to \infty, \tag{19}$$

where $\alpha_1$ is dimensionless distance, $N_c$ is the ratio due to buoyancy force, $Ra_b$ is the bioconvection Rayleigh number, $\beta$ is the ferrohydrodynamic interaction parameter, $\varepsilon$ the Curie temperature, $\lambda$ is the heat dissipation parameter, $S$ is the heat generation/absorption parameter ($S > 0$ for suction and $S < 0$ for injection), $Nb$ and $Nt$ are the Brownian motion and thermophoresis parameters, the Prandtl number is $Pr$, the Eckert number is $Ec$, the radiation parameter is $Rd$, the local inertia parameter is $F_r$, the chemical reaction parameter is $\delta$, the porosity parameter is $P_1$, the Schmidt number is $Sc$, the Lewis number of bioconvection is $Lb$, the Peclet number $Pe$, $\delta_n$ is the bioconvection constant, and quantities are defined by

$$S = \frac{W_w x}{v_f Ra^{\frac{1}{2}}}, \, N_c = \frac{g\beta_C(C_w - C_\infty)x^3 cos(\alpha_1)}{v_f^2 Ra_x}, \, Ra_b = \frac{g\beta_N(N_w - N_\infty)x^3 cos(\alpha)}{v_f^2 Ra_x}, \, P_1 = \frac{\mu_f}{k_o^*},$$

$$L_i = \frac{C_b}{\sqrt{k_o^*}}, \, \beta = \frac{\gamma\lambda_o K \rho_f(T_w - T_\infty)}{2\pi\mu_f^2}, \, Pr = \frac{\mu_f c_p}{k_f}, \, Ec = \frac{W_w^2}{c_p(T_w - T_\infty)}, \, \lambda = \frac{\mu_f^2}{\rho_f(T_w - T_\infty)Ra_x^{\frac{3}{4}}}, \tag{20}$$

$$\delta = \frac{K_1 x^2}{v_f Ra_x^{\frac{1}{2}}}, \, Rd = \frac{16\sigma^* T_\infty^3}{3k^* k_f}, \, Sc = \frac{v_f}{D_B}, \, Pe = \frac{bW_c}{D_n}, \, Lb = \frac{v_f}{D_n}, \, \epsilon = \frac{T_\infty}{T_\infty - T_w},$$

$$d = \frac{aRa_x}{x}, \, Nt = \frac{\tau D_T(T_w - T_\infty)}{T_\infty \alpha_f}, \, Nb = \frac{\tau D_B(C_w - C_\infty)}{\alpha_f}, \, \delta_n = \frac{N_\infty}{N_w - N_\infty}.$$

The local Nusselt, Sherwood, and local Density expressions, as well as the coefficient of skin friction, will be computed by

$$C_f = \frac{2\tau_w}{\rho_f W_w^2}, \, Nu_x = \frac{q_h x}{k_f(T_w - T_\infty)}, \, Sh_x = \frac{q_m x}{D_B(C_w - C_\infty)}, \, Sn_x = \frac{q_n x}{D_n(N_w - N_\infty)}, \tag{21}$$

$$\tau_w = \mu_f u_s|_{s=0}, \, q_h = [-k_f T_s + q_r]|_{s=0}, \, q_m = -D_B C_s|_{s=0}, \, q_n = -D_n N_s|_{s=0}, \tag{22}$$

$$Ra_x^{\frac{1}{4}} C_f = 2f''(0), \, Nu = -Ra_x^{\frac{1}{4}}(1 + Rd)\theta'(0), \, Sh = -Ra_x^{\frac{1}{4}}\phi'(0), \, Sn = -Ra_x^{\frac{1}{4}}\chi'(0). \tag{23}$$

## 3. HAM Solutions Methodology

The homotopy analysis method (HAM) was applied to solve Equations (15)–(18). Shijun Liao developed this technique in 1992. It is often valid, regardless of whether there are a limited number of parameters or otherwise. It can be used to solve both weakly and strongly nonlinear problems. It offers a wide range of options for selecting the base functions of solutions, as well as discretion in choosing the linear operators. However, it provides a convenient method for ensuring the convergence of series solutions. Therefore, this method differs from other techniques, with examples like Adomain decomposition and the delta expansion methods. In the introduction section, some studies on the approach were presented.

Taking the initial guesses of the $f(\zeta), \theta(\zeta), \phi(\zeta)$, and $\chi(\zeta)$ with the auxiliary linear operators respectively as

$$f_0(\zeta) = 1 - e^{-\zeta}, \, \theta_0(\zeta) = \left(\frac{B_i}{1 + B_i}\right)e^{-\zeta}, \, \phi_0(\zeta) = -\left(\frac{Nt}{Nb}\right)e^{-\zeta}, \, \chi_0(\zeta) = e^{-\zeta}. \tag{24}$$

and

$$\mathcal{L}_f = f''' - f', \, \mathcal{L}_\theta = \theta'' - \theta, \, \mathcal{L}_\phi = \phi'' - \phi, \, \mathcal{L}_\chi = \chi'' - \chi, \tag{25}$$

the properties are satisfied as given below

$$\mathcal{L}_f(\Lambda_1 + \Lambda_2 e^\zeta + \Lambda_3 e^{-\zeta}) = 0, \quad \mathcal{L}_\theta(\Lambda_4 e^\zeta + \Lambda_5 e^{-\zeta}) = 0,$$

$$\mathcal{L}_\phi(\Lambda_6 e^\zeta + \Lambda_7 e^{-\zeta}) = 0, \quad L_\chi(\Lambda_8 e^\zeta + \Lambda_9 e^{-\zeta}) = 0 \tag{26}$$

with arbitrary constants $\Lambda_i, i \in [1,9]$. The Zeroth order form of the problem is given by

$$(1-p)\mathcal{L}_f[f(\zeta;p) - f_0(\zeta)] = ph_f \mathbf{N}_f[f(\zeta,p), \theta(\zeta,p), \phi(\zeta,p), \chi(\zeta,p)], \tag{27}$$

$$(1-p)\mathcal{L}_\theta[\theta(\zeta;p) - \theta_0(\zeta)] = ph_\theta \mathbf{N}_\theta[\theta(\zeta,p), f(\zeta,p), \phi(\zeta,p)], \tag{28}$$

$$(1-p)\mathcal{L}_\phi[\phi(\zeta,p) - \phi_0(\zeta)] = ph_\phi \mathbf{N}_\phi[\phi(\zeta,p), \theta(\zeta,p), f(\zeta,p)], \tag{29}$$

$$(1-p)\mathcal{L}_\chi[\chi(\zeta,p) - \chi_0(\zeta)] = ph_\chi \mathbf{N}_\chi[\chi(\zeta,pp), \phi(\zeta,p), f(\zeta,p)], \tag{30}$$

with $p \in [0,1]$ as the embedded parameter, and nonlinear operators $\mathbf{N}_f$, $\mathbf{N}_\theta$, $\mathbf{N}_\phi$, and $\mathbf{N}_\chi$ obtained by using Equations (15)–(18).

The problems' equivalent $m$ order of the deformation are

$$L_f[f_m(\zeta,p) - \eta_m f_{m-1}(\zeta)] = h_f \mathcal{R}_{f,m}(\zeta), \tag{31}$$

$$\mathcal{L}_\theta[\theta_m(\zeta,p) - \eta_m \theta_{m-1}(\zeta)] = h_\theta \mathcal{R}_{\theta,m}(\zeta), \tag{32}$$

$$\mathcal{L}_\phi[\phi_m(\zeta,p) - \eta_m \phi_{m-1}(\zeta)] = h_\phi \mathcal{R}_{\phi,m}(\zeta), \tag{33}$$

$$\mathcal{L}_\chi[\chi_m(\zeta,p) - \eta_m \chi_{m-1}(\zeta)] = h_\chi \mathcal{R}_{\chi,m}(\zeta), \tag{34}$$

$$f_m = S, f'_m = 0, \theta'_m - B_i\theta_m = 0, Nb\phi'_m + Nt\theta'_m = 0, \chi_m = 0, at\ \zeta = 0$$

$$f'_m = 0, \theta_m = 0, \phi_m = 0, \chi_m = 0 as \zeta \to \infty. \tag{35}$$

$$\eta_m = \begin{cases} 0, & \text{if } m \leq 1 \\ 1, & \text{if } m > 1, \end{cases} \tag{36}$$

where $\mathcal{R}_f^m(\zeta), \mathcal{R}_\theta^m(\zeta), \mathcal{R}_\phi^m(\zeta), \mathcal{R}_\chi^m(\zeta)$ can be obtained using Equations (15)–(18).

The general solutions are given by

$$f_m(\zeta) = f_m^s(\zeta) + \Lambda_1 + \Lambda_2 e^\zeta + \Lambda_3 e^{-\zeta}, \tag{37}$$

$$\theta_m(\zeta) = \theta_m^s(\zeta) + \Lambda_4 e^\zeta + \Lambda_5 e^{-\zeta}, \tag{38}$$

$$\phi_m(\zeta) = \phi_m^s(\zeta) + \Lambda_6 e^\zeta + \Lambda_7 e^{-\zeta}, \tag{39}$$

$$\chi_m(\zeta) = \chi_m^s(\zeta) + \Lambda_8 e^\zeta + \Lambda_9 e^{-\zeta}, \tag{40}$$

where $(f_m^s(\zeta), \theta_m^s(\zeta), \phi_m^s(\zeta), \chi_m^s(\zeta))$ are special solutions.

## 4. Analysis of Convergence of the Solutions

The convergence Table 1 is organized for each profile up towards the 35th order of approximation. Table 2 compares the current work to the published work and reveals that there is very close agreement.

**Table 1.** Convergence of HAM solutions with order approximations.

| Order of Approximations | $-f''(0)$ | $-\theta'(0)$ | $-\phi'(0)$ | $-\chi'(0)$ |
|---|---|---|---|---|
| 1 | 1.87633 | 0.6333 | 1.12222 | 0.69596 |
| 5 | 1.93521 | 0.76748 | 1.12327 | 0.69583 |
| 10 | 1.93385 | 0.76804 | 1.12349 | 0.69575 |
| 15 | 1.93135 | 0.7658 | 1.12354 | 0.69575 |
| 25 | 1.93026 | 0.7658 | 1.12354 | 0.69575 |
| 30 | 1.93026 | 0.7658 | 1.12354 | 0.69575 |
| 35 | 1.93026 | 0.7658 | 1.12354 | 0.69575 |

**Table 2.** Comparison of $-\theta'(0)$ and $-\phi'(0)$ by alternating buoyancy ratio parameter $N_c$ with published work.

| $N_c$ | $-\theta'(0)$ Reddy et al. [12] | Present | $-\phi'(0)$ Reddy et al. [12] | Present |
|---|---|---|---|---|
| 0.1 | 0.32598 | 0.32601 | 1.48394 | 1.48381 |
| 0.2 | 0.32405 | 0.32411 | 1.46789 | 1.46801 |
| 0.3 | 0.32229 | 0.32231 | 1.45214 | 1.45215 |
| 0.4 | 0.32125 | 0.32129 | 1.43598 | 1.43594 |
| 0.5 | 0.31868 | 0.31867 | 1.41938 | 1.41940 |

## 5. Results and Discussion

The role of $\beta$ on velocity profile is detected in Figure 2. The ferromagnetic parameter emphasizes the effect of the magnetic dipole's external magnetic field on fluid dynamics. As the magnetic field acts as a deforming force, the axial velocity decreases. Figure 3 depicts the important properties of the $Fr$ on the dimensionless velocity. Obviously, increasing $Fr$ causes the velocity of fluid layers to decrease. From a physical standpoint, the local inertia parameter generates resistance forces against the motion of fluid particles. Moreover, as the local inertia parameter increases, so does the velocity. Figure 4 depicts the impact of increasing $P_1$ on the velocity. The porosity parameter is defined as the kinematic viscosity to permeability strength of porous space ratio. As the porosity parameter increases, the velocity curves definitely decrease. Figure 5 depicts the effect of the $Ra_b$ on the velocity profile. It was discovered that when the values of $Ra_b$ rise, the velocity accelerates. As a result of the buoyancy forces caused by bioconvection, the fluid velocity increases by increasing the bioconvection Raleigh number. Figure 6 shows that the temperature on the boundary layers enhances as a result of increasing the values of $\beta$. This is due to an interaction of the fluid's movement and the interference of the ferromagnetic particles. The interplay reduces the velocity while frictional heating increases between the fluid layers, resulting in an increase in the thermal boundary layer thickness. The effect of $\lambda$ on the temperature profile is depicted in Figure 7. In this case, temperature is displayed as an increasing function of $\lambda$. Usually, as the values of $\lambda$ increase, so does the thermal conductivity, and thus the temperature. Figure 8 displays the role of $Fr$ on temperature profile. A rise in $Fr$ results in a rise in temperature and the thermal boundary layer thickness. Figure 9 reveals that as $B_i$ increases, so does the thickness of the thermal boundary layer, and the temperature also enhances. A higher Biot number contributes to more convection, which leads to the enhancement of the temperature and thermal boundary layer thickness. Figure 10 shows the effect of $S < 0$ on the dimensionless temperature. This figure reveals that when increasing $S < 0$, the temperature and the thickness of the thermal boundary layers both decrease. As an outcome, suction is removed from the warm fluid in the boundary layer region to a large extent. Moreover, the opposite trend is observed in the temperature profile, as shown in Figure 11 with $S > 0$. This is attributable to the fact that the temperature of the fluid is raised by injecting warm fluid into the boundary layer region. Figure 12 depicts the effects of $Rd$ on the temperature profile for various $Rd$ values.

An increase in the $Rd$ causes a rise in the temperature, and the effect of thermal radiation improves the medium's thermal diffusive. Besides that, for higher $Nt$ values, temperatures rise in the boundary layer region (Figure 13). The thermophoretic force developed in the boundary layer regime is an outcome of the temperature gradient; such forces entail the diffusion of nanoparticles out of the higher temperature area to a lower temperature area, leading to a thermal boundary layer thickness enhancement. Figure 14 depicts the $Nb$ characteristics on a temperature profile. Usually, a rise in $Nb$ improves the motion of fluid particles randomly, resulting in more heat generation. As a result, the temperature rises. Figure 15 depicts the temperature distribution by raising $Ec$ values. The $Ec$ defines the connection among the flow of kinetic energy and heat enthalpy variation. As a consequence, raising $Ec$ also raises the kinetic energy. Moreover, temperature is well understood to be defined as average kinetic energy. As a result, the fluid's temperature rises. This graph shows that as $Ec$ increases, so does the temperature. Figure 16 shows the role of $Pr$ on the temperature profile. It has been discovered that raising the $Pr$ lowers the temperature of the fluid flow. Large $Pr$ values clearly result in the thinning of thermal boundary layers. As $\delta$ increases, so does the concentration profile. The concentration becomes less effective as the delta values increase, as shown in Figure 17. The physical effect of $Nb$ on a concentration profile is depicted in Figure 18. Brownian motion does play a role in determining the efficiency of heat transfer during nanofluid flow. The nanoparticles collide with one another and transfer energy due to the random motion of a nanofluid. As a result, as $Nb$ levels rise, the concentration profile falls. The effect of $Nt$ on nanoparticle concentration is depicted in Figure 19. The concentration field rises in this case due to an increase in $Nt$. Larger $Nt$ causes an increase in thermophoresis forces, which further frequently carries nanomaterials from higher to lower temperature regions. As a result, the concentration decreases. The effect of $Sc$ on concentration is depicted in Figure 20. Sc denotes the momentum-to-mass diffusivity ratio, which measures the relative efficacy of momentum and mass transport through diffusion within concentration boundary layers. Figures 21 and 22 show the effects of $Pe$ and $Lb$ on the microorganism profile. According to these figures, the microorganism field decreases with increase of both numbers. According to Table 3, as $Pr$ estimates increase, so does the Nusselt number. Table 4 shows that as $\delta$ and $Sc$ increase, so does the Sherwood number. The results of motile microorganism density are increased by increasing $Lb$ and $Ra_b$, as shown in Table 5.

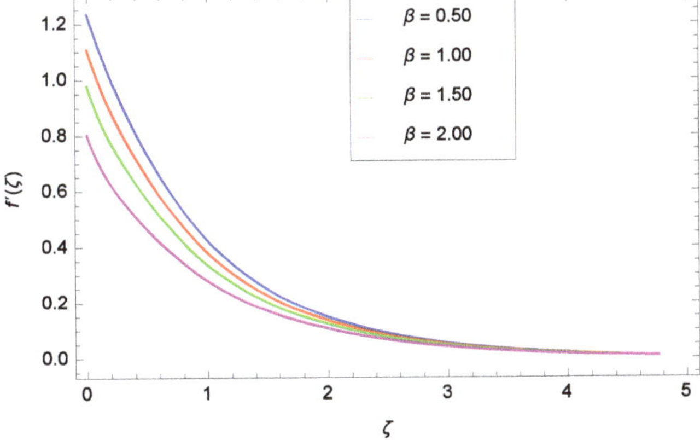



**Figure 2.** Influence of $\beta$ on $f'(\zeta)$.

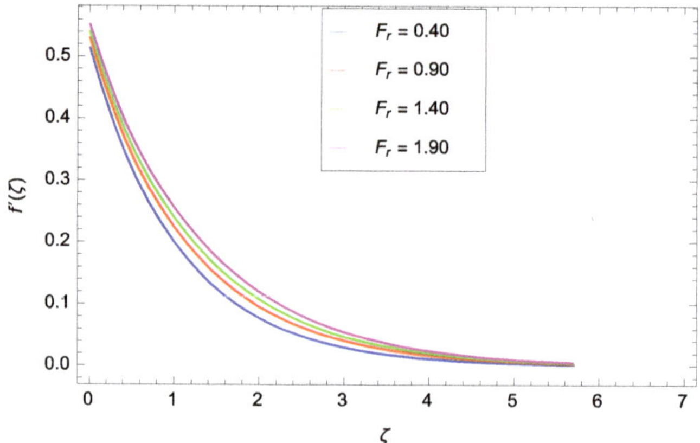

**Figure 3.** Influence of $Fr$ on $f'(\zeta)$.

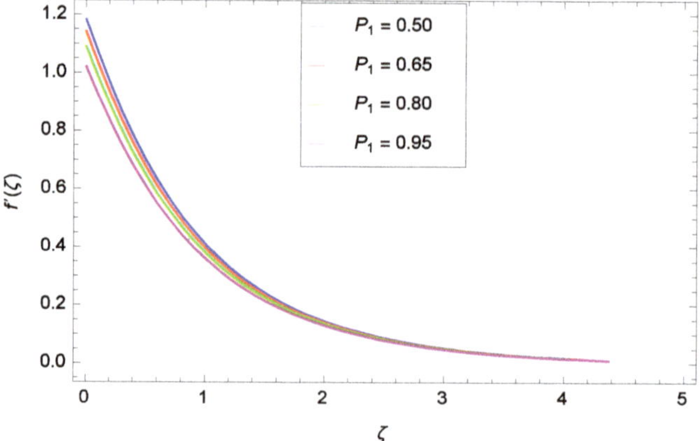

**Figure 4.** Influence of $P_1$ on $f'(\zeta)$.

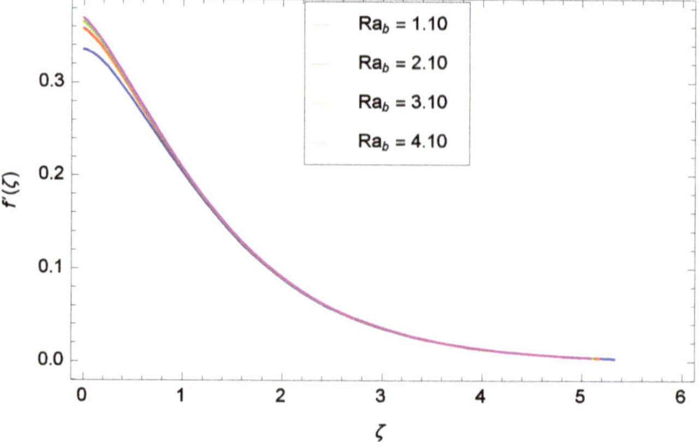

**Figure 5.** Influence of $Ra_b$ on $f'(\zeta)$.

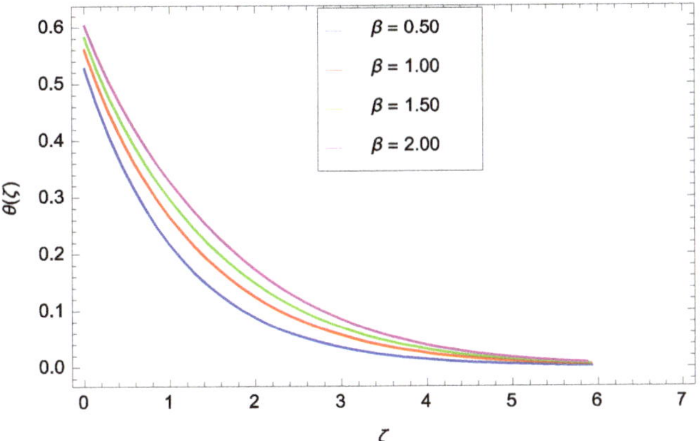

**Figure 6.** Influence of $\beta$ on $\theta(\zeta)$.

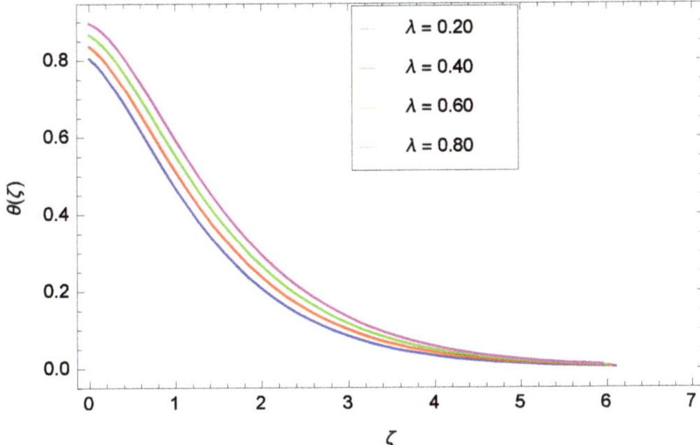

**Figure 7.** Influence of $\lambda$ on $\theta(\zeta)$.

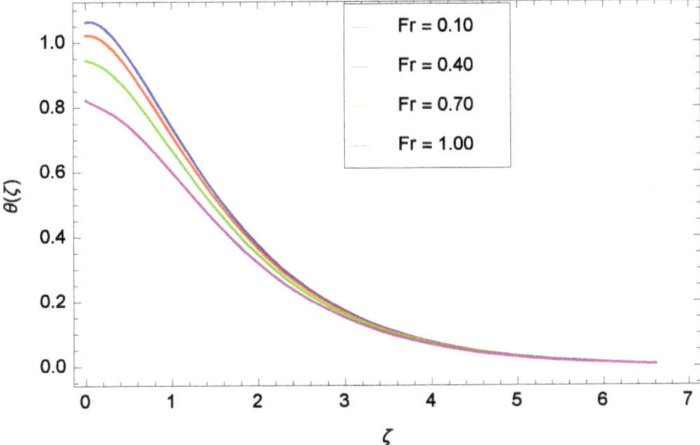

**Figure 8.** Influence of $Fr$ on $\theta(\zeta)$.

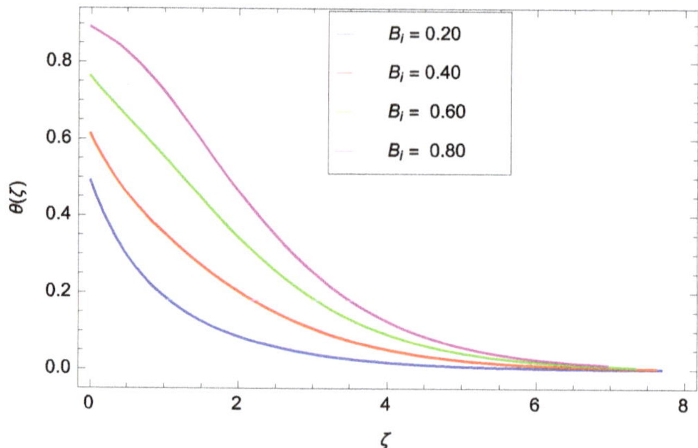

**Figure 9.** Influence of $B_i$ on $\theta(\zeta)$.

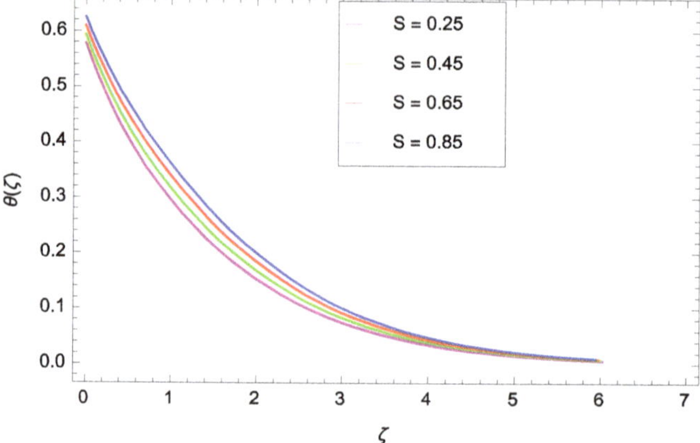

**Figure 10.** Influence of $S < 0$ on $\theta(\zeta)$.

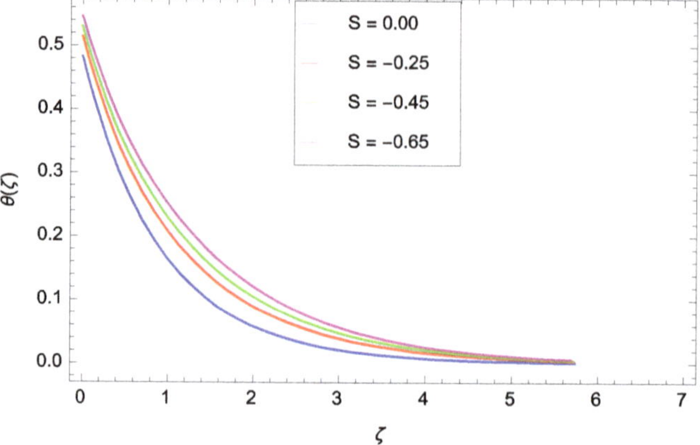

**Figure 11.** Influence of $S > 0$ on $\theta(\zeta)$.

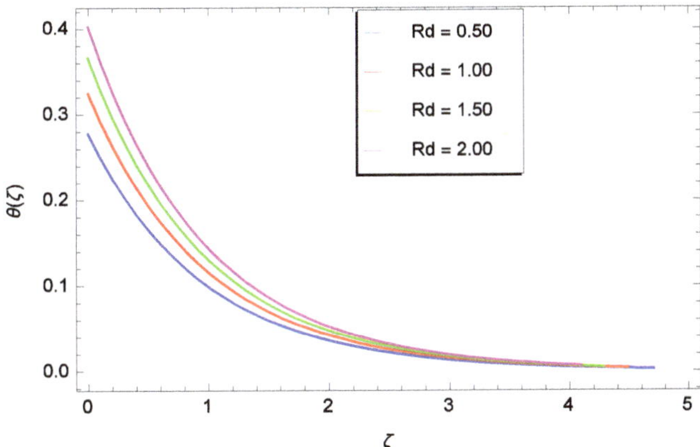

**Figure 12.** Influence of $Rd$ on $\theta(\zeta)$.

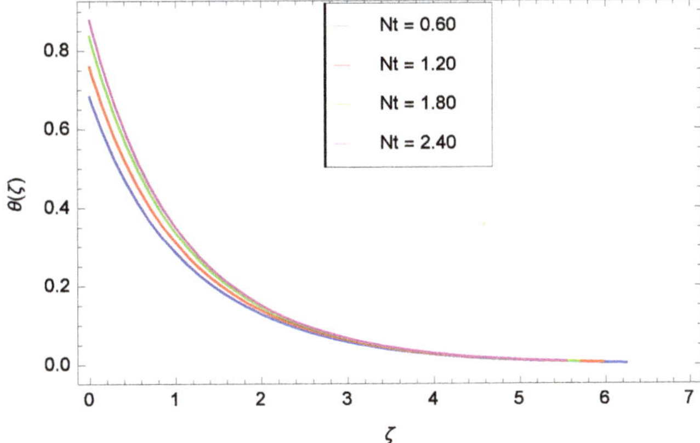

**Figure 13.** Influence of $Nt$ on $\theta(\zeta)$.

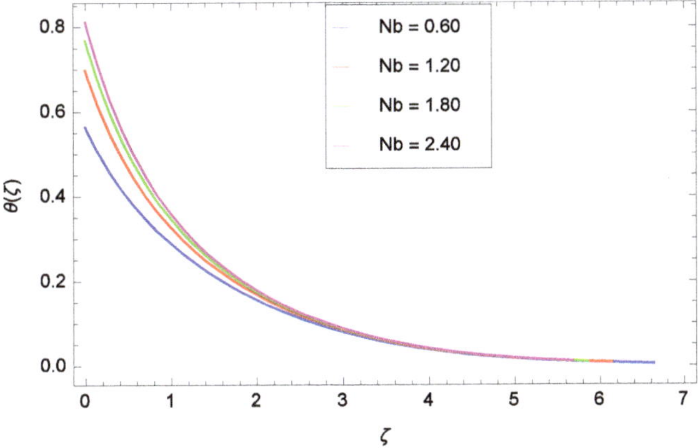

**Figure 14.** Influence of $Nb$ on $\theta(\zeta)$.

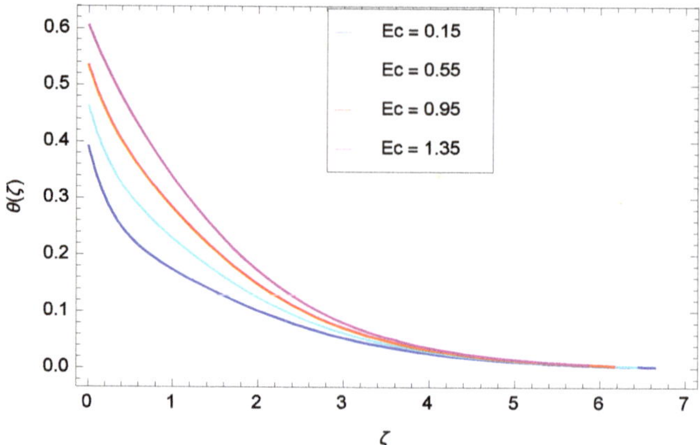

**Figure 15.** Influence of $Ec$ on $\theta(\zeta)$.

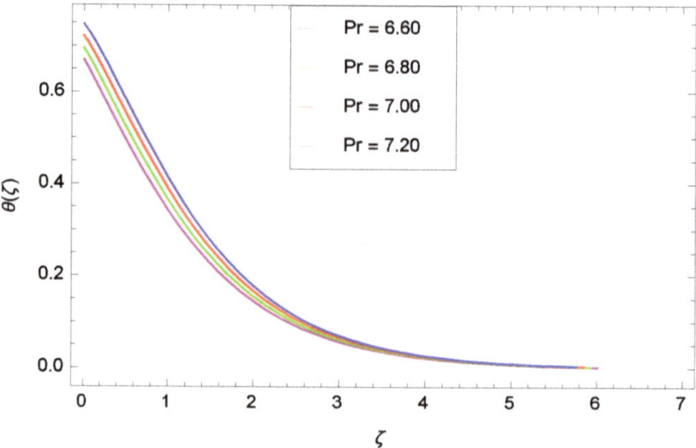

**Figure 16.** Influence of $Pr$ on $\theta(\zeta)$.

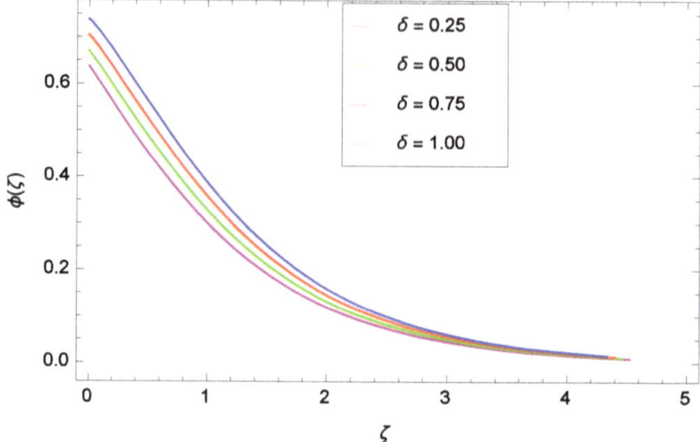

**Figure 17.** Influence of $\delta$ on $\phi(\zeta)$.

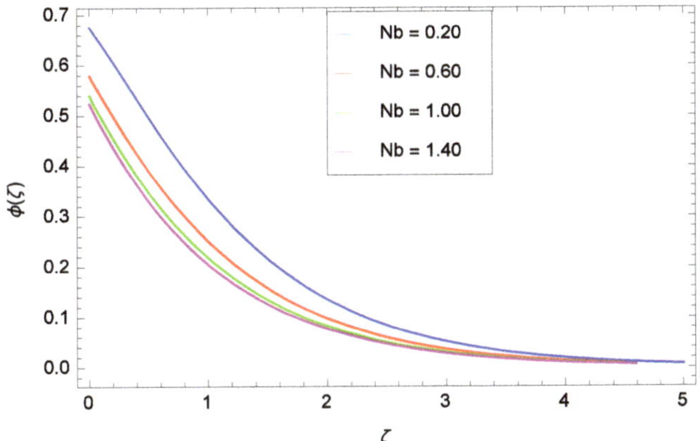

**Figure 18.** Influence of $Nb$ on $\phi(\zeta)$.

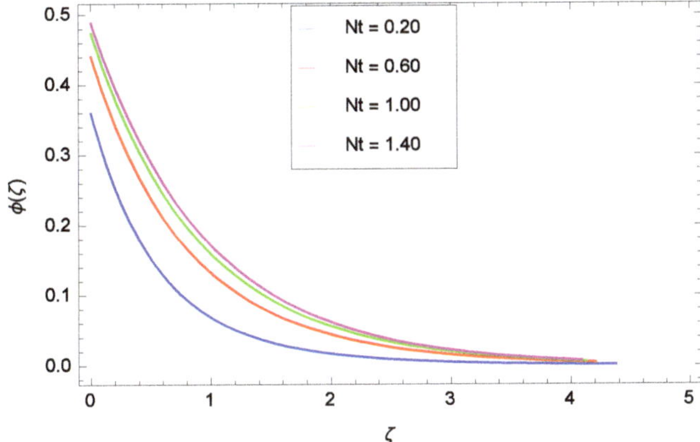

**Figure 19.** Influence of $Nt$ on $\phi(\zeta)$.

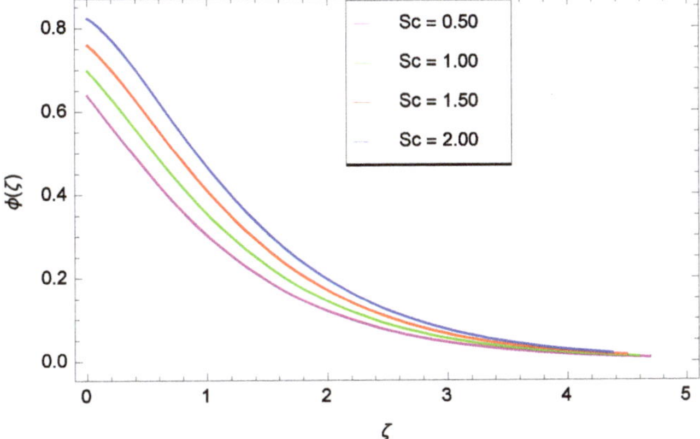

**Figure 20.** Influence of $Sc$ on $\phi(\zeta)$.

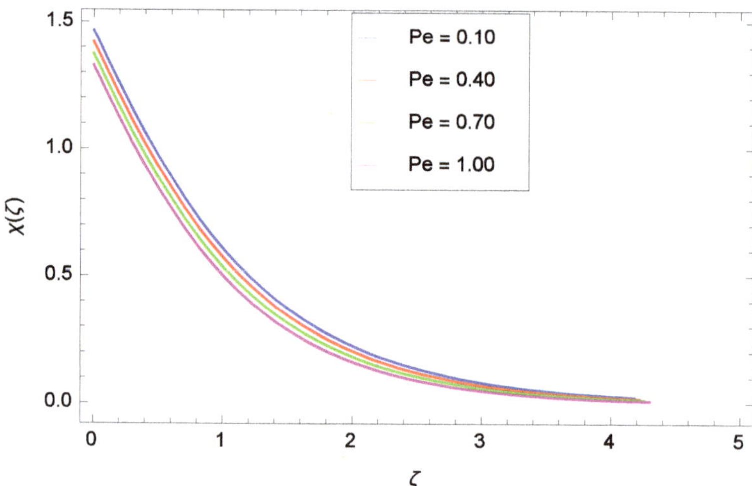

Figure 21. Influence of $Pe$ on $\chi(\zeta)$.

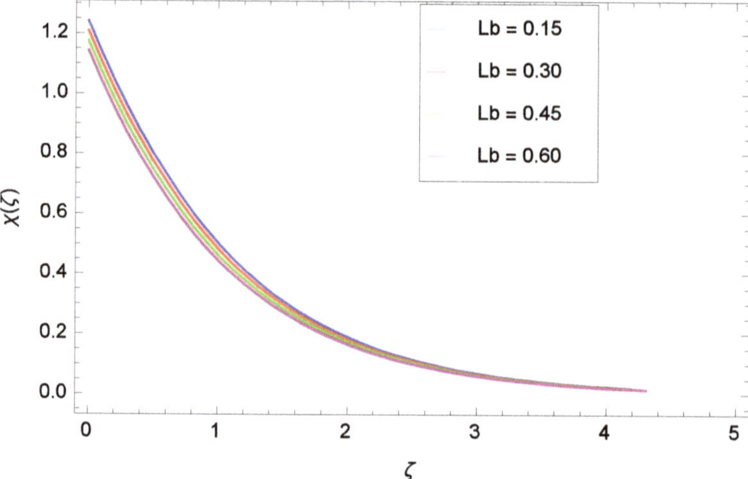

Figure 22. Influence of $Lb$ on $\chi(\zeta)$.

**Table 3.** Influence of $\beta$, $\varepsilon$, $\lambda$, $S$, $Rd$, $Fr$, $Ec$, and $Pr$ on $\theta'(0)$.

| $\beta$ | $\varepsilon$ | $\lambda$ | $S$ | $Rd$ | $Fr$ | $Ec$ | $Pr$ | $\theta'(0)$ |
|---|---|---|---|---|---|---|---|---|
| 0.2 | 0.3 | 0.2 | 0.4 | 0.3 | 0.2 | 0.1 | 0.7 | 0.44701 |
| 0.3 | | | | | | | | 0.43167 |
| 0.5 | | | | | | | | 0.40752 |
| | 0.4 | | | | | | | 0.39766 |
| | 0.5 | | | | | | | 0.38914 |
| | 0.6 | | | | | | | 0.37245 |
| | | 0.5 | | | | | | 0.45806 |
| | | 0.8 | | | | | | 0.45739 |
| | | 1.1 | | | | | | 0.45623 |
| | | | 0.8 | | | | | 1.03123 |
| | | | 1.2 | | | | | 0.87433 |
| | | | 1.6 | | | | | 0.64876 |
| | | | | 0.5 | | | | 0.45342 |
| | | | | 0.7 | | | | 0.44534 |
| | | | | 0.9 | | | | 0.43998 |
| | | | | | 0.6 | | | 0.64554 |
| | | | | | 1.0 | | | 0.63854 |
| | | | | | 1.4 | | | 0.61291 |
| | | | | | | 0.3 | | 0.38453 |
| | | | | | | 0.5 | | 0.38123 |
| | | | | | | 0.7 | | 0.37941 |
| | | | | | | | 6.7 | 0.75651 |
| | | | | | | | 7.7 | 0.80612 |
| | | | | | | | 8.7 | 0.86432 |

**Table 4.** Influence of $\delta$, $Sc$, $Nb$ on $\phi'(0)$.

| $\delta$ | $Sc$ | $Nb$ | $\phi'(0)$ |
|---|---|---|---|
| 0.2 | 0.3 | 0.5 | 1.28733 |
| 0.4 | | | 1.26931 |
| 0.6 | | | 1.24887 |
| | 0.5 | | 1.32742 |
| | 0.7 | | 1.31022 |
| | 0.9 | | 1.30271 |
| | | 1.0 | 1.33075 |
| | | 1.5 | 1.41186 |
| | | 2.0 | 1.47572 |

**Table 5.** Influence of $N_\delta$, $Pe$, $Lb$, $Ra_b$ on $\chi'(0)$.

| $\delta_n$ | $Pe$ | $Lb$ | $Ra_b$ | $\chi'(0)$ |
|---|---|---|---|---|
| 0.1 | 0.3 | 0.3 | 0.2 | 0.57238 |
| 0.2 | | | | 0.56103 |
| 0.3 | | | | 0.55271 |
| | 0.5 | | | 0.42714 |
| | 1 | | | 0.45102 |
| | 1.5 | | | 0.49327 |
| | | 0.5 | | 0.53185 |
| | | 1 | | 0.62386 |
| | | 1.5 | | 0.66671 |
| | | | 0.3 | 0.75408 |
| | | | 0.4 | 0.76965 |
| | | | 0.5 | 0.77121 |

## 6. Conclusions

A vertical cone has been used to study the flow of ferromagnetic nanofluid with bio-convection and magnetic dipole elements. There is also evidence of the Darcy–Forchheimer flow model. Viscosity dissipation, Brownian motion, chemical reaction, and thermophoresis are involved. Using appropriate transformations, nonlinear PDEs were reduced to a set of nonlinear ODEs. The important outcomes are listed below:

- The velocity decreases as the ferromagnetic interaction parameter, porosity parameter, and local inertia parameter increase.
- The temperature rises as the ferromagnetic interaction, heat dissipation, injection, thermal radiation parameters and Eckert number are raised, and reduces when the Prandtl number and sunction parameter are raised.
- When the Brownian motion, chemical reaction parameters and Schmidt number increase, the concentration decreases, while it increases when the thermophoresis parameter increases.
- The motile microorganism density decreases through raising the Peclet number and bioconvection Lewis number.

**Author Contributions:** A.H.U. and W.K. formulated and solved the problem. A.H.U. and Z.S. wrote the draft and the revised manuscript. A.H.U. and Z.S. sketched the graphs. U.W.H. and P.K. performed the investigations, supervision, and funding. All authors have read and agreed to the published version of the manuscript.

**Funding:** This research received no external funding.

**Institutional Review Board Statement:** Not applicable.

**Informed Consent Statement:** Not applicable.

**Data Availability Statement:** Availability exists for entirety of the data.

**Acknowledgments:** This research project is supported by Thailand Science Research and Innovation (TSRI) Basic Research Fund: Fiscal year 2022. The authors are thankful to KMUTT for providing the technical and financial support. The first author is supported by the Petchra Pra Jom Klao Doctoral Scholarship Academic for Ph.D. program of King Mongkut's University of Technology Thonburi (KMUTT) [Grant No. 13/2562].

**Conflicts of Interest:** The authors have no conflict of interests.

## References

1. Raj, K.; Moskowitz, R. Commercial applications of ferrofluids. *J. Magn. Magn. Mater.* **1990**, *85*, 233–245. [CrossRef]
2. Goldman, A. *Handbook of Modern Ferromagnetic Materials*; Springer Science and Business Media: Berlin, Germany, 2012.
3. Hathway, D.B. dB-sound. *Eng. Mag.* **1979**, *13*, 42–44.
4. Andersson, H.I.; Valnes, O.A. Flow of a heated ferrofluid over a stretching sheet in the presence of a magnetic dipole. *Acta Mech.* **1998**, *128*, 39–47. [CrossRef]
5. Zeeshan, A.; Majeed, A.; Ellahi, R.; Zia, Q.M.Z. Mixed convection flow and heat transfer in ferromagnetic fluid over a stretching sheet with partial slip effects. *Therm. Sci.* **2018**, *22*, 2515–2526. [CrossRef]
6. Hayat, T.; Ahmad, S.; Khan, M.I.; Alsaedi, A. Exploring magnetic dipole contribution on radiative flow of ferromagnetic Williamson fluid. *Results Phys.* **2018**, *8*, 545–551. [CrossRef]
7. Choi, S.U.; Eastman, J.A. *Enhancing Thermal Conductivity of Fluids with Nanoparticles*; Argonne National Lab: Lemont, IL, USA, 1995.
8. Ellahi, R. The effects of MHD and temperature dependent viscosity on the flow of non-Newtonian nanofluid in a pipe: Analytical solutions. *Appl. Math. Model.* **2013**, *37*, 1451–1467. [CrossRef]
9. Ellahi, R.; Raza, M.; Akbar, N.S. Study of peristaltic flow of nanofluid with entropy generation in a porous medium. *J. Porous Media* **2017**, *20*, 461–478. [CrossRef]
10. Hayat, T.; Ahmad, S.; Khan, M.I.; Alsaedi, A. Modeling and analyzing flow of third grade nanofluid due to rotating stretchable disk with chemical reaction and heat source. *Phys. B Condens. Matter.* **2018**, *537*, 116–126. [CrossRef]
11. Awais, M.; Shah, Z.; Perveen, N.; Ali, A.; Kumam, P.; Thounthong, P. MHD effects on ciliary-induced peristaltic flow coatings with rheological hybrid nanofluid. *Coatings* **2020**, *10*, 186. [CrossRef]

12. Reddy, P.S.; Sreedevi, P.; Chamkha, A.J. MHD natural convection boundary layer flow of nanofluid over a vertical cone with chemical reaction and suction/injection. *Comput. Therm. Sci. Int. J.* **2017**, *9*, 165–182. [CrossRef]
13. Alsaedi, A.; Khan, M.I.; Farooq, M.; Gull, N.; Hayat, T. Magnetohydrodynamic (MHD) stratified bioconvective flow of nanofluid due to gyrotactic microorganisms. *Adv. Powder Technol.* **2017**, *28*, 288–298. [CrossRef]
14. Hayat, T.; Bashir, Z.; Qayyum, S.; Alsaedi, A. Nonlinear radiative flow of nanofluid in presence of gyrotactic microorganisms and magnetohydrodynamic. *Int. J. Numer. Methods Heat Fluid Flow* **2019**, *29*, 3039–3055. [CrossRef]
15. Nadeem, S.; Ijaz, M.; El-Kott, A.; Ayub, M. Rosseland analysis for ferromagnetic fluid in presence of gyrotactic microorganisms and magnetic dipole. *Ain Shams Eng. J.* **2020**, *11*, 1295–1308. [CrossRef]
16. Bhatti, M.M.; Michaelides, E.E. Study of Arrhenius activation energy on the thermo-bioconvection nanofluid flow over a Riga plate. *J. Therm. Anal. Calorim.* **2021**, *143*, 2029–2038. [CrossRef]
17. Waqas, H.; Khan, S.A.; Alghamdi, M.; Alqarni, M.S.; Muhammad, T. Numerical simulation for bio-convection flow of magnetized non-Newtonian nanofluid due to stretching cylinder/plate with swimming motile microorganisms. *Eur. Phys. J. Spec. Top.* **2021**, *230*, 1239–1256. [CrossRef]
18. Ingham, D.B.; Pop, I. (Eds.) *Transport Phenomena in Porous Media III*; Elsevier: Oxford, UK, 2005; Volume 3.
19. Zeng, Z.; Grigg, R. A criterion for non-Darcy flow in porous media. *Transp. Porous Media* **2006**, *63*, 57–69. [CrossRef]
20. Kumar, B.R.; Sivaraj, R. MHD viscoelastic fluid non-Darcy flow over a vertical cone and a flat plate. *Int. Commun. Heat Mass Transf.* **2013**, *40*, 1–6. [CrossRef]
21. Chamkha, A.; Abbasbandy, S.; Rashad, A.M. Non-Darcy natural convection flow for non-Newtonian nanofluid over cone saturated in porous medium with uniform heat and volume fraction fluxes. *Int. J. Numer. Method Heat Fluid Flow* **2015**, *25*, 422–437. [CrossRef]
22. Mallikarjuna, B.; Rashad, A.M.; Hussein, A.K.; Raju, S.H. Transpiration and thermophoresis effects on Non-Darcy convective flow past a rotating cone with thermal radiation. *Arab. J. Sci. Eng.* **2016**, *41*, 4691–4700. [CrossRef]
23. Durairaj, M.; Ramachandran, S.; Mehdi, R.M. Heat generating/absorbing and chemically reacting Casson fluid flow over a vertical cone and flat plate saturated with non-Darcy porous medium. *Int. J. Numer. Method Heat Fluid Flow* **2017**, *27*, 156–173. [CrossRef]
24. Pătrulescu, F.O.; Groşan, T.; Pop, I. Natural convection from a vertical plate embedded in a non-Darcy bidisperse porous medium. *J. Heat Transf.* **2020**, *142*, 012504. [CrossRef]
25. Liao, S.J. The Proposed Homotopy Analysis Method for the Solution of Nonlinear Problems. Ph.D. Thesis, Shanghai Jiao Tong University, Shanghai, China, 1992.
26. Liao, S.J. An explicit, totally analytic approximate solution for Blasius viscous flow problems. *Int. J. Non-Linear Mech.* **1999**, *34*, 759–778. [CrossRef]
27. Liao, S. *Beyond Perturbation: Introduction to the Homotopy Analysis Method*; CRC Press: Boca Raton, FL, USA, 2003.
28. Irfan, M.; Khan, M.; Khan, W.A.; Sajid, M. Thermal and solutal stratifications in flow of Oldroyd-B nanofluid with variable conductivity. *Appl. Phys. A* **2018**, *124*, 674. [CrossRef]
29. Tlili, I.; Waqas, H.; Almaneea, A.; Khan, S.U.; Imran, M. Activation energy and second order slip in bioconvection of Oldroyd-B nanofluid over a stretching cylinder: A proposed mathematical model. *Processes* **2019**, *7*, 914. [CrossRef]
30. Usman, A.H.; Khan, N.S.; Humphries, U.W.; Ullah, Z.; Shah, Q.; Kumam, P.; Thounthong, P.; Khan, W.; Kaewkhao, A.; Bhaumik, A. Computational optimization for the deposition of bioconvection thin Oldroyd-B nanofluid with entropy generation. *Sci. Rep.* **2021**, *11*, 11641. [CrossRef] [PubMed]
31. Hayat, T.; Waqas, M.; Shehzad, S.A.; Alsaedi, A. Mixed convection flow of viscoelastic nanofluid by a cylinder with variable thermal conductivity and heat source/sink. *Int. J. Numer. Method Heat Fluid Flow* **2016**, *26*, 214–234. [CrossRef]

*Article*

# Evaluation of Coating Film Formation Process Using the Fluorescence Method

**Ayako Yano *, Kyoichi Hamada and Kenji Amagai**

Department of Mechanical Science and Technology, Gunma University, Gunma 376-8515, Japan;
t171b070@gunma-u.ac.jp (K.H.); amagai@gunma-u.ac.jp (K.A.)
* Correspondence: yano@gunma-u.ac.jp

**Abstract:** In this paper, we invented a novel observation method of the coating film formation process using the fluorescence method. With this method, the temporal change in the coating film thickness can be evaluated quantitatively. In addition, since the thickness and flow of the coating film can be measured simultaneously, the detailed coating film formation process was clarified. In the experiment, the adhesion behavior of the spray-paint droplets when applied to a wall was investigated. The characteristics of coating films formed by the spray droplets, particularly the influence of injection pressure on the coating film formation, were determined using the fluorescence method. At the initial stage of the coating process, the coating area increased linearly. When the ratio of the coating area to the measurement range reached about 80%, the rate at which the coating area increased slowed down, and an overlap began. The amount of paint that adhered to the coating film formation could be estimated by calculating the overlap ratio. Moreover, the thickness and smoothness of the coating film were evaluated using the histogram data of the fluorescence intensity. The leveling process was discussed in relation to the standard deviation of the histogram data. In addition, the flow of the paint during the coating film formation was investigated using tracer particles, and the effect of the spray gun injection pressure on the leveling process was investigated. Changes in the film thickness and flow during the coating film formation process could be evaluated through fluorescence observation.

**Keywords:** spray coating; coating film formation; leveling of coating surface; fluorescence method; visualization

**Citation:** Yano, A.; Hamada, K.; Amagai, K. Evaluation of Coating Film Formation Process Using the Fluorescence Method. *Coatings* **2021**, *11*, 1076. https://doi.org/10.3390/coatings11091076

Academic Editor: Cecilia Bartuli

Received: 10 August 2021
Accepted: 30 August 2021
Published: 6 September 2021

**Publisher's Note:** MDPI stays neutral with regard to jurisdictional claims in published maps and institutional affiliations.



## 1. Introduction

Reducing the emission of volatile organic compounds (VOCs) generated in various industrial fields is crucial, due to its adverse effects on humans and the atmosphere. The World Health Organization (WHO) has defined organic compounds with a boiling point between 50 and 260 °C, such as toluene, xylene, and formaldehyde, as VOCs. Air pollution caused by suspended particulate matter (SPM) and photochemical oxidants is the problem caused by VOCs, which is still a serious situation. In addition, VOCs are involved in the generation of fine particles (PM 2.5) as a precursor. VOCs are primarily released during painting processes, because organic solvents, such as thinner, are used to dilute the paint. Thinner contains xylene, toluene, etc., which can easily be released into the atmosphere. In particular, spray coating is performed as a coating method for large machines, such as automobiles; however, paint consumption is enormous, and a large amount of VOCs are generated. In addition, VOC emissions from shipbuilding, building construction, and automotive production have been reported [1–3], and the effect on humans has been discussed [4,5].

To reduce VOC emission during the painting process, it is important to improve the adhesion efficiency of the paint and optimize the coating time. Therefore, clarifying the relationship between the characteristics of the paint spray and the paint surface quality is necessary. Coating process optimization technology could reduce paint consumption while maintaining surface quality [6–8]. Additionally, spray-coating modeling has been investigated as a means of predicting coating thickness [9]. Research has been conducted to understand the mechanism behind coating film formation using spray coating [10,11]. However, in conventional studies, observations have been made with the naked eye and with a camera, and only the area and shape of the paint covering the observation range have been measured. The coating thickness and paint flow during the coating film formation process have not been reported in previous studies. In this study, we investigated the coating process in which the atomized paint adheres to the coated surface. Further, we propose a methodology to determine the change in thickness and smoothness of the coating film quantitatively using the fluorescence method. Fluorescence has been used in various fields, such as biology, medicine, and combustion diagnostics in engineering [12]. However, fluorescence has not been previously applied as an evaluation method to elucidate the paint film formation mechanisms. In this study, the fluorescence method is newly applied to the coating technology.

## 2. Experimental Methods

Figure 1 shows an outline of the experimental setup. The experimental equipment consisted of a paint sprayer that sprayed paint and an observation unit that observed how paint adheres to and forms a coating film. The gravity feed spray gun (CREAMY(KP)5A-12, Kinki Factory Co., Ltd., Osaka, Japan) was used. The nozzle diameter of the spray gun was 1.2 mm, and the injection pressure was set to $P_{in}$ = 0.1, 0.2, and 0.3 MPa. The paints used in the experiment were colorless acrylic resin paints (Econet EB Two Pack Paints, Origin Electric Co., Ltd., Saitama, Japan), and their components are 40%–45% for acrylic polyol, 35%–40% for butyl acetate, and 15%–20% for diisobutyl ketone. Three types of viscosities were prepared by changing the amount of thinner. As a result of measuring each paint viscosity with an Iwata type viscosity cup, $\mu$ = 0.023, 0.033, and 0.037 Pa·s, respectively. Coumarin 153, which is a fluorescent agent, was used in order to assess the droplet adhesion behavior and the paint film formation. It was mixed with each paint at a ratio of 2.0 g/L. The absorption wavelength range of Coumarin 153 is 360 to 480 nm, the maximum absorption wavelength is 423 nm, and the maximum fluorescence wavelength is 530 nm. Phosphorescent powder was also mixed with the paint to measure the flow characteristics of the paint film. The wavelengths of lights used were 420 and 480 nm. The 420 nm light is near the absorption wavelength range of Coumarin 153. The phosphorescence from the powder was emitted by the 420 and 480 nm lights.

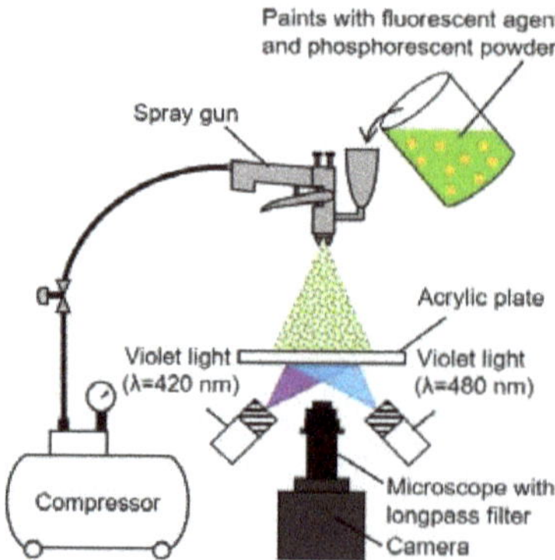


**Figure 1.** Experimental setup.

In the observation section, a transparent acrylic plate, with a length of 100 mm, a width of 100 mm, and a thickness of 3 mm, was installed as a painted surface at 300 mm downstream from the tip of the spray gun. Since the wavelength of light that can be transmitted through the transparent acrylic plate is about 400 nm or more, it was able to transmit the 420 and 480 nm lights used here. The paint mixed with the fluorescent agent and the phosphorescent powder was measured for fluorescence and phosphorescence by shining light of a specific wavelength from the underside of this acrylic plate.

The coating film formation process for the paint sprayed on the acrylic plate was photographed using a digital camera (DMC-GH4, Panasonic Corporation, Osaka, Japan) fitted with a microscope (VZM1000, Edmund Optics, Tokyo, Japan). In addition, a long pass filter (Φ 25 mm 500 nm High-Performance Longpass Filter, Edmund Optics) for blocking light at a wavelength of 490 nm or less was inserted between the microscope and camera to block the violet light, and only the light of fluorescence and phosphorescence wavelengths was photographed. In order to measure the paint film thickness quantitatively, the fluorescence intensities were calibrated. Paint containing a fluorescent agent was sandwiched between two glass plates, as shown in Figure 2, and the fluorescence intensities with against the various gap distances between the glass and acrylic plate were obtained. Figure 3 shows the calibrated data. From these data, the relationship between the fluorescence intensity and the thickness of the paint film was ascertained.

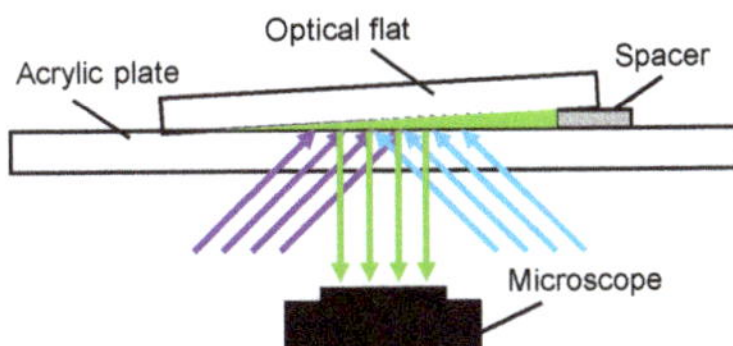


**Figure 2.** Experimental setup for calibrations of the relationship between fluorescence intensity and the coating film thickness.

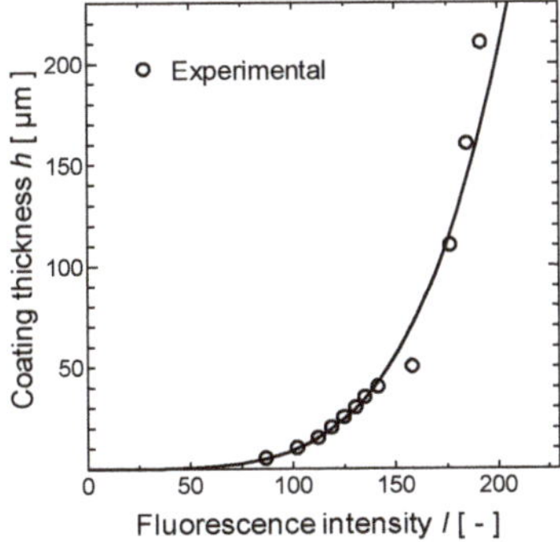


**Figure 3.** Calibration curve showing the coating film thickness in relation to the fluorescence intensity.

The obtained image was binarized to determine the adhesion area of the paint and the size distribution of the adherent droplets. In addition, to evaluate the film formation process, we defined the area ratio, $\alpha(t)$, and the overlap ratio, $\beta(t)$, from a previous study [12]. Figure 4 shows an example of the time dependence of the area and overlap ratios. The area ratio is the ratio of the droplet adhesion area to the measurement range. The overlap ratio is the value obtained by subtracting $\alpha(t)$ from the value that multiplies the increase ratio, $\alpha'_0$, of the area ratio at the initial stage of the droplet adhesion by the elapsed time $t$[s]. If the droplets do not overlap, the area ratio is considered to increase as $\alpha'_0 \cdot t$, and the overlap ratio indicates how much extra paint is attached as compared with do not overlap case.

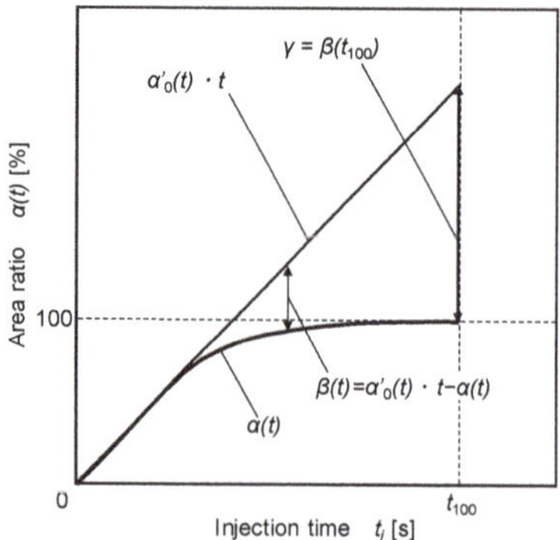

**Figure 4.** Model of coating formation [12] (adapted with permission from The Japanese Society for Experimental Mechanics).

In addition, the temporal change in the coating film thickness was determined by measuring the fluorescence intensity distribution on across-section of the fluorescence image. Additionally, the smoothness of the coating film was evaluated by examining the standard deviation of the fluorescence intensity distribution over the entire fluorescence image.

### 3. Experimental Results

*3.1. Fluorescent Images and Thickness of Adherent Paint Droplets*

Figure 5 shows fluorescent and pseudo-color images of an adherent paint droplet, measured from time-lapse photography taken under the conditions of $\mu = 0.037$ Pa·s and $P_{in} = 0.3$ MPa. It is shown that the area where the paint adheres becomes green. The red solid line represents the measurement position of the cross-sectional fluorescence intensity. Figure 6 shows the time change in the cross-sectional thickness of the droplets when adhering to the acrylic plate. To derive the film thickness, the aforementioned calibration data were used. The coating thickness, $h$, was calculated from the average fluorescence intensity using the calibration data. The horizontal axis is distance, $X$, and the vertical axis is coating thickness, $h$. $T_D = 0$ s was defined as one frame before droplet impact, and the droplet behavior was measured at $T_D = 1/90, 3/90$, and $6/90$ s. It was confirmed that the center part of the droplet is thinner than the outer parts at $T_D = 1/90$s. After adhesion, the center part becomes thick at $T_D = 3/90$ s. This behavior is caused by the impact when the droplet adheres. Here, the high, spike-like values seen in the cases of $T_D = 3/90$ s and $6/90$ s are the parts of the bubbles existing in the adherent droplet. The presence of air bubbles resulted in a high value due to the increase in local thickness, as shown in the fluorescence images. Finally, the droplet thickness converged to about 50 μm at $T_D = 6/90$ s. In addition, there was almost no difference in position between $T_D = 1/90$ and $T_D = 6/90$ s. Therefore, it was confirmed that the paint droplet adhering to the acrylic plate did not spread on the painted surface.

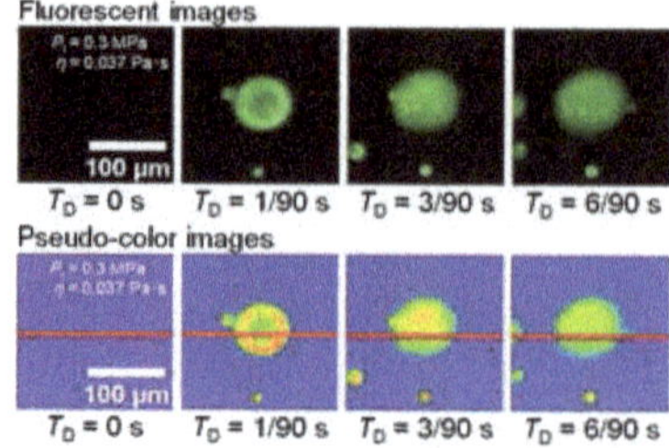

**Figure 5.** Fluorescent and pseudo-color images of the paint droplet.

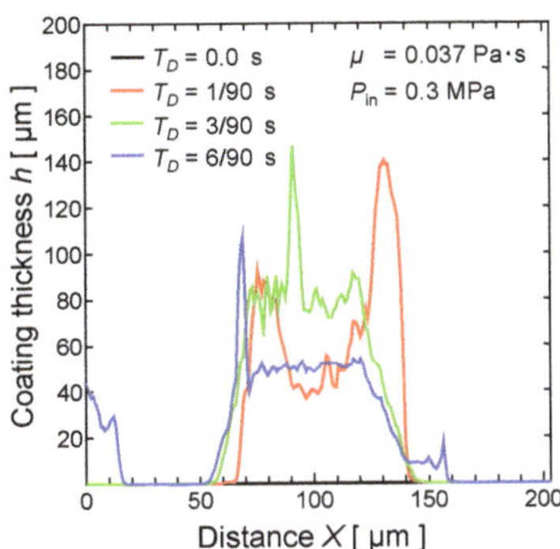

**Figure 6.** Coating thickness on a red line in Figure 5.

Figure 7 shows fluorescent and pseudo-color images of the droplet on the formed coating surface. Other conditions are the same as in Figure 5. Figure 8 represents the time change in the cross-sectional thickness of the droplet. It was confirmed that both ends of the droplet show a high value when the droplet adheres to the coating at $T_D = 1/90$ s. However, the center part of the droplet did not become thick after adhesion. The spike-shaped part seen in the $T_D = 3/90$ s data indicated the presence of air bubbles. The coating thickness before ($T_D = 0/90$ s) and after ($T_D = 6/90$ s) the droplet adhesion was compared. The coating thickness did not change in the range where the droplet adhered ($70 < X < 170$). However, the coating thickness increased in the range where the droplet did not adhere ($1 < X < 70$, $170 < X < 205$). This result shows that the adherent droplet spread around and integrated with the coating. Using the fluorescence method, it is possible to analyze the change in the coating thickness in a short time (within minutes) by examining the fluorescence intensity distribution in the cross-section.

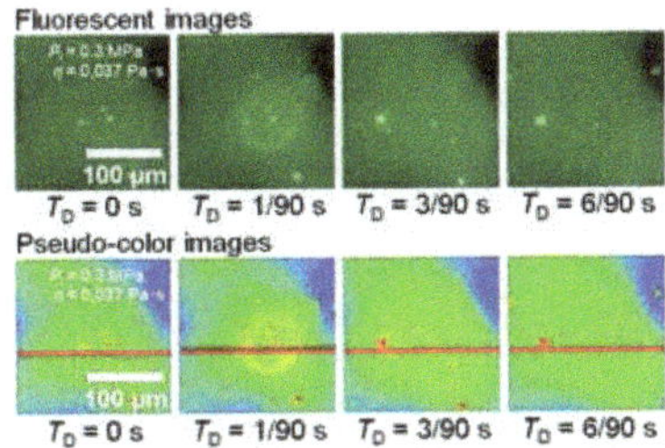


**Figure 7.** Fluorescent and pseudo-color images of the paint droplet.

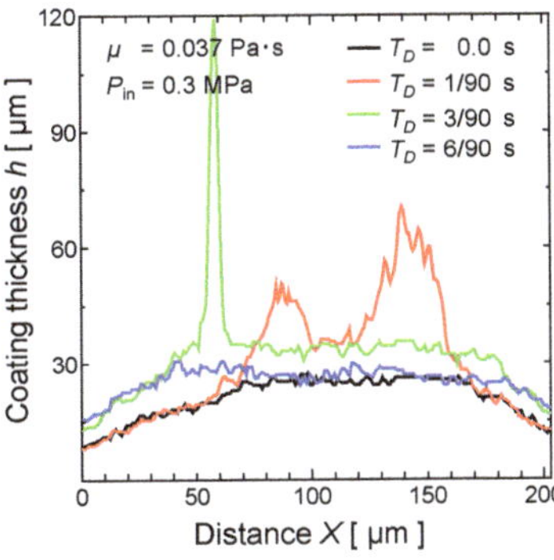

**Figure 8.** Coating thickness on a red line in Figure 7.

### 3.2. Observation of Coating Film Formation Process

In the above section, we focused on one droplet, but in the following, we observe a wider range of coating films using fluorescence. Figure 9 shows a fluorescence image of a painted surface, taken under conditions of injection pressure, $P_{in}$ = 0.2 MPa, and paint viscosity, $\mu$ = 0.033 Pa·s. The area where the paint adheres becomes green, the adhesion area increases with time, and the overall fluorescence intensity increases. The red solid line shown in the image at $t$ = 0.2 s represents the measurement position of the cross-sectional fluorescence intensity.

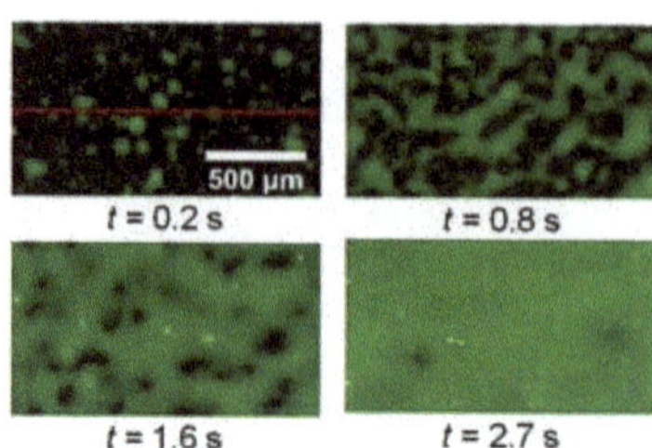


**Figure 9.** Fluorescence image of spray coating.

In the past, the authors changed the injection pressure $P_{in}$ to arbitrary values and examined the time change in the area and overlap ratios [12]. In the time change graph of the area and overlap ratios when changing the injection pressure, the increase ratio of the area ratio grows with the increase in the injection pressure. The higher the injection pressure, the shorter the time until $\alpha(t)$ = 100%. This was due to the increase in the paint flow rate per unit of time caused by the increase in injection pressure. The area ratio $\alpha(t)$ increases immediately after adhesion for any injection pressure, but increases gradually from around 50%. This was caused by the fact that the paint overlapped and began to adhere. The overlap ratio, $\beta(t)$, shows a value close to 0, because there was almost no overlap of droplets at the beginning of the paint adhesion. However, it increased with time. In addition, the overlap ratio in the time ($t = T_{100}$) it took for area ratio to reach 100%, with an injection pressure of $P_{in}$ = 0.2, 0.3, and 0.4 MPa, was about 200%, 300%, and 350%, respectively. It was found that the paint, which was about 2.3 to 2.8 times the measurement range, had adhered prior to the film formation.

Figure 10 shows the cross-sectional film thickness distribution at each stage, in the case of injection pressure $P_{in}$ = 0.2 MPa and viscosity $\mu$ = 0.033 Pa·s. The cross-sectional coating-thickness distribution was obtained by measuring the fluorescence intensity distribution on the line shown by the red solid line in Figure 9 and calculating the distribution of thickness from the calibration data of the fluorescence intensity and the coating thickness. It can be seen that the variation in coating thickness decreases with time and that a uniform and thick coating is formed. Thus, the temporal change in the coating film thickness can be examined from the distribution of fluorescence intensity.

Figure 11 shows the temporal change in the coating thickness under the conditions of $P_{in}$ = 0.2 MPa and $\mu$ = 0.033 Pa·s. The coating film thickness increased from the start of the spray coating. At the end of the spray coating ($t = T_{fin}$), the change in the coating thickness became flat.

### 3.3. Evaluation of Coating Surface Smoothness and Leveling Process

Leveling phenomena in the spray coating process were analyzed using the intensity distribution of the fluorescence image. To evaluate the surface smoothness, a histogram, or frequency distribution, of the tone values of the fluorescence image was produced. Figure 12 shows the histograms produced under the conditions of injection pressure $P_{in}$ = 0.2 MPa and paint viscosity $\mu$ = 0.033 Pa·s. The horizontal axis of Figure 12 indicates the tone values of fluorescence intensity, which was changed from 0 to 256. The vertical axis indicates the probability density.

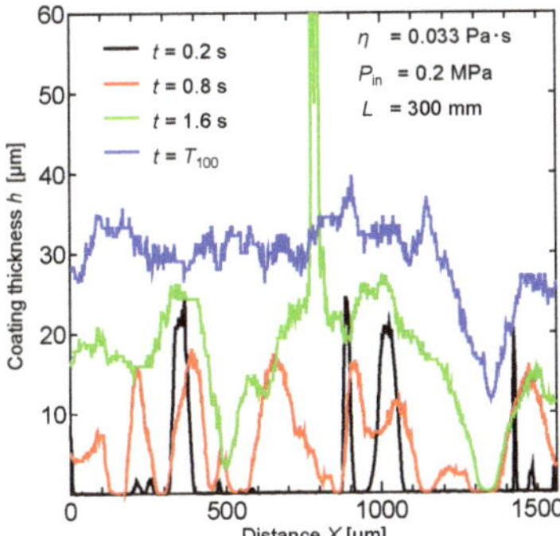

**Figure 10.** Fluorescence intensity distribution on a line graph.

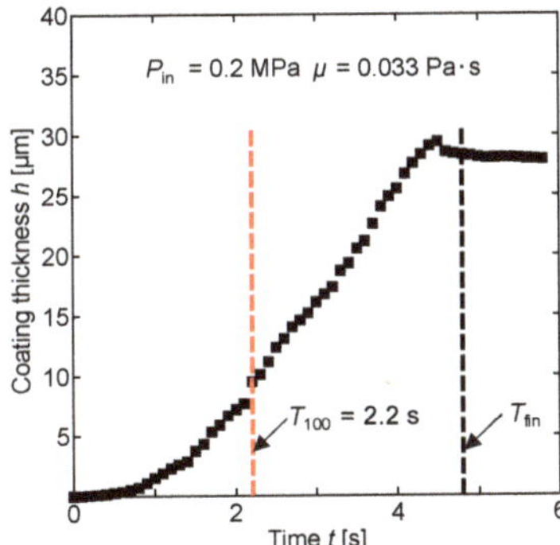

**Figure 11.** Evolution of the coating thickness over time.

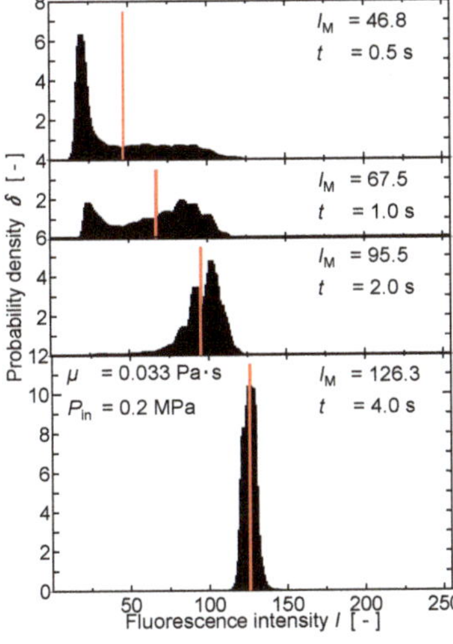

**Figure 12.** Fluorescence intensity distribution.

At the first stage of spray coating, the histogram has a wide range distribution. This result shows that fluorescence intensities have various tone values because the paint droplets adhered to the coating surface individually. The distribution becomes gradually narrower with an increase in the coating time, $t$. At the final stage, the distribution settles into a monodispersed shape. This means that the coating surface became flat. As described here, leveling phenomena can evaluate the width of the histogram. In this study, the standard deviation, $\sigma$ [-], was used for estimating the coating surface smoothness.

The center of the distribution, $I_M$, is shifted to the higher intensity side. This result shows that the average thickness of the coating film increased with the change in distribution shape. Therefore, the change in the average thickness of the coating film can be evaluated using the position of the center value.

The standard deviation of the intensity distribution in the histogram was investigated. Figure 13 shows the temporal change in the standard deviation. As shown in Figure 11, the standard deviation increased after the start of the spray-droplet adhesion (part of (A) in Figure 13) and decreased in part (B) of Figure 13. In this part, the leveling phenomena between adherent droplets occurred via their coalescence. $T_{100}$ indicates a point in time when the entire surface was covered by paint droplets. At the end of part (B), the coating surface became flat by the finish of leveling. However, in part (C), the standard deviation increased because the paint droplets overlapped on the coating surface. At the final stage of spray coating (in part (D)), the standard deviation decreased, gradually, with the progress of the leveling. In the figure, $T_{fin}$ indicates the stop time of spray coating. As described above, the average thickness and smoothness of the coating film could be easily estimated using the fluorescence method. In particular, it was clearly shown that the standard deviation data from the fluorescence intensity distribution provided important information about the smoothness of the coating surface.

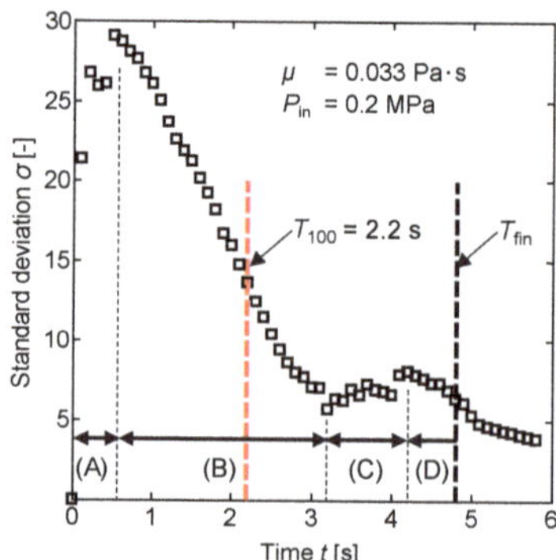


**Figure 13.** Standard deviation.

### 3.4. Influence of Injection Pressure and Paint Viscosity on Film Formation

Figure 14 shows the average thickness of the coating film, $h$ [mm], under various amounts of injection pressure $P_{in}$. The $P_{in}$ was changed from 0.1 to 0.5. It can be seen that the coating thickness increased uniformly until the end of injection, $t = T_{fin}$. The film thickness of $P_{in} = 0.2$ MPa was thicker than that of $P_{in} = 0.1$ MPa, because the injection flow rate of the former was greater than that of the latter.

In the case of $P_{in} = 0.5$ MPa, a thinner thickness value was maintained, even though the spray coating was sustained. The reason for this characteristic can be explained with reference to the airflow effect. In this study, the two-fluid atomizer was used for the spray formation. Therefore, strong airflow was generated at the coating surface under the high injection pressure condition. When atomized paint is stuck to the paint surface, paint flow is caused by the impact of adhesion and the contact of paint droplets. This flow affected the coating film formation and the quality of the paint surface. In this study, paint flow on the coating surface was visualized using the tracer particles in the paint.

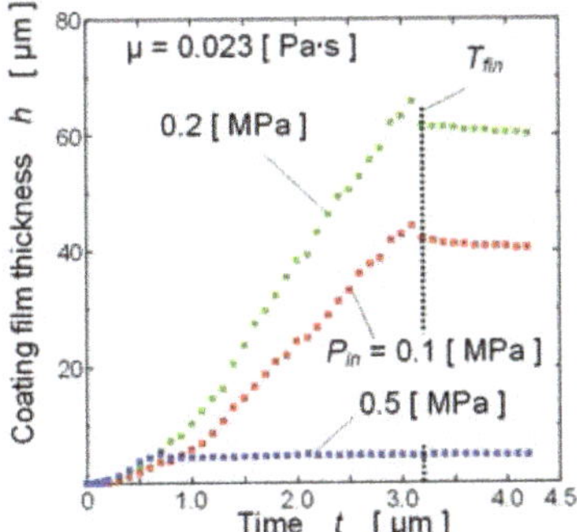

Figure 14. Coating film thickness under various injection pressures $P_{in}$.

The first point where paint adhesion was defined as the origin $(X_0, Y_0) = (0, 0)$. The length of the particle trajectory, $L_p$, and the absolute displacement, $L_d$, were measured under several experimental conditions. Figure 15 shows a schematic image of $L_p$ (the black line) and $L_d$ (the dashed line). $L_p$ is the total distance the particle moved, and $L_d$ is the linear distance from the original point to the end point after being moved by the flow. It was predicted that the flow would show random movement caused by the paint impact and leveling. Therefore, it was expected that $L_p$ and $L_d$ would become different values. If these values indicated almost the same number, then the flow caused by other forces would be generated. In our experiments, the time evolution of $L_p$ and $L_d$ was measured.

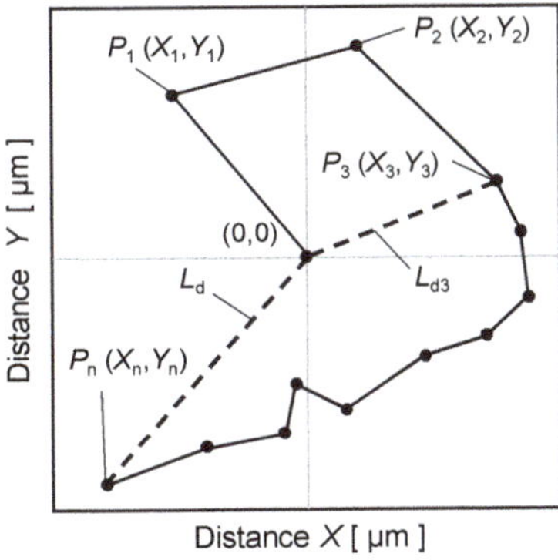

Figure 15. Definition of $L_p$ and $L_d$.

Figure 16, Figure 17, Figure 18 show the time evolution of the absolute displacement, $L_d$, using paint with a viscosity of $\mu = 0.023$ Pa·s and an injection pressure of $P_{in} = 0.1$, 0.2, and 0.5 Pa, respectively. The horizontal axis represents the injection time, $t$ s, and the vertical axis represents $L_d$. The point in time when the droplet starts to adhere is $t = 0$. Each plotline corresponds to one tracer particle.

As shown in Figure 16, the $L_d$ of most droplets that adhered at a relatively early time became 100–150 μm, and, after that, $L_d$ was maintained or decreased. This indicates the flow when the droplets that initially adhered began to coalesce to form a coating film. The paint droplet that first adhered to the surface did not move much. However, it moved when the droplet coalesced with other droplets.

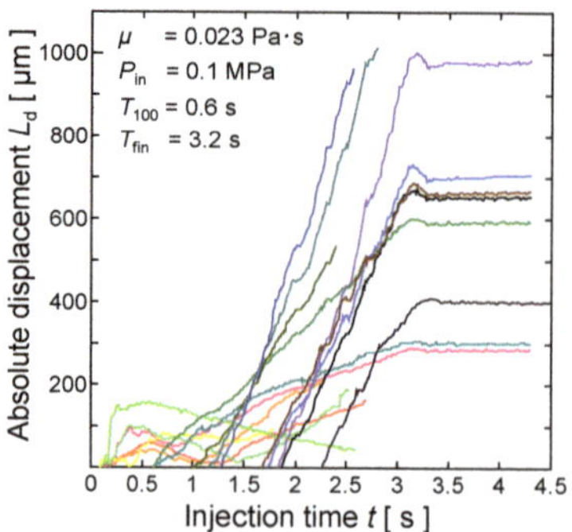

**Figure 16.** Time evolution of $L_d$ with injection pressure of $P_{in} = 0.1$ Pa.

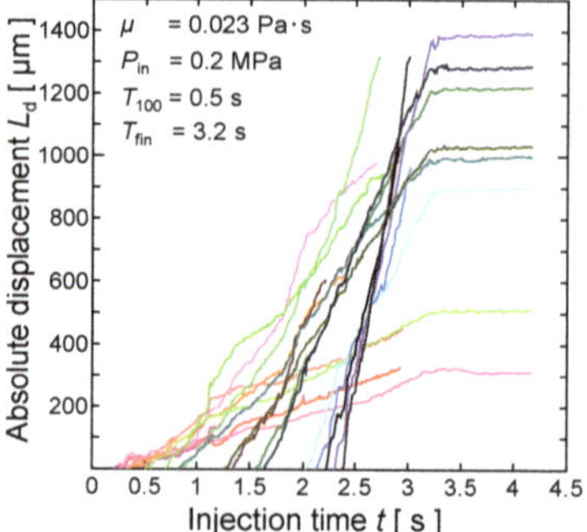

**Figure 17.** Time evolution of $L_d$ with injection pressure of $P_{in} = 0.2$ Pa.

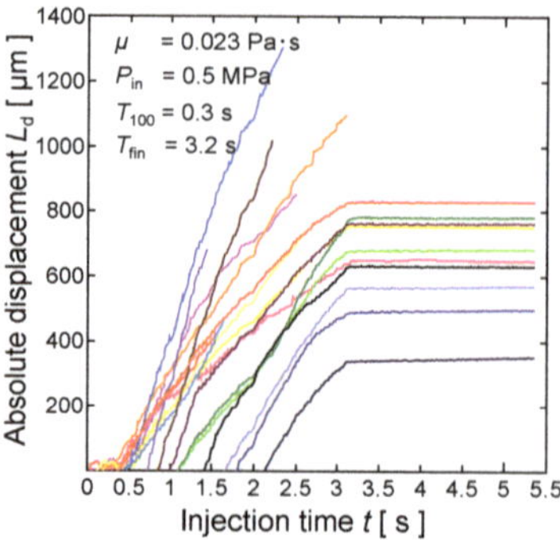

**Figure 18.** Time evolution of $L_d$ with injection pressure of $P_{in} = 0.5$ Pa.

However, the $L_d$ of the droplet that adhered after $t = 1.5$ s increased steadily until $t = T_{fin}$. In addition, the $L_d$ of some droplets that initially adhered increased to around $t = 1.0$ s, although the rate of increase was lower than that of the droplets that adhered later. These are the effects of the flow in the coating film. The $L_d$ of the droplets that adhered later increased at the same rate, and a uniform flow occurred on the surface of the entire area.

The fastest flow was generated on the surface of the coating film because the droplets that adhered later had a higher rate of increase of $L_d$. In addition, the particles that initially adhered were closer to the wall surface than the particles that adhered later. Therefore, the particles moved at a lower velocity than the surface velocity. Some of the droplets that adhered at the initial stage, whose $L_d$ had a low rate of increase from around $t = 1.0$ s, were caused by this movement.

As shown in Figure 17, the $L_d$ of all the droplets increased steadily until $t = T_{fin}$. However, there was a difference in the rate of increase of $L_d$. The droplets that adhered later showed a higher rate of increase, and it was confirmed that the particles moved to a point away from the initial position—above $L_d = 1200$ μm. In accordance with this, a flow faster than $P_{in} = 0.1$ Pa was expected to occur. In addition, the slow flow was generated near the wall because the $L_d$ of all particles increased, regardless of the time of adherence. The initially deposited paint was replaced by the latterly deposited paint as a result of this slow flow.

As shown in Figure 18, the $L_d$ of all the droplets increased at the same rate. Therefore, the $L_d$ of the droplets that initially adhered became larger than that of the droplets that adhered later. According to this result, a uniform flow was generated in the coating film. Even near the wall, the flow moved at the same velocity as the paint on the surface, if it was outside the boundary layer. Moreover, in this case, the replacement of the paint that adhered in the initial stage was performed more actively. The above results suggest that under the condition of a paint viscosity of $\mu = 0.023$ Pa·s, a flow occurs in the coating film, and the higher the injection pressure, the greater the replacement of the paint that initially adheres to it. Therefore, we consider that the coating film flowed out of the observation range before it hardened, and the coating thickness did not increase. In addition, we consider this to be because the average film thickness does not increase under the conditions of $\mu = 0.023$ Pa·s and $P_{in} = 0.5$ MPa, as shown in Figure 14.

## 4. Conclusions

In this study, we investigated the process of forming a coating film by changing the paint injection pressure. In addition, the thickness and smoothness of the coating film were investigated using the fluorescence method. The coating could be realized using the fluorescence method by mixing a fluorescent agent with the paint. The higher the injection pressure, the shorter the time until the coating was formed. The area ratio rises, linearly, immediately after the start of droplet deposition, but reached 100%, while decreasing the increase rate of the area ratio by overlapping. It was possible to estimate the amount of paint that overlapped and adhered using the overlapping ratio. The thickness of the coating could be calculated from the fluorescence intensity. Additionally, the smoothness of the coating film could be evaluated by examining the standard deviation of the fluorescence intensity. The thickness of the coating increased, linearly, from the start of droplet deposition to the end of spraying. The standard deviation increased near the onset of droplet deposition and then decreased until the time of film formation. At the end of the injection, the standard deviation decreased, and a very smooth coating was formed. The flow of the coating film formation was observed by tracking the fluorescent particles in the paint. We found that the higher the injection pressure, the more active the flow of the coating film. In this study, the thickness of the coating film and the flow of the formation process could be observed in detail at the same time using fluorescence observation. By using the methodology of this paper, it is possible to quantitatively analyze the coating film formation process. This method is applicable for various industrial processes, such as the painting of automobiles, ships and buildings. The conditions for forming a smooth and high-quality coating surface with a smaller amount of paint can be determined on the basis of data rather than empirical rules.

**Author Contributions:** Conceptualization, K.A.; investigation, A.Y., K.H. and K.A.; writing—original draft preparation, K.H.; writing—review and editing, A.Y. and K.A.; supervision, K.A. All authors have read and agreed to the published version of the manuscript.

**Funding:** This research received no external funding.

**Institutional Review Board Statement:** Not applicable.

**Informed Consent Statement:** Not applicable.

**Data Availability Statement:** Data is contained within the article.

**Conflicts of Interest:** The authors declare no conflict of interest.

## References


1. Celebi, U.B.; Vardar, N. Investigation of VOC Emissions from Indoor and Outdoor Painting Processes in Shipyards. *Atmos. Environ.* **2008**, *42*, 5685–5695. [CrossRef]
2. Li, F.; Niu, J.; Zhang, L. A Physically-based Model for Prediction of VOCs Emissions from Paint Applied to an Absorptive Substrate. *Build. Environ.* **2006**, *41*, 1317–1325. [CrossRef]
3. Julio, L.; Rivera, T.R. A Framework for Environmental and Energy Analysis of the Automobile Painting Process. *Procedia CIRP* **2014**, *15*, 171–175.
4. Kim, B.; Yoon, J.; Choi, B.; Shin, Y.C. Exposure Assessment Suggests Exposure to Lung Cancer Carcinogens in a Painter Working in an Automobile Bumper Shop. *Saf. Health Work* **2013**, *4*, 216–220. [CrossRef] [PubMed]
5. Mølgaard, B.; Viitanen, A.; Kangas, A.; Huhtiniemi, M.; Larsen, S.T.; Vanhala, E.; Hussein, T.; Boor, B.E.; Hämeri, K.; Koivisto, A.J. Exposure to Airborne Particles and Volatile Organic Compounds from Polyurethane Molding, Spray Painting, Lacquering, and Gluing in a Workshop. *Int. J. Environ. Res. Public Health* **2015**, *12*, 3756–3773. [CrossRef] [PubMed]
6. Nwaogu, U.C.; Tiedje, N.S. Foundry Coating Technology: A Review. *Mater. Sci. Appl.* **2011**, *2*, 1143–1160. [CrossRef]
7. Akafuah, N.K.; Poozesh, S.; Salaimeh, A.; Patrick, G.; Lawler, K.; Saito, K. Evolution of the Automotive Body Coating Process-A Review. *Coatings* **2016**, *6*, 24. [CrossRef]
8. Slegers, S.; Linzas, M.; Drijkoningen, J.; D'Haen, J.; Reddy, N.K.; Deferme, W. Surface Roughness Reduction of Additive Manufactured Products by Applying a Functional Coating Using Ultrasonic Spray Coating. *Coatings* **2017**, *7*, 208. [CrossRef]
9. Luangkularb, S.; Prombanpong, S.; Tangwarodomnukun, V. Material Consumption and Dry Film Thickness in Spray Coating Process. *Procedia CIRP* **2014**, *17*, 789–794. [CrossRef]
10. Tanno, S.; Ohani, S. Mechanism of Paint Film Formation in Spray Coating. *Kagaku Kogaku Ronbunshu* **1977**, *3*, 593–599. [CrossRef]
11. Tachi, K.; Okuda, C.; Oyama, Y. Shouichi Suzuki: Paint Film Formation Process in Spray Coating. *J. Jpn. Soc. Color Mater.* **1986**, *59*, 711–718. [CrossRef]
12. Yano, A.; Oe, T.; Takaishi, K.; Amagai, K. Visualization of the Paint Film Formation Process During Spray Coating. *Adv. Exp. Mech.* **2019**, *4*, 43–48.

*Article*

# Numerical Investigation on the Evaporation Performance of Desulfurization Wastewater in a Spray Drying Tower without Deflectors

Debo Li [1,2,*], Ning Zhao [2], Yongxin Feng [2] and Zhiwen Xie [2]

[1] Guangdong Electric Power Research Institute, Guangzhou 510663, China
[2] China Southern Power Grid Technology Co., Ltd., Guangzhou 510080, China; 15088050825@163.com (N.Z.); yongxingf@126.com (Y.F.); betty8228@126.com (Z.X.)
* Correspondence: ldbyx@126.com

**Abstract:** The desulfurization wastewater evaporation technology with flue gas has been widely applied to dispose of desulfurization wastewater. This paper investigates the effect of flue gas flow rate and temperature, wastewater flow rate and initial temperature, and droplet size on the evaporation performance of the desulfurization wastewater in a spray drying tower without deflectors. The results show that the flue gas flow rate and temperature affect the evaporation performance of desulfurization wastewater. The larger flow rate and higher temperature of flue gas correspond to the faster evaporation speed and the shorter complete evaporation distance of the wastewater droplet. Decreasing the flow rate and increasing the initial temperature of the desulfurization wastewater is advantageous to enhance the evaporation speed and shorten the complete evaporation distance of the wastewater droplet. Reducing the droplet size is beneficial to improve the evaporation performance of the desulfurization wastewater. The orthogonal test results show that the factors affecting droplet evaporation performance are ranked as follows: flue gas flow rate > wastewater flow rate > flue gas temperature > wastewater initial temperature > droplet size. Considering the evaporation ratio and the complete evaporation distance, the optimal setting is 14.470 kg/s for flue gas flow rate, 385 °C for flue gas temperature, 0.582 kg/s for wastewater flow rate, 25 °C for wastewater initial temperature, and 60 μm for droplet size. These studied results can provide valuable information to improve the operational performance of the desulfurization wastewater evaporation technology with flue gas.

**Keywords:** desulfurization wastewater evaporation technology; evaporation performance; orthogonal test; simulation

**Citation:** Li, D.; Zhao, N.; Feng, Y.; Xie, Z. Numerical Investigation on the Evaporation Performance of Desulfurization Wastewater in a Spray Drying Tower without Deflectors. *Coatings* **2021**, *11*, 1022. https://doi.org/10.3390/coatings11091022

Academic Editors: Eduardo Guzmán and Alexandru Enesca

Received: 20 July 2021
Accepted: 24 August 2021
Published: 26 August 2021



## 1. Introduction

In China, wet flue gas desulfurization technology is widely applied to coal-fired thermal power plants to remove $SO_2$ in the flue gas because it has the advantages of high efficiency, low operating cost, and high reliability [1–4]. However, this technology produces large quantities of desulfurization wastewater, which contains many acidic ions, heavy metal ions, and suspended solids [5–7]. Releasing desulfurization wastewater into the environment is strictly prohibited [8,9]. Therefore, the methods for desulfurization wastewater disposal have gained extensive research interest in recent years [10].

Some technologies have been proposed to dispose of the desulfurization wastewater, such as chemical precipitation, membrane separation, evaporative crystallization, electrodialysis technology, etc. [11–14]. Of these desulfurization wastewater disposal technologies, desulfurization wastewater evaporation technology is an effective method to achieve zero-emission of desulfurization wastewater [14–17]. Significantly, flue gas after the air preheater is the best choice as a heating source to evaporate the desulfurization wastewater, as shown in Figure 1 [18]. The flue gas is injected into the spray drying tower through a special-designed channel, and the desulfurization wastewater is sprayed into a spray drying tower through a high-speed rotating atomizer, and then evaporated under the preheating effect of the flue gas. The residual solid particles after evaporation are expected to be captured [19]. To improve the evaporation performance of desulfurization wastewater, some efforts have been performed. Liang et al. [20] investigated the evaporation and crystallization behaviors of the desulfurization wastewater droplet using thermogravimetric analysis. They found that the increase in heating rate can promote evaporation and crystallization rates simultaneously, while the final temperature has a limited effect on these rates. Deng et al. [21] numerically studied the effect of the position and number of nozzles, droplet size, and flue gas temperature on evaporation performance. Ma et al. [8] simulated the evaporation behavior of desulfurization wastewater and found that smaller droplet size, and higher flue gas flow rate and temperature could benefit the complete evaporation of the desulfurization wastewater. Zheng et al. [22] explored the chlorine migration of various chlorine salt solutions and typical desulfurization wastewater at high temperatures during the evaporation process of concentrated wastewater by a laboratory-scale tube furnace and a pilot-scale system. Although numerous efforts have been devoted to studies of the desulfurization wastewater evaporation, there are still few works that comprehensively investigate the desulfurization wastewater evaporation.

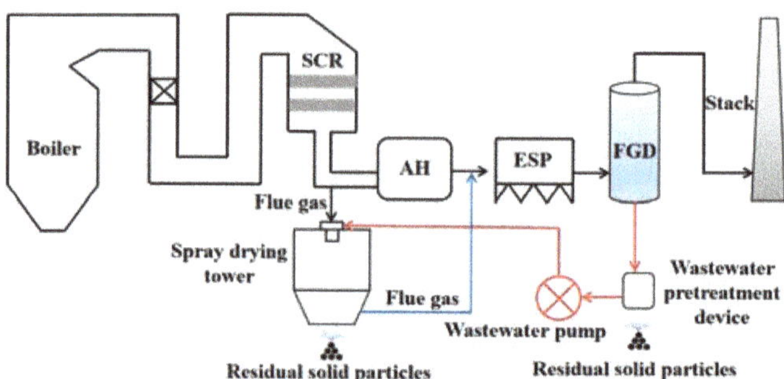


**Figure 1.** The desulfurization wastewater evaporation technology with flue gas.

This work is aimed to comprehensively investigate the effects of flue gas flow rate, flue gas temperature, wastewater flow rate, initial wastewater temperature, and droplet size on the desulfurization wastewater evaporation performance. These studied results can give more helpful information to guide the desulfurization wastewater evaporation technology with flue gas.

## 2. Physical Model and Numerical Method

### 2.1. Spray Drying Tower Description

The schematic diagram of a spray drying tower without deflectors is shown in Figure 2. The height is 17.9 m, and the diameter is 7.2 m. The flue gas is injected into the spray drying tower through a specially-designed volute structure. In the atomization process, the desulfurization wastewater is firstly sent to the atomizer by the metering pump. The motor drives the atomizing disc to rotate at high speed, and the desulfurization wastewater is thrown out from the channels, thus atomizing into droplets. The droplets are then evaporated under the preheating effect of the flue gas. In this study, the complete evaporation distance (as shown in Figure 2) is defined as the maximum distance from the atomizer in the vertical direction when the droplet particle mass is zero. Table 1 lists the physical parameters of flue gas under full load.

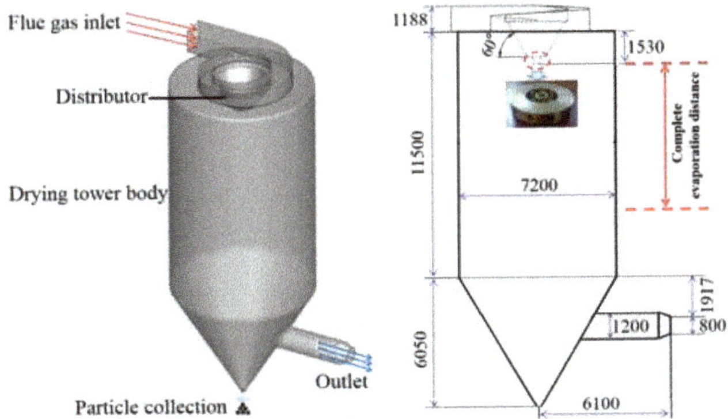


**Figure 2.** Schematic diagram of spray drying tower without deflectors.

**Table 1.** Physical parameters of flue gas in actual operation under full load.

| Item | Value |
| --- | --- |
| Volume flow of flue gas ($Nm^3/h$) | 32,315 |
| Temperature (K) | 638 |
| Density ($kg/m^3$) | 1.24 |
| viscosity ($kg/m \cdot s$) | $3.22 \times 10^{-5}$ |
| Specific heat capacity at constant pressure ($J/kg \cdot k$) | 1159.2 |
| Components (%) | |
| $N_2$ | 73.00 |
| $O_2$ | 3.00 |
| $CO_2$ | 13.00 |
| $H_2O_{(g)}$ | 11.00 |

### 2.2. Numerical Method

In this work, commercial software (Ansys Fluent version 16.0) is adopted to perform this simulation. The continuous phase of drying gas is treated by an Eulerian approach, and a standard $k$-$\varepsilon$ model is utilized for the turbulence description.

Continuity and momentum equations [23]:

$$\frac{\partial \rho}{\partial t} + \nabla \left( \rho \vec{v} \right) = S_m \tag{1}$$

$$\frac{\partial}{\partial t} \left( \rho \vec{v} \right) + \nabla \cdot \left( \rho \vec{v} \vec{v} \right) = -\nabla p + \nabla \cdot \left[ \mu \left( \nabla \vec{v} + \nabla \vec{v}^{\mathrm{T}} \right) - \frac{2}{3} \mu \nabla \cdot \vec{v} I \right] + \rho \vec{g} + \vec{F} \tag{2}$$

where $\rho$ and $v$ are drying gas density and velocity, and $S_m$ is the mass source term. $\rho \vec{g}$ is the gravitational force, $\vec{F}$ is the sum forces exerted by particles on the gas phase, $\mu$ is the drying gas effective viscosity, $I$ is the unit tensor.

The standard $k$-$\varepsilon$ is adopted to describe the flow. The turbulence kinetic energy and its rate of dissipation are obtained from the following transport equations [24]:

$$\frac{\partial}{\partial t} (\rho k) + \frac{\partial}{\partial x_i} (\rho k u_i) = \frac{\partial}{\partial x_j} \left[ \left( \mu + \frac{\mu_t}{\sigma_k} \right) \frac{\partial k}{\partial x_j} \right] + G_k + G_b - \rho \varepsilon - Y_M + S_k \tag{3}$$

$$\frac{\partial}{\partial t} (\rho \varepsilon) + \frac{\partial}{\partial x_i} (\rho \varepsilon u_i) = \frac{\partial}{\partial x_j} \left[ \left( \mu + \frac{\mu_t}{\sigma_\varepsilon} \right) \frac{\partial \varepsilon}{\partial x_j} \right] + C_{1\varepsilon} \frac{\varepsilon}{k} (G_k + C_{3\varepsilon} G_b) - C_{2\varepsilon} \rho \frac{\varepsilon^2}{k} + S_\varepsilon \tag{4}$$

where $G_k$ represents the generation of turbulence kinetic energy due to the mean velocity gradients, $G_b$ is the generation of turbulence kinetic energy due to buoyancy, $Y_M$ is the generation of turbulence kinetic energy due to buoyancy. $C_{1\varepsilon}$, $C_{2\varepsilon}$, and $C_{3\varepsilon}$ are constants. $\sigma_k$ and $\sigma_\varepsilon$ are the turbulent Prandtl numbers for $k$ and $\varepsilon$. $S_k$ and $S_\varepsilon$ are user-defined source terms.

The trajectory of a discrete phase droplet integrates the force balance on the particle, which is written in a Lagrangian reference frame [25,26]. This force balance equates the droplet inertia with the forces acting on the droplet and can be written as:

$$\frac{d\vec{u}_p}{dt} = F_D(\vec{u} - \vec{u}_p) + \frac{\vec{g}(\rho_p - \rho)}{\rho_p} + \vec{F} \tag{5}$$

$$F_D = \frac{18\mu}{\rho_p d_p^2}\frac{C_d \mathrm{Re}_p}{24} \tag{6}$$

$$\mathrm{Re}_p = \frac{\rho d_p\left|\vec{u}_p - \vec{u}\right|}{\mu} \tag{7}$$

where $\vec{u}_p$ is the droplet velocity, $F_D(\vec{u} - \vec{u}_p)$ is the drag force per unit droplet mass, $\vec{u}$ is the fluid phase velocity, $\rho_p$ is the density of the droplet, $\rho$ is the fluid density, $\vec{F}$ is an additional acceleration (force/unit droplet mass) term, $\mu$ is the molecular viscosity of the fluid, $Re$ is the relative Reynolds number of gas and liquid. $C_d$ is the drag coefficient.

The droplet temperature is updated according to a heat balance that relates the sensible heat change in the droplet to the convective and latent heat transfer between the droplet and the continuous phase [27]:

$$m_p c_p \frac{dT_p}{dt} = hA_p(T_\infty - T_p) \qquad T_p < T_{vap} \tag{8}$$

$$m_p c_p \frac{dT_p}{dt} = hA_p(T_\infty - T_p) + \frac{dm_p}{dt}h_{lg} \qquad T_{vap} \leq T_p < T_{bp} \tag{9}$$

$$\frac{d(d_p)}{dt} = \frac{4k_g}{\rho_p c_{p,g} d_p}\left(1 + 0.23\sqrt{\mathrm{Re}_p}\right)\ln\left[1 + \frac{c_{p,g}(T_\infty - T_p)}{h_{lg}}\right] \qquad T_{bp} \leq T_p \tag{10}$$

where $m_p$, $c_p$, $T_p$, and $A_p$ are the mass, specific heat capacity at constant pressure, temperature, and surface area of droplet particles, $T_\infty$ is the temperature in the flue gas, $T_{vap}$ is the droplet vaporization temperature, $T_{bp}$ is the boiling temperature of the droplet, $h_{lg}$ is the latent heat of the droplet vaporization, $c_{p,g}$ is the heat capacity of the gas, $\rho_p$ is the droplet density, $k_g$ is the thermal conductivity of the gas. $h$ is the convective heat transfer coefficient, which is calculated with a modified $Nu$ number as follows [28]:

$$Nu = \frac{hd_p}{k_\infty} = 2.0 + 0.6\mathrm{Re}_d^{1/2}Pr^{1/3} \tag{11}$$

where $k_\infty$ is the thermal conductivity of the continuous phase, $Re_d$ is the Reynolds number, and $P_r$ is the Prandtl number of the continuous phase.

The gas phase is described using the species transport model. The atomization model of the droplet adopts the hollow cone model. The atomization angle and inner diameter are 89° and 0.23 m, respectively. The droplet after atomizing follows the Rosin–Rammler distribution with the constant distribution coefficient (1.2). Different parts are meshed separately to attain a high-quality grid, as shown in Figure 3. After performing grid-independence tests, the mesh with 650,000 cells was finally used in the simulation study.

The simulated cases are listed in Table 2. Cases 1, 2, and 3 are used to study the influence of flue gas flow rate on evaporation performance. Cases 1, 4, and 5 are used to discuss the effect of flue gas temperature on evaporation performance. Cases 1, 6, and 7 are used to discuss the effect of wastewater flow rate on evaporation performance. Cases 1, 8, and 9 are used to discuss the effect of the initial temperature of wastewater on evaporation performance. Cases 10, 11, and 12 are used to discuss the effect of the droplet size of wastewater on evaporation performance. Cases 13–30 are the orthogonal test cases of evaporation performance.

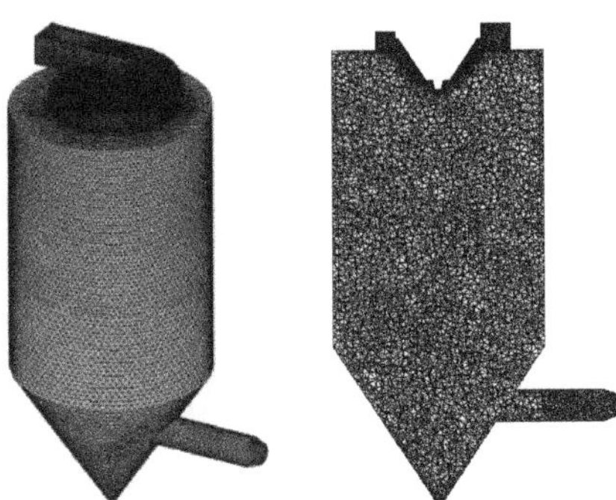

**Figure 3.** Grid division of this spray drying tower without deflectors.

**Table 2.** Case setting.

| Cases | Flow Rate of Flue Gas | Temperature of Flue Gas | Flow Rate of Wastewater | Initial Temperature of Wastewater | Droplet Size |
|---|---|---|---|---|---|
| | kg/s | K | kg/s | K | μm |
| 1 | 11.131 | 638 | 0.832 | 298 | |
| 2 | 14.470 | 638 | 0.832 | 298 | |
| 3 | 7.792 | 638 | 0.832 | 298 | |
| 4 | 11.131 | 658 | 0.832 | 298 | size range: 10–200 μm, |
| 5 | 11.131 | 618 | 0.832 | 298 | an average diameter: 60 μm, |
| 6 | 11.131 | 638 | 1.082 | 298 | spread parameter: 1.2 |
| 7 | 11.131 | 638 | 0.582 | 298 | |
| 8 | 11.131 | 638 | 0.832 | 323 | |
| 9 | 11.131 | 638 | 0.832 | 348 | |
| 10 | 11.131 | 638 | 0.832 | 298 | 5 μm |
| 11 | 11.131 | 638 | 0.832 | 298 | 60 μm |
| 12 | 11.131 | 638 | 0.832 | 298 | 150 μm |
| 13–30 | Orthogonal test of evaporation performance | | | | |

### 2.3. Validation of the Simulated Results

To verify the model's reliability, the measured and simulated temperatures, $H_2O_{(g)}$ concentration at the outlet are compared, as shown in Table 3. The simulated results are consistent with the measured, proving that the model has high reliability and is acceptable in engineering.

**Table 3.** Comparison between the simulated and measured results.

| | Item | 100% Load (600 MW) | 66% Load (400 MW) |
|---|---|---|---|
| Operation parameter | Flue gas temperature (K) | 638 | 610 |
| | Flow rate of wastewater (kg/s) | 0.832 | 0.582 |
| | Inlet wastewater temperature (K) | 298 | 298 |
| Temperature at the outlet (K) | Simulated | 460 | 446 |
| | Measured | 453 | 432 |
| $H_2O_{(g)}$ concentration at the outlet (%) | Simulated | 17.5 | 16.9 |
| | Measured | 17.1 | 16.3 |

## 3. Results and Discussion

### 3.1. Effect of Flue Gas Flow Rate on the Evaporation Performance

Figure 4 shows the turbulence kinetic energy under different flue gas flow rates. By increasing the flue gas flow rate, the turbulence kinetic energy in the mixing zone of the flue gas and the droplets gradually increases, which indicates more vigorous mixing between the flue gas and the droplets. Figure 5 displays temperature under different flue gas flow rates. A larger flue flow rate corresponds to more heat, which is advantageous to droplet evaporation.

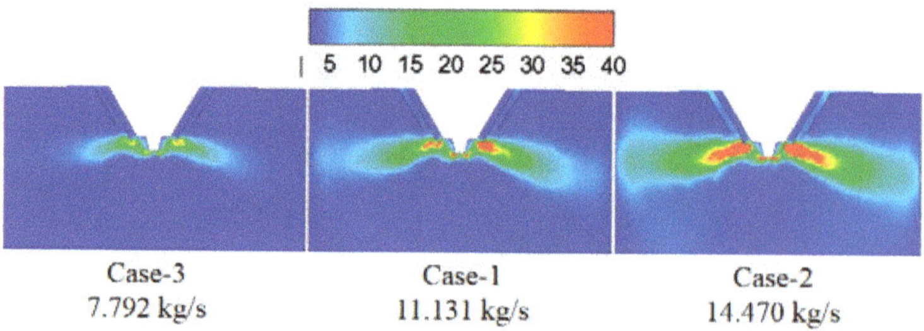


**Figure 4.** Turbulence kinetic energy under different flue gas flow rates ($m^2/s^2$).

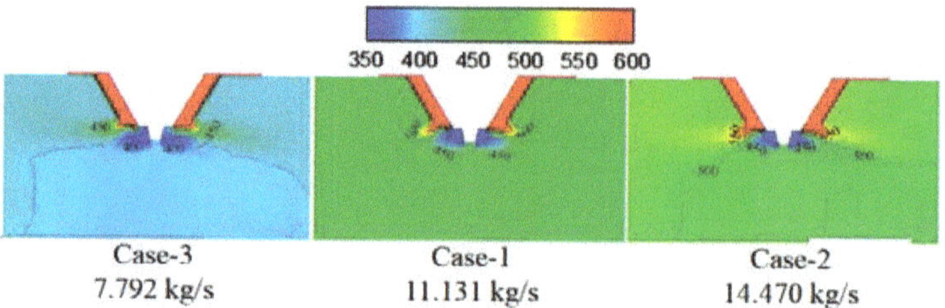


**Figure 5.** Temperature under different flue gas flow rates (K).

Figure 6 shows the effect of flue gas flow rate on the evaporation performance of the droplets in the spray drying tower. It can be seen that as increasing the flue gas flow rate from 7.792 (case-3) to 11.131 kg/s (case-1), the droplet evaporation time in the spray drying tower significantly decreases, and the distance of complete evaporation greatly decreases from 6.8 to 4.8 m. This can be explained by considering that the increase of the flue gas flow rate introduces more heat into the drying tower, which enhances the droplets evaporation performance. By further increasing the flue gas flow rate from 11.131 to 14.470 kg/s, the droplet evaporation time changes slightly, and the distance of complete evaporation only decreases from 4.8 to 4.7 m. Therefore, higher flue gas flow rate corresponds to shorter droplet evaporation time and shorter complete evaporation distance. Still, if the flue gas flow rate exceeds a certain level, the improvement of the droplet evaporation performance is not apparent. In actual operation, taking the evaporation performance and the safety of boiler operation into consideration, an appropriate amount of flue gas flow should be chosen.

### 3.2. Effect of Flue Gas Temperature on the Evaporation Performance

Figure 7 shows the turbulence kinetic energy under different flue gas temperatures. By increasing the flue gas temperature, the turbulence kinetic energy in the mixing zone of the flue gas and the droplets changes slightly, which indicates the slight difference in the mixing intensity between flue gas and droplets. Figure 8 displays temperature under different flue gas flow rates. Higher flue gas temperature corresponds to higher temperature and more heat in the mixing zone of flue gas and droplets, which is advantageous to evaporate the droplets.

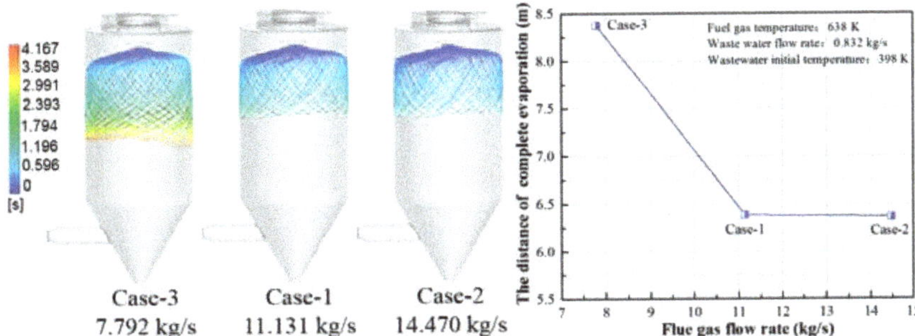



Case-3    Case-1    Case-2
7.792 kg/s   11.131 kg/s   14.470 kg/s

**Figure 6.** Effect of flue gas flow rate on evaporation performance.

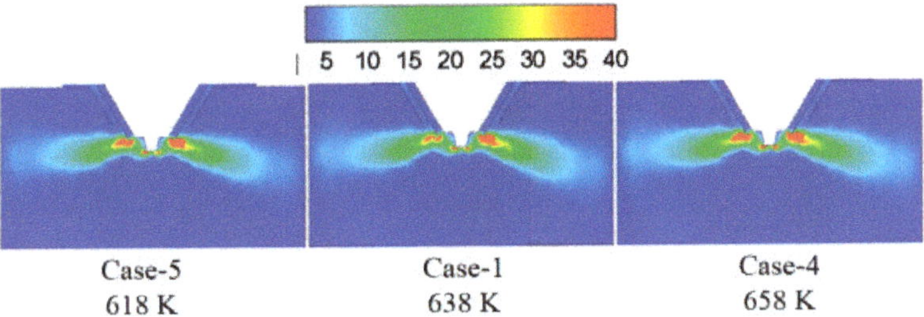



Case-5    Case-1    Case-4
618 K    638 K    658 K

**Figure 7.** Turbulence kinetic energy under different flue gas temperatures ($m^2/s^2$).

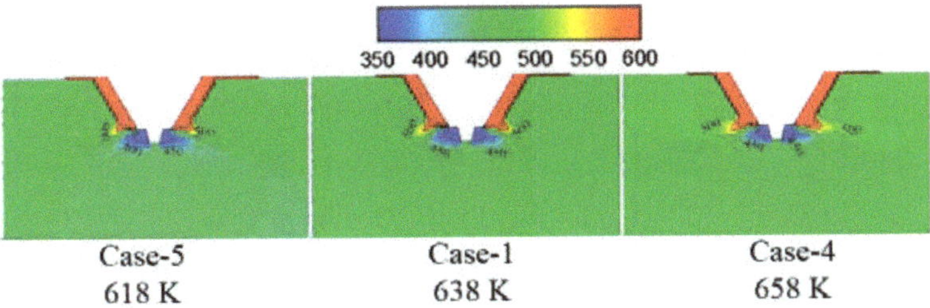



Case-5    Case-1    Case-4
618 K    638 K    658 K

**Figure 8.** Temperature under different flue gas temperatures (K).

Figure 9 shows the effect of flue gas temperature on the droplet evaporation performance in the spray drying tower. It can be seen that increasing the flue gas temperature is beneficial for reducing the droplet residence time in the spray drying tower and improving droplet evaporation performance. This is because the higher flue gas temperature increases the temperature difference between the flue gas and the droplets, enhancing the diffusion and thermophoretic force effects. As a result, heat and mass transfer become more vital, and the droplet evaporation speed is accelerated and reduces the complete evaporation distance.

### 3.3. Effect of Wastewater Flow Rate on The Evaporation Performance

Figure 10 shows the turbulence kinetic energy under different wastewater flow rates. By increasing the wastewater flow rate, the turbulence kinetic energy in the mixing zone of flue gas and droplet changes slightly, which indicates the slight difference in the mixing intensity between flue gas and droplet. Figure 11 displays temperature under different wastewater flow rates. The larger wastewater flow rate corresponds with the lower temperature in the mixing zone of flue gas and droplets because the evaporation process of larger wastewater flow rate absorbs more heat.

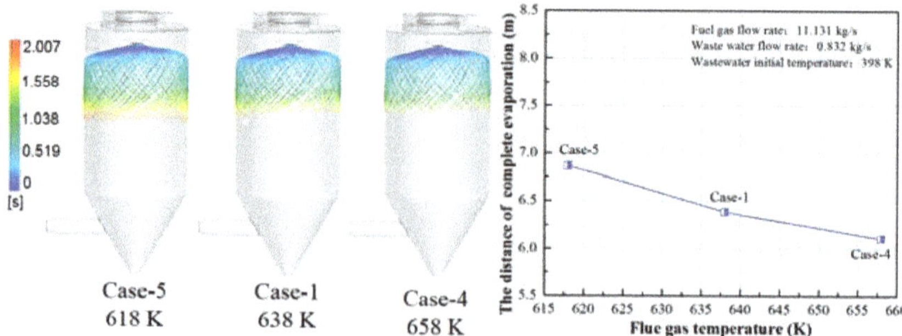

Case-5 618 K  Case-1 638 K  Case-4 658 K

**Figure 9.** Effect of flue gas temperature on evaporation performance.

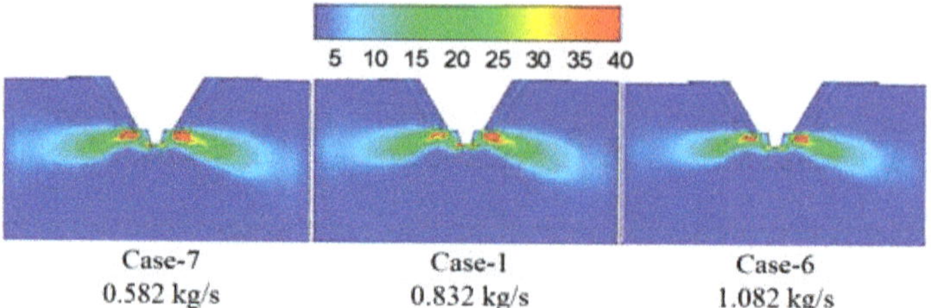

Case-7 0.582 kg/s  Case-1 0.832 kg/s  Case-6 1.082 kg/s

**Figure 10.** Turbulence kinetic energy under different waste water flow rates ($m^2/s^2$).

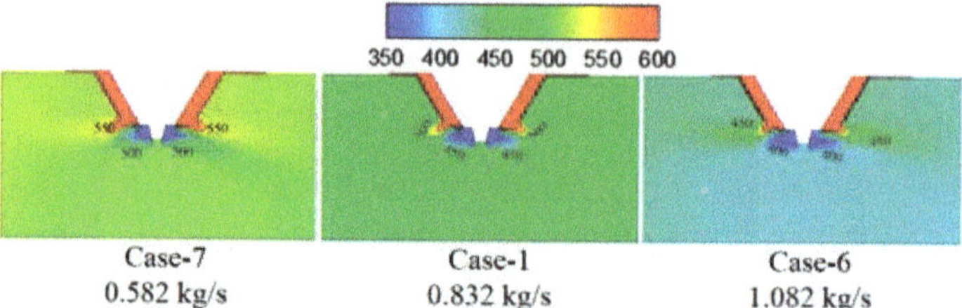

Case-7 0.582 kg/s  Case-1 0.832 kg/s  Case-6 1.082 kg/s

**Figure 11.** Temperature under different wastewater flow rates (K).

Figure 12 shows the effect of wastewater flow rate on the droplet evaporation performance in the spray drying tower. With the increase of the wastewater flow rate, the droplet residence time in the spray drying tower becomes longer and the complete evaporation distance increases, thus, reducing the wastewater flow rate is beneficial to the evaporation performance of the wastewater. On the one hand, under the same flue gas flow rate, the input heat is constant. The larger wastewater that needs to be processed, the more time it takes to evaporate. On the other hand, the larger wastewater flow rate contains more large-diameter droplets. For these large-diameter droplets, more time is required for them to evaporate completely, and the complete evaporation distance is longer. Therefore, it is necessary to design the wastewater flow rate in a spray drying tower.

### 3.4. Effect of Wastewater Initial Temperature on the Evaporation Performance

Figure 13 shows the turbulence kinetic energy under different initial wastewater temperatures. By increasing the initial wastewater temperature, the turbulence kinetic energy in the mixing zone of the flue gas and the droplets changes slightly. Figure 14 displays temperature under different initial wastewater temperatures. Higher initial wastewater temperature corresponds to the higher temperature of the flue gas and the droplets in the mixing zone because the evaporation process of the wastewater droplet absorbs less heat under the constant input heat.

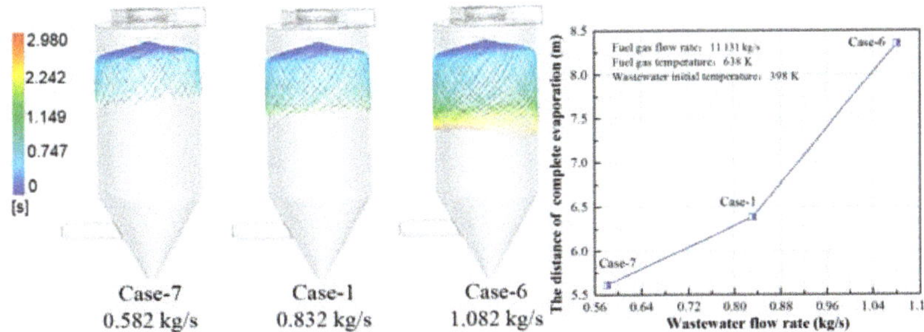

Case-7
0.582 kg/s

Case-1
0.832 kg/s

Case-6
1.082 kg/s

**Figure 12.** Effect of wastewater flow rate on evaporation performance.

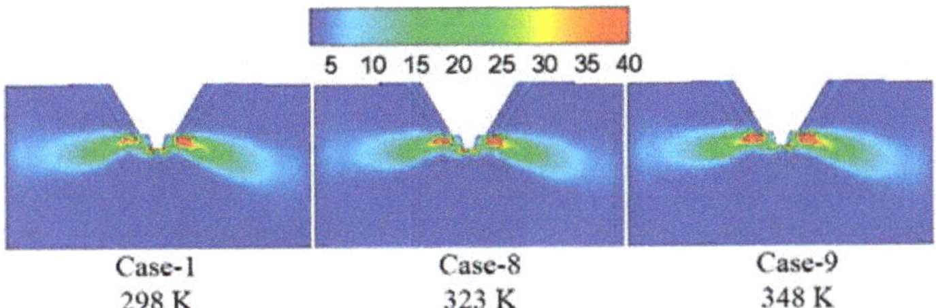


Case-1
298 K

Case-8
323 K

Case-9
348 K

**Figure 13.** Turbulence kinetic energy under different wastewater initial temperatures ($m^2/s^2$).

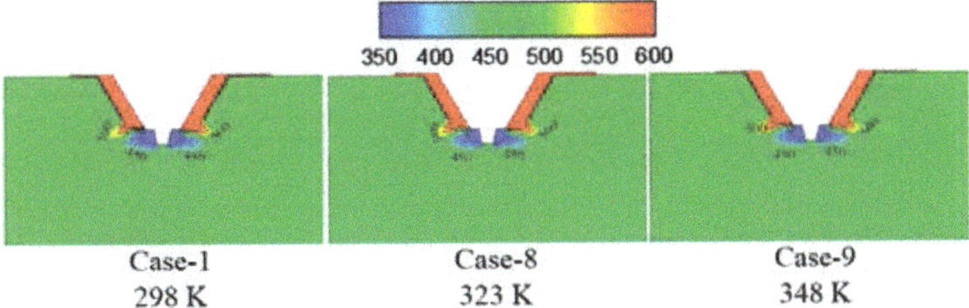


Case-1
298 K

Case-8
323 K

Case-9
348 K

**Figure 14.** Temperature under different wastewater initial temperatures (K).

Figure 15 shows the effect of wastewater flow rate on the evaporation performance of the droplet in the spray drying tower. It can be seen that as the initial temperature of the wastewater increases from 298 to 348 K, the complete evaporation distance is reduced from 4.8 to 4.6 m, which is only 0.2 m. Additionally, there is no significant difference in the trajectory of the droplets under different initial wastewater temperatures. It can be concluded that the complete evaporation distance gradually decreases with the increase of the initial temperature of the wastewater, but the reduction degree is minimal. This is because if the initial temperature of the wastewater is higher, less heat is needed, and the evaporation temperature is reached quicker. However, this part of the heat only accounts for a small proportion of the heat carried by the flue gas, thus, the change of the initial temperature of the droplet has little effect on its complete evaporation distance. In the actual operation of the power plant, although the evaporation process of the desulfurization wastewater can be improved by increasing the initial temperature of the wastewater, its effect is minimal.

### 3.5. Effect of Droplet Size on the Evaporation Performance

Figure 16 shows the turbulence kinetic energy under different droplet sizes. By increasing the droplet size from 5 to 60 μm, the turbulence kinetic energy in the mixing zone of the flue gas and the droplets changes slightly, which indicates the slight difference in the mixing intensity between flue gas and droplet. By further increasing the droplet size from 60 to 150 μm, the turbulence kinetic energy in the mixing zone of the flue gas and the droplets becomes significantly stronger, and this is because the larger droplet size has a more substantial rigidity and turbulence. Figure 17 displays temperature under different droplet sizes, and smaller droplet size corresponds with the higher temperature in the mixing zone of the flue gas and the droplets because the smaller size droplet has a larger specific surface area and is sufficiently heated to evaporate.

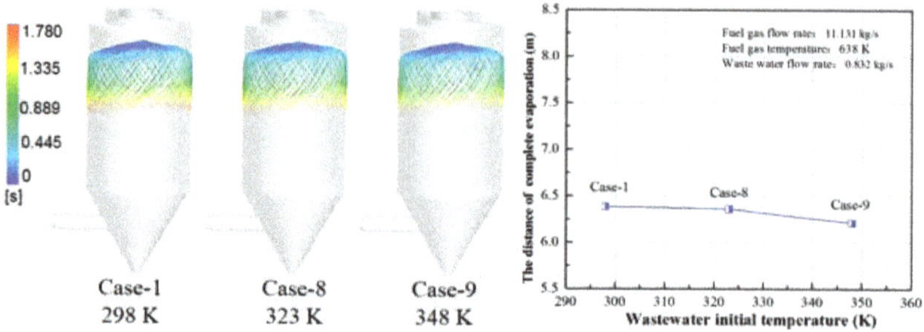


**Figure 15.** Effect of initial wastewater temperature on evaporation performance.

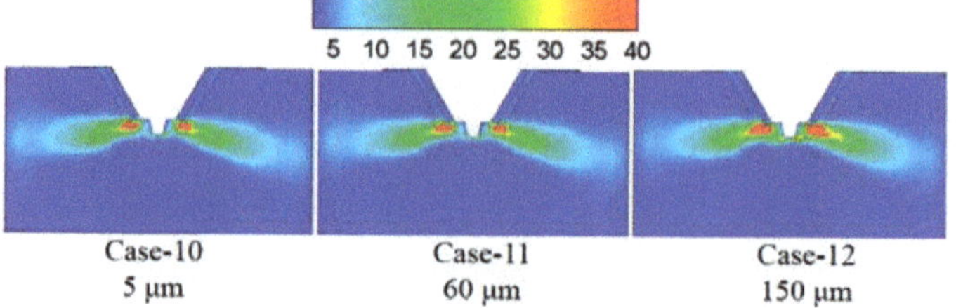


**Figure 16.** Turbulence kinetic energy under different droplet sizes ($m^2/s^2$).

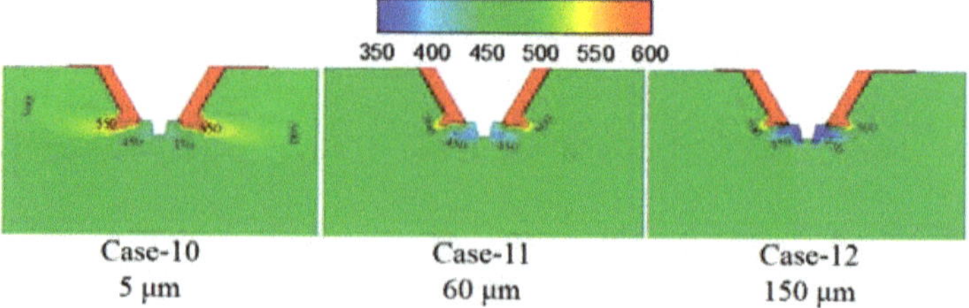


**Figure 17.** Temperature under different droplet sizes (K).

Figure 18 shows the effect of wastewater flow rate on the evaporation performance of the droplet in the spray drying tower. As the droplet size increases, it can be seen that the trajectory of the droplets will diffuse toward the tail of the drying tower, the complete evaporation distance is extended, and the atomization evaporation effect is reduced. When the droplet sizes are 5 and 60 μm, the difference in the evaporation performance is slight. Considering the atomization cost and other factors, it is more economical and practical to choose a droplet size of 60 μm.

### 3.6. Orthogonal Test of Evaporation Performance

The orthogonal method is adopted to compare the degree of influence of various conditions on the evaporation performance, which includes five factors: flue gas flow rate, flue gas temperature, wastewater flow rate, initial wastewater temperature, and droplet size. Each factor is set at three levels with a total of 18 cases. The orthogonal results of each factor on the droplet evaporation performance are shown in Table 4. The orthogonal test results show that the evaporation ratio of cases 19, 21, 24, 27, 28, and 29 cannot reach 100%, and the evaporation ratio of case-21 is the lowest, only 90.39%. Compared to other cases, this case has the lowest flue gas flow, the largest wastewater flow rate, the largest flue gas/wastewater ratio, and the lowest initial temperature of the wastewater. Therefore, the input heat is insufficient, the evaporation performance is poor, and the desulfurization wastewater cannot be completely evaporated. According to the comparison of the R-value of the complete evaporation distance of each case, the factors affecting the droplet evaporation performance are ranked as follows: flue gas flow rate > wastewater flow rate > flue gas temperature > wastewater initial temperature > droplet size. Considering the evaporation ratio and the complete evaporation distance, the optimal setting is 14.470 kg/s for flue gas flow rate, 385 °C for flue gas temperature, 0.582 kg/s for wastewater flow rate, 25 °C for wastewater initial temperature, and 60 μm for droplet size.

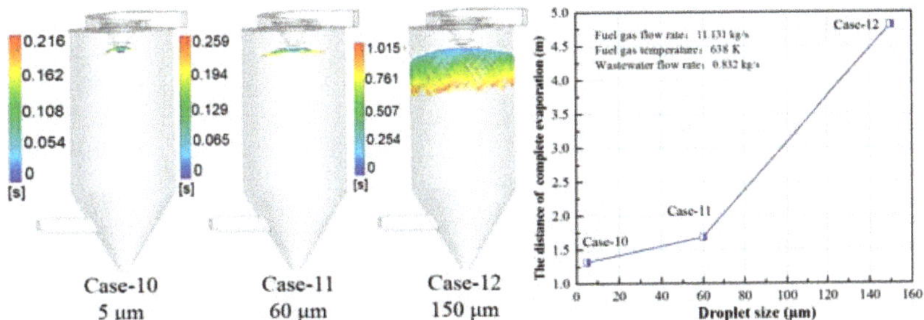


**Figure 18.** Effect of droplet sizes on evaporation performance.

**Table 4.** Orthogonal table of each factor.

| Cases | Flue Gas Flow Rate | Flue Gas Temperature | Wastewater Flow Rate | Wastewater Initial Temperature | Droplet Size | Evaporation Ratio * | The Distance of Complete Evaporation |
|---|---|---|---|---|---|---|---|
| | kg/s | °C | kg/s | °C | μm | % | m |
| 13 | 14.470 | 385 | 1.082 | 75 | 150 | 100 | 4.774 |
| 14 | 14.470 | 365 | 0.832 | 50 | 60 | 100 | 1.888 |
| 15 | 14.470 | 345 | 0.582 | 25 | 5 | 100 | 2.504 |
| 16 | 11.131 | 385 | 1.082 | 50 | 60 | 100 | 2.021 |
| 17 | 11.131 | 365 | 0.832 | 25 | 5 | 100 | 2.664 |
| 18 | 11.131 | 345 | 0.582 | 75 | 150 | 100 | 3.854 |
| 19 | 7.792 | 385 | 0.832 | 75 | 5 | 96.51 | - |
| 20 | 7.792 | 365 | 0.582 | 50 | 150 | 100 | 3.687 |
| 21 | 7.792 | 345 | 1.082 | 25 | 60 | 90.39 | - |
| 22 | 14.470 | 385 | 0.582 | 25 | 60 | 100 | 1.780 |
| 23 | 14.470 | 365 | 1.082 | 75 | 150 | 100 | 5.107 |
| 24 | 14.470 | 345 | 0.832 | 50 | 5 | 98.68 | - |
| 25 | 11.131 | 385 | 0.832 | 25 | 150 | 100 | 4.480 |
| 26 | 11.131 | 365 | 0.582 | 75 | 60 | 100 | 1.783 |
| 27 | 11.131 | 345 | 1.082 | 50 | 5 | 98.34 | - |
| 28 | 7.792 | 385 | 0.582 | 50 | 5 | 99.71 | - |
| 29 | 7.792 | 365 | 1.082 | 25 | 150 | 97.04 | - |
| 30 | 7.792 | 345 | 0.832 | 75 | 60 | 100 | 2.153 |
| $K_1$ | 598.68 | 596.22 | 585.77 | 596.51 | 597.04 | - | - |
| $K_2$ | 598.34 | 597.04 | 595.19 | 596.73 | 590.39 | - | - |
| $K_3$ | 583.65 | 587.41 | 599.71 | 587.43 | 593.24 | - | - |
| $k_1$ | 99.78 | 99.37 | 97.63 | 99.42 | 99.51 | - | - |
| $k_2$ | 99.72 | 99.51 | 99.20 | 99.46 | 98.40 | - | - |
| $k_3$ | 97.28 | 97.90 | 99.95 | 97.91 | 98.87 | - | - |
| $R$ | 2.50 | 1.61 | 2.32 | 1.55 | 1.11 | - | - |

* The evaporation ratio is defined as: $(\dot{m}_{out} - \dot{m}_{in-flue\ gas})/\dot{m}_{wastewater}$, where $\dot{m}_{in-flue\ gas}$ indicates the mass flow of the flue gas injected into the spray drying tower, $\dot{m}_{out}$ indicates the mass flow of the mixture gas flowing out of the spray drying tower, $\dot{m}_{wastewater}$ is the mass flow of the wastewater (liquid) injected into the spray drying tower. $K_i$ ($i$ = 1, 2, 3) indicates the test index sum of each factor at the same level; $k_i$ ($i$ = 1, 2, 3) indicates the test index average value of each factor at the same level; $R$ indicates the range, which refers to the difference between the maximum value and the minimum value of the sum of each level index in the same factor, $R = \max\{k_1, k_2, k_3\} - \min\{k_1, k_2, k_3\}$ in any column. For more details about the orthogonal test please refer to [29].

## 4. Conclusions

This paper investigates the effect of flue gas flow rate, flue gas temperature, wastewater flow rate, initial wastewater temperature, and droplet size on the desulfurization wastewater evaporation performance in a spray drying tower without deflectors. These studied results can provide valuable information to improve the operational performance of the desulfurization wastewater evaporation technology with flue gas. Some conclusions are reached:

(1) The flue gas flow rate and temperature affect the evaporation performance of the desulfurization wastewater. The larger the flue gas flow rate and the higher the flue gas temperature, the faster the wastewater droplet evaporation and the shorter the complete evaporation distance. However, if the flue gas flow rate exceeds a certain level, the improvement of the droplet evaporation performance is not apparent. In actual operation, the effect of atomization and the safety of boiler operation should be considered simultaneously, an appropriate amount of flue gas flow should be extracted, and the wastewater flow of the spray drying tower should be practically designed.

(2) The smaller the wastewater flow rate, the higher the initial wastewater temperature, the faster the wastewater droplet evaporation, and the shorter the complete evaporation distance.

(3) Reducing the droplet size is beneficial to improving the evaporation effect of the desulfurization wastewater. Considering all factors, it is more economical and practical to choose a droplet size of 60 μm.

(4) The orthogonal test results of various factors show that the factors affecting droplet evaporation performance are as follows: flue gas flow rate > wastewater flow rate > flue gas temperature > initial wastewater temperature > droplet size. Considering the evaporation ratio and the distance of complete evaporation, the optimal setting is 14.470 kg/s for flue gas flow rate, 385 °C for flue gas temperature, 0.582 kg/s for wastewater flow rate, 25 °C for initial wastewater temperature, and 60 μm for droplet size.

**Author Contributions:** Writing—original draft, D.L.; investigation, N.Z.; writing—review and editing, Y.F.; project administration, Z.X. All authors have read and agreed to the published version of the manuscript.

**Funding:** This research was funded by Critical Technology Research and Integrated Application for Wastewater Resources and Zero-Emission in High Water Consumption Enterprises (GD-KJXM20183546).

**Institutional Review Board Statement:** Not applicable.

**Informed Consent Statement:** Not applicable.

**Data Availability Statement:** Data is contained within the article.

**Conflicts of Interest:** The authors declare no conflict of interest.

## References

1. Córdoba, P. Status of flue gas desulphurisation (FGD) systems from coal-fired power plants: Overview of the physic-chemical control processes of wet limestone FGDs. *Fuel* **2015**, *144*, 274–286. [CrossRef]
2. Guo, Y.; Xu, Z.; Zheng, C.; Shu, J.; Dong, H.; Zhang, Y.; Weng, W.; Gao, X. Modeling and optimization of wet flue gas desulfurization system based on a hybrid modeling method. *J. Air Waste Manag.* **2019**, *69*, 565–575. [CrossRef] [PubMed]
3. Gutiérrez Ortiz, F.J.; Vidal, F.; Ollero, P.; Salvador, L.; Cortés, V.; Gimenez, A. Pilot-plant technical assessment of wet flue gas desulfu-rization using limestone. *Ind. Eng. Chem. Res.* **2006**, *45*, 1466–1477. [CrossRef]
4. Srivastava, R.K.; Jozewicz, W. Flue gas desulfurization: The state of the art. *J. Air Waste Manag.* **2001**, *51*, 1676–1688. [CrossRef]
5. Ma, S.; Chai, J.; Chen, G.; Wu, K.; Xiang, Y.; Wan, Z.; Zhang, J.; Zhu, H. Partitioning characteristic of chlorine ion in gas and solid phases in process of desulfurization waste water evaporation: Model development and calculation. *Environ. Sci. Pollut. Res.* **2019**, *26*, 8257–8265. [CrossRef] [PubMed]

6. Shafiq, I.; Hussain, M.; Shehzad, N.; Maafa, I.M.; Akhter, P.; Amjad, U.; Razzaq, S.S.A.; Yang, W.; Tahir, M.; Russo, N. The effect of crystal facets and induced porosity on the performance of monoclinic BiVO$_4$ for the enhanced visible-light driven photocatalytic abatement of methylene blue. *J. Environ. Chem. Eng.* **2019**, *7*, 103265. [CrossRef]
7. Azam, K.; Raza, R.; Shezad, N.; Shabir, M.; Yang, W.; Ahmad, N.; Shafiq, I.; Akhter, P.; Razzaq, A.; Hussain, M. Development of recoverable magnetic mesoporous carbon ad-sorbent for removal of methyl blue and methyl orange from waste water. *J. Environ. Chem. Eng.* **2020**, *8*, 104220. [CrossRef]
8. Ma, S.; Chai, J.; Chen, J.; Wu, K.; Wan, Z.; Zhang, J. Numerical simulation of bypass evaporation system treating FGD waste water using high temperature flue gas. *Environ. Technol.* **2018**, *5*, 1–24.
9. Guo, J.; Shu, S.; Liu, X.; Wang, X.; Yin, H.; Chu, Y. Influence of Fe loadings on desulfurization performance of activated carbon treated by nitric acid. *Environ. Technol.* **2017**, *38*, 266–276. [CrossRef]
10. Ma, S.; Chai, J.; Chen, G.; Yu, W.; Zhu, S. Research on desulfurization waste water evaporation: Present and future perspectives. *Renew. Sust. Energ. Rev.* **2016**, *58*, 1143–1151.
11. Cui, L.; Li, G.; Li, Y.; Yang, B.; Zhang, L.; Dong, Y.; Ma, C. Electrolysis-electrodialysis process for removing chloride ion in wet flue gas desulfurization waste water (DW): Influencing factors and energy consumption analysis. *Chem. Eng. Res. Des.* **2017**, *123*, 240–247. [CrossRef]
12. Enoch, G.D.; Spiering, W.; Tigchelaar, P.; de Niet, J.; Lefers, J.B. Treatment of waste water from wet lime (stone) flue gas desulfurization plants with aid of crossflow microfiltration. *Sep. Sci. Technol.* **1990**, *25*, 1587–1605. [CrossRef]
13. Ye, C.S.; Huang, J.W.; Liu, T. Technology progress and treatment methods of flue Gas desulfurization waste water in coal-fired plants. *Environ. Eng.* **2017**, *35*, 10–13.
14. Liu, Z.; Wey, M.; Lin, C. Reaction characteristics of Ca(OH)$_2$, HCl and SO$_2$ at low temperature in a spray dryer integrated with a fabric filter. *J. Hazard. Mater.* **2002**, *95*, 291–304. [CrossRef]
15. Ukai, N.; Nagayasu, T.; Kamiyama, N.; Fukuda, T. Spray-Drying Device for Dehydrated Filtrate from Desulfurization Waste Water, Air Pollution Control System and Flue Gas Treatment Method. U.S. Patent No. 9,555,341, 31 January 2017.
16. Shaw, W.A. Fundamentals of zero liquid discharge system design. *Power* **2011**, *155*, 56–58.
17. Shaw, W.A. Benefits of evaporation FGD purge water. *Power* **2008**, *152*, 60–63.
18. Mosti, C.; Cenci, V. ZLD systens applied to ENEL coal-fired power plants. *Power Tech.* **2012**, *92*, 69–73.
19. Xu, Y.; Jin, B.S.; Zhou, Z.; Fang, W. Experimental and numerical investigations of desulfurization waste water evaporation in a lab-scale flue gas duct: Evaporation and HCl release characteristics. *Environ. Technol.* **2021**, *42*, 1411–1427. [CrossRef]
20. Liang, Z.; Zhang, L.; Yang, Z.; Qiang, T.; Pu, G.; Ran, J. Evaporation and crystallization of a droplet of desulfurization waste water from a coal-fired power plant. *Appl. Therm. Eng.* **2017**, *119*, 52–62. [CrossRef]
21. Deng, J.J.; Pan, L.M.; Chen, D.Q.; Dong, Y.Q.; Wang, C.M.; Liu, H.; Kang, M.Q. Numerical simulation and field test study of desulfuriza-tion waste water evaporation treatment through flue gas. *Water Sci. Technol.* **2014**, *70*, 1285–1291. [CrossRef]
22. Zheng, C.; Zheng, H.; Yang, Z.; Liu, S.; Li, X.; Zhang, Y.; Weng, W.; Gao, X. Experimental study on the evaporation and chlorine migration of desulfurization waste water in flue gas. *Environ. Sci. Pollut. Res.* **2019**, *26*, 4791–4800. [CrossRef] [PubMed]
23. Jen, T.C.; Li, L.; Cui, W.; Chen, Q.; Zhang, X. Numerical investigations on cold gas dynamic spray process with nano- and microsize particles. *Int. J. Heat Mass Tran.* **2005**, *48*, 4384–4396. [CrossRef]
24. Launder, B.E.; Spalding, D.B. *Lectures in Mathe-Matical Models of Turbulence*; Von Karman In-stitutefor Fluid Dynamics: Sint-Genesius-Rode, Belgium, 1972.
25. ANSYS. *Ver A F 16.0—User's Guide*; ANSYS Inc.: Canonsburg, PA, USA, 2015.
26. Orourke, P.J.; Amsden, A.A. The TAB method for numerical calculation of spray droplet breakup. In Proceedings of the International Fuels and Lubricants Meeting and Exposition, Toronto, ON, Canada, 2 November 1987.
27. Miura, K.; Miura, T.; Ohtani, S. Heat and mass transfer to and from droplets. *Chem. Eng. Prog. Sym. Ser.* **1977**, *163*, 95–102.
28. Sazhin, S.S. Advanced models of fuel droplet heating and evaporation. *Prog. Energ. Combust. Sci.* **2006**, *32*, 162–214. [CrossRef]
29. Zhou, J.; Zhao, M.; Wang, C.; Gao, Z. Optimal design of diversion piers of lateral intake pumping station based on orthogonal test. *Shock Vib.* **2021**, *5*, 1–9.

Article

# On the Analysis of the Non-Newtonian Fluid Flow Past a Stretching/Shrinking Permeable Surface with Heat and Mass Transfer

Shahid Khan [1], Mahmoud M. Selim [2,3], Aziz Khan [4], Asad Ullah [5], Thabet Abdeljawad [4,6,7,*], Ikramullah [8], Muhammad Ayaz [9] and Wali Khan Mashwani [1]

[1] Institute of Numerical Sciences, Kohat University of Science & Technology, Kohat 26000, Khyber Pakhtunkhwa, Pakistan; shahidkhan@kust.edu.pk (S.K.); mashwanigr8@gmail.com (W.K.M.)
[2] Department of Mathematics, Al-Aflaj College of Science and Humanities Studies, Prince Sattam Bin Abdulaziz University, Al-Aflaj 710-11912, Saudi Arabia; m.selim@psau.edu.sa
[3] Department of Mathematics, Suez Faculty of Science, Suez University, Suez 34891, Egypt
[4] Department of Mathematics and General Sciences, Prince Sultan University, P.O. Box 66833, Riyadh 11586, Saudi Arabia; akhan@psu.edu.sa
[5] Department of Mathematical Sciences, The University of Lakki Marwat, Lakki Marwat 28420, Khyber Pakhtunkhwa, Pakistan; asad@ulm.edu.pk
[6] Department of Medical Research, China Medical University, Taichung 40402, Taiwan
[7] Department of Computer Science and Information Engineering, Asia University, Taichung 40402, Taiwan
[8] Department of Physics, Kohat University of Science & Technology, Kohat 26000, Khyber Pakhtunkhwa, Pakistan; ikramullah@kust.edu.pk
[9] Department of Mathematics, Abdul Wali Khan University, Mardan 23200, Khyber Pakhtunkhwa, Pakistan; mayazmath@awkum.edu.pk
* Correspondence: tabdeljawad@psu.edu.sa

Citation: Khan, S.; Selim, M.M; Khan, A.; Ullah, A.; Abdeljawad. T.; Ikramullah; Ayaz. M.; Mashwani, W.K. On the Analysis of the Non-Newtonian Fluid Flow Past a Stretching/Shrinking Permeable Surface with Heat and Mass Transfer. *Coatings* **2021**, *11*, 566. https://doi.org/10.3390/coatings11050566

Academic Editor: Eduardo Guzmán

Received: 4 April 2021
Accepted: 29 April 2021
Published: 12 May 2021

Publisher's Note: MDPI stays neutral with regard to jurisdictional claims in published maps and institutional affiliations.



**Abstract:** The 3D Carreau fluid flow through a porous and stretching (shrinking) sheet is examined analytically by taking into account the effects of mass transfer, thermal radiation, and Hall current. The model equations, which consist of coupled partial differential equations (PDEs), are simplified to ordinary differential equations (ODEs) through appropriate similarity relations. The analytical procedure of HAM (homotopy analysis method) is employed to solve the coupled set of ODEs. The functional dependence of the hydromagnetic 3D Carreau fluid flow on the pertinent parameters are displayed through various plots. It is found that the x-component of velocity gradient ($f^{'}(\eta)$) enhances with the higher values of the Hall and shrinking parameters ($m, \varrho$), while it reduces with magnetic parameter and Weissenberg number ($M, We$). The y-component of fluid velocity ($g(\eta)$) rises with the augmenting values of $m$ and $M$, while it drops with the augmenting viscous nature of the Carreau fluid associated with the varying Weissenberg number. The fluid temperature $\theta(\eta)$ enhances with the increasing values of radiation parameter ($Rd$) and Dufour number ($Du$), while it drops with the rising Prandtl number ($Pr$). The concentration field ($\phi(\eta)$) augments with the rising Soret number ($Sr$) while drops with the augmenting Schmidt number ($Sc$). The variation of the skin friction coefficients ($C_{fx}$ and $C_{fz}$), Nusselt number ($Nu_x$) and Sherwood number ($Sh_x$) with changing values of these governing parameters are described through different tables. The present and previous published results agreement validates the applied analytical procedure.

**Keywords:** thermal radiations; magnetic field; Carreau fluid; stretching/shrinking surface; Hall effect; nonlinear radiations; HAM

## 1. Introduction

The thermal energy transportation and the fluid boundary layer motion over stretching (shrinking) sheets are the areas of immense importance due to its broad range industrial and technological applications. Some of the applications consist of: growing crystals structures, plastic sheets preparation, manufacturing of electronic chips and materials, paper industry, cooling process, and so on [1,2]. The basic work in this regard was started by Crane [3]. Andersson et al. [4], and Vajravelu [5] discussed the different aspects of fluids flowing over stretching surfaces. It is important to mention here that the gradients' existence are essential for the growth of various fluxes and flows. In fluids, there are two important effects named as Soret and Dufor effects. In Soret effect, the existence of temperature gradient results in thermal diffusion which governs the thermal energy flow. The mass transfer is mainly governed by Dufor effect, which gives rise to the diffusion-thermo effect. These effects have an influential role in governing the natural convective flow, which is one of the modes by which thermal energy can transfer due to the aggregate motion of the heated fluid. The term cross diffusion refers to the process in which the existence of concentration gradient of one specie develops the flux of the other. This means that the cross diffusion is associated with both thermal and mass diffusion. The heat energy exchangers, steel processing, cooling of nuclear power plant, etc. are the well-known technological sectors in which the convection thermal energy transportation plays an important role. The different aspects during the heat energy flow over a 3D exponentially stretching surface are investigated by Liu et al. [6]. Hayat et al. [7] examined the boundary layer Carreau fluid motion and obtained that the presence of suction depreciates (enhances) the Carreau fluid speed (boundary layer thickness). Further detail analysis about the boundary layer flow can be accessed in refs [8–10].

The Magneto-Hydro-dynamics (MHD) studies the evolution of the macroscopic behaviors of fluids in the ambient magnetic field presence. The MHD flow finds its applications in Astrophysics and Astronomy, nuclear reactors cooling, engineering and technology, Plasma Physics, etc. Nazar et al. [11] investigated the thermal energy transformation during the magnetized flow over a vertical and stretchable surface. During their investigation, they found that the enhancing B-field magnitude reduces the coefficient of skin friction and the thermal energy loss. The analytic investigation of heat energy transfer during the 3D MHD migration over a stretchable plate is carried out by Xu et al. [12] using the series solution approach. The MHD stagnation flow toward an extendable surface is examined by Ishak et al. [13]. A more recent study on stagnation point flow can be found in references [14,15]. The heat energy transfer through convection during the magnetized 3D motion on an extendable surface is worked out by Vajravelu et al. [16]. Pop and Na [17] examined the impacts of B-field on the fluid flowing through a porous and stretchable surface. The recent developments on the magnetized boundary-layer motion can be found in references [18–21].

The thermal energy radiations and its analysis are extremely important in the solar energy, fission reactors, engines, propulsion equipment for speedy aircrafts, and various chemical phenomena operating at extreme temperatures. Gnaneswara Reddy [22] studied the magnetized nanofluid motion by incorporating the impact of thermal energy radiation. The mixed convection MHD fluid flow through a perforated enclosure is examined by Gnaneswara Reddy [23]. He investigated the effects produced due to chemical reaction, Ohmic dissipation, and heat energy source. Emad [24] investigated the different impacts that arose due to the inclusion of thermal radiations in a conducting fluid flow. The influence of thermal radiations on the thermal energy transfer through convection in an electrically conducting fluid of varying viscosity moving over an extending surface is worked out by Abo-Eldahab and Elgendy [25]. Gnaneswara Reddy [26] investigated the various impacts arose due to Joule-heating, thermo-phoresis and viscous nature of a magnetized fluid flowing over an isothermal, perforated and inclined surface. A more recent and detailed investigation of the MHD flow can be found in references [27–30]. Yulin et al. [31] numerically studied the natural convection flow of nanofluid over inclined enclosure. They investigated the different impacts due to constant heat energy source and temperature. Zhe et al. [32] performed an experimental analysis of water, ethylene glycol, and copper oxides mixture by employing statistical techniques for the multi-walled-carbon-nanotubes (MWCNTs). Shah et al. [33–35] analytically scrutinized the micropolar fluid in different frames. Hayat et al. [36] analyzed the Cu-water MHD nanofluids flow in the rotating disks. Dat et al. [37] have recently studied numerically the $\gamma-$AlOOH nanoliquid by using different shaped nanoparticles within a wavy container. The recent studies about the nanaofluids along with different advantages can be studied in refs. [38–46].

Fluids are categorized broadly as non-Newtonian and Newtonian. The Newton viscosity relation which shows that the shear stress and strain are directly related, is applicable in the Newtonian fluid. The non-Newtonian fluid can not be described by this simple direct relation between stress and strain. The non-Newtonian fluids, for example manufactured and genetic liquid organisms, blood, polymers, liquids, etc., have central importance in this advance technological world. The non-Newtonian fluids are very hard to be analytically and numerically treated, as compared to the Newtonian fluids, due to its nonlinear behavior. The credit goes to Carreau [47], who developed a relation that describes both, the viscoelastic and nonlinear properties of of such type of complex fluids. Ali and Hayat [48] worked out the Carreau fluid peristaltic motion through an asymmetrical enclosure. Goodarzi et al. [49] analyzed the simultaneous impact of slip and temperature jump over the Non-Newtoinian nanofluid (alumina + carboxy-methyl cellulose) motion through microtube, and investigated the impacts of pertinent parameters over the nanofluid state variables. Maleki et al. [50] analyzed the impacts of heat generation (absorption), suction (injection), nanoparticles type (volume fraction), thermal and velocity slip parameter, and radiation on the temperature and velocity fields of four different types of nanofluids moving over a perforated flat surface. Hayat et al. [51] studied the impacts of induced magnetic field on the flow of Carreau fluid. Tshehla [52] examined the Carreau fluid migration past an inclined surface. Elahi et al. [53] analyzed the Carreau fluid 3D migration from a duct. Gnaneswara Reddy et al. [54] studied the effects due to Ohmic heating during the MHD viscous nanofluid motion through a nonlinear, permeable, and extending surface. Jiaqiang et al. [55] employed the wetting models in order to explain the working procedures of different surfaces found in nature. Khan et al. [56,57] employed the fractional model to Casson and Brinkman types fluids. The impacts produced due to the incorporation of thermal radiations in the presence of suction (injection) on the MHD flow of fluids are investigated by researchers [58–60]. Maleki et al. [61] analyzed the impact of heat generation (absorption) and viscous dissipation on the heat transfer during the non-Newtonian pseudoplastic nanomaterial motion over a perforated flate. Gheynani et al. [62] examined the turbulent motion of a non-Newtonian Carboxymethyl cellulose copper oxide nanofluid in a 3D microtube by investigating the impacts of nanoparticle concentration and diameter over the temperature and velocity fields. Maleki et al. [63] studied the heat transfer characteristics of pseudo-plastic non-Newtonian nanofluid motion over a permeable surface in the presence of suction and injection. The system of governing PDEs is converted to ODEs by using similarity solution technique, and then solved numerically by employing Runge–Kutta–Fehlberg fourth–fifth order (RKF45) method. The numerical investigation of (water + alumina) nanofluid mixed flow through a 2D square cavity having porous medium is carried out by Nazari et al. [64] employing a Fortran Code.

The phenomenon in which the application of an external magnetic-field to a conducting fluid produces potential difference, is termed the Hall effect. The impacts due to the inclusion of Hall effect are examined by various researchers due to its relevance with a variety of technological and industrial applications. Biswal and Sahoo [65] investigated the impacts of Hall current on the magnetized fluid motion over a vertical, permeable and oscillating surface. Raju et al. [66] worked out the Hall current impacts on the MHD flow over an oscillatory surface having porous upper wall. Datta and Janna [67] analyzed the magnetized and oscillatory fluid motion on a flat surface in the presence of Hall current. Aboeldahab and Elbarbary [68] analyzed the impacts due to Hall current during the MHD fluid dynamics through a semi-infinite and perpendicular plate. Khan et al. [69] used the finite element method for the Newtonian fluid past a semi-circular cylinder. The variation in temperature and mass diffusion in the MHD fluid flow considering the inclusion of Hall effect is examined by Rajput and Kanaujia [70]. Further studies on similar footings are performed by Shah et al. [71–73] employing semi-analytical calculations. The magnetized and peristaltic fluid dynamics of Carreau–Yasuda fluid through a channel is numerically investigated by Abbasi et al. [74] taking into account the Hall effect impact. Abdeljawad et al. [74] investigated the 3D magnetite Carreau fluid migration through a surface of parboloid of revolution by incorporating mass transfer and thermal radiations. The impacts of Hall current and cross diffusion on the two dimensional (2D) MHD Carreau fluid flow through a perforated and stretchable (shrinkable) surface is recently investigated in [75].

Here, we extend the previous work [75] to 3-dimensional space in order to analyze what actually happens in the most general situation. The novelty of the current investigation is to examine analytically the thermal energy and mass transfer properties of the MHD Carreau fluid 3D motion through a perforated stretching sheet by considering the effects of Hall current and cross diffusion. This research work has potential applications in problems involving motion of the non-Newtonian fluid over perforated stretching (shrinking) surfaces. The research work carried out is organized in the following manner:

The geometrical description and model equations of the current investigation are presented in Section 2. The obtained results are discussed and explained by plotting various graphs in Section 3. The comparison and the computation of engineering-based related quantities are discussed through different tables in Section 4. The work is finally concluded in Section 5.

## 2. Mathematical Modeling

The 3D magnetized Carreau fluid is considered along a linear stretching and contracting permeable sheet by incorporating the impacts of thermal radiations and Hall current. The flow is assumed to be incompressible, laminar, steady, and electrically conducting. The external magnetic field $B_0$ is applied in the $y$-direction. The thermal energy and mass diffusion impacts due to the existence of temperature gradient and concentration gradient are considered as well. The geometry is chosen in such a way that the sheet velocities along $x$-and $y-$axis are respectively $u_w$ and $v_w$, whereas the flow is restricted to the positive $z-$axis, as can be seen in Figure 1. Furthermore, convective heat energy flow and mass transfer are considered on the sheet, such that the assumed liquid below the sheet has temperature $T_f$ and concentration $C_f$ in order to make them consistent with the heat and mass conversion coefficients $h_1$ and $h_2$.

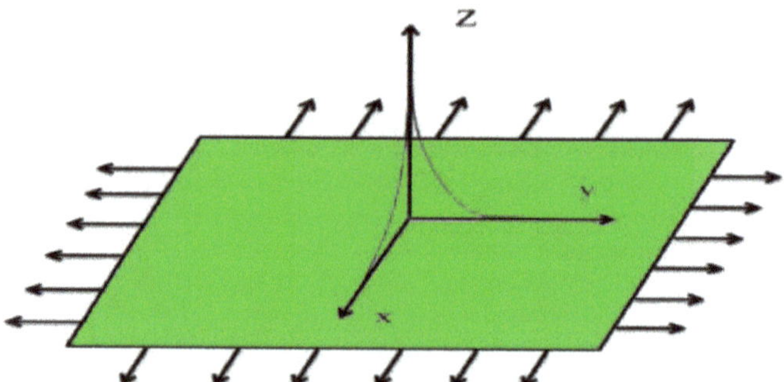


**Figure 1.** Geometrical description of the study.

The Carreau fluid flow is governed by the relation [47,76]:

$$\eta = \left[ \eta_\infty + (\eta_0 - \eta_\infty)\left(1 + (\lambda\dot{\gamma})^2\right)^{\frac{n-1}{2}} \right],$$

(1)

where $\eta_0$ $(\eta_\infty)$ denotes the zero (infinite) shear-rate viscosity, $n$ is the index of power law, $\lambda$ denotes the time constant of the material. The symbol $\dot\gamma$ is given by [76]:

$$\dot\gamma = \sqrt{\frac{1}{2}\sum_i\sum_j \dot\gamma_{ij}\dot\gamma_{ji}} = \sqrt{\frac{1}{2}\Pi}, \tag{2}$$

where $\Pi$ is the strain-rate tensor second invariant. Hall effect arises when magnetic field is applied externally to the conducting fluid which can modify the flow pattern. This phenomenon can be studied with the help of Ohm's law [75,77] given as:

$$\vec{j} + \frac{w_e t_e}{B_0} \times (\vec{j}\times\vec{B}) + \frac{\sigma P_e}{e n_e} = \sigma(\vec{V}\times\vec{B}+\vec{E}), \tag{3}$$

where $\vec{j}$ is the current density, $w_e$ $(t_e)$ is the angular frequency (collision time interval) of electrons, $\sigma$ denotes the conductivity, $\vec{E}$ $(\vec{B})$ is the electric field (magnetic field), $n_e(\ e)$ is the number density (charge) of electrons, and $P_e$ is the pressure of electrons. The $y-$ component of $\vec{j}$ is zero due to the application of external magnetic field in this direction. The $x$ and $z-$components of $\vec{j}$ are expressed in the chosen geometry as:

$$j_x = \frac{\sigma B_0^2(mu - w)}{1 + m^2}, \tag{4}$$

$$j_z = \frac{\sigma B_0^2(mw + u)}{1 + m^2}, \tag{5}$$

where $m = w_e t_e$ is the Hall parameter. Using Equations (1)–(5) at $\eta_\infty = 0$, the Carreau fluid equations are written respectively as [75]:

$$\frac{\partial u}{\partial x} + \frac{\partial v}{\partial y} + \frac{\partial w}{\partial z} = 0, \tag{6}$$

$$u\frac{\partial u}{\partial x} + v\frac{\partial u}{\partial y} + w\frac{\partial u}{\partial z} - v\frac{\partial^2 u}{\partial y^2}\left(1 + \left(\frac{n-1}{2}\right)\lambda^2\left(\frac{\partial u}{\partial y}\right)^2\right) = v(n-1)\lambda^2\frac{\partial^2 u}{\partial y^2}\left(\frac{\partial u}{\partial y}\right)^2 \times \left(1 + \left(\frac{n-3}{2}\right)\lambda^2\left(\frac{\partial u}{\partial y}\right)^2\right)$$
$$- \frac{\sigma B_0^2(mw + u)}{\rho(1 + m^2)} - \frac{vu}{k}, \tag{7}$$

$$u\frac{\partial w}{\partial x} + v\frac{\partial w}{\partial y} + w\frac{\partial w}{\partial z} - v\frac{\partial^2 w}{\partial y^2}\left(1 + \left(\frac{n-1}{2}\right)\lambda^2\left(\frac{\partial w}{\partial y}\right)^2\right) = v(n-1)\lambda^2\frac{\partial^2 w}{\partial y^2}\left(\frac{\partial w}{\partial y}\right)^2 \times \left(1 + \left(\frac{n-3}{2}\right)\lambda^2\left(\frac{\partial w}{\partial y}\right)^2\right)$$
$$+ \frac{\sigma B_0^2(mu - w)}{\rho(1 + m^2)} - \frac{vw}{k}, \tag{8}$$

$$u\frac{\partial T}{\partial x} + v\frac{\partial T}{\partial y} + w\frac{\partial T}{\partial z} = -\frac{1}{\rho c_p}\frac{\partial q_r}{\partial y} + \alpha\frac{\partial^2 T}{\partial y^2} + \frac{D_m K_T}{c_s c_p}\frac{\partial^2 C}{\partial y^2}, \tag{9}$$

$$u\frac{\partial C}{\partial x} + v\frac{\partial C}{\partial y} + w\frac{\partial C}{\partial z} = \frac{D_m K_T}{T_m}\frac{\partial^2 T}{\partial y^2} + D_m\frac{\partial^2 C}{\partial y^2}. \tag{10}$$

The system boundary restrictions are the following:

$$u = u_w(x) + L_1\frac{\partial u}{\partial y}, v = v_w, \frac{\partial T}{\partial y} = -\frac{h_1}{k}(T_f - T), \frac{\partial T}{\partial y} = -\frac{h_2}{D_m}(C_f - C), w = 0 \text{ at } y = 0, \tag{11}$$

$$u \to 0, w \to 0, T \to T_\infty, C \to C_\infty \text{ as } y \to \infty,$$

where $B_0$ is the magnetic field magnitude, $T$ $(\rho)$ is the Carreau fluid temperature (density), $k$ $(c_s)$ is the Carreau fluid thermal conductivity (susceptibility of concentration), $u_w(x) = ax$ $(v)$ is the fluid velocity $x$ $(y)$ component, $K_T$ is the thermal diffusion ratio, $C$

is the concentration of the fluid, $L_1$ is the factor of the velocity slip, and $D_m$ is the mass diffusivity. Furthermore, $v_w$ is the mass flow velocity, and $C_f$ $(T_f)$ is the convective fluid concentration (temperature).

The flux of the radiations $q_r$ is [75,78]:

$$\frac{\partial q_r}{\partial y} = -\frac{16\sigma^s T_0^3}{3k_1}\frac{\partial^2 T}{\partial y^2}, \tag{12}$$

where $\sigma^s$ and $k_1$ are respectively the Stefan constant and average absorption coefficient. Applying Equation (12) to Equation (9), we will get the

$$u\frac{\partial T}{\partial x} + v\frac{\partial T}{\partial y} - \frac{D_m K_T}{c_s c_p}\frac{\partial^2 C}{\partial y^2} = \frac{\partial}{\partial y}\left[\left(\alpha + \frac{16\sigma^s}{3k_1}T_\infty^3\right)\frac{\partial T}{\partial y}\right]. \tag{13}$$

Using the similarity variables as below [75]:

$$\psi = \sqrt{av}f(\eta)x, \eta = \sqrt{\frac{a}{v}}y, T - T_\infty = (T_f - T_\infty)\theta(\eta), w = axg(\eta),$$

$$C - C_\infty = (C_f - C_\infty)\phi(\eta), T - T_\infty = T_\infty(\theta_w - 1)\theta, \theta_w = \frac{T_f}{T_\infty}. \tag{14}$$

Here, $a$ is constant. The symbols $f$, $\theta$, and $\phi$ represent the non-dimensional fluid velocity, temperature, and concentration, respectively. The symbol $\psi$ denotes the stream function satisfying $u = \frac{\partial \psi}{\partial y}$ and $v = -\frac{\partial \psi}{\partial x}$.

Applying these transformations in Equations (6), (7), (10), (13), and (14), we obtain

$$f'''\left[1 + \left(\frac{n-1}{2}\right)Wef''^2\right] + 2f'''\left[\left(\frac{n-1}{2}\right)Wef''^2\right]\left[1 + \left(\frac{n-3}{2}\right)Wef''^2\right]$$
$$+ ff'' - f'^2 - \frac{M}{1+m^2}(f' + mg) - rf' = 0, \tag{15}$$

$$g''\left[1 + \left(\frac{n-1}{2}\right)Weg'^2\right] + 2g''\left[\left(\frac{n-1}{2}\right)Weg'^2\right]\left[1 + \left(\frac{n-3}{2}\right)Weg'^2\right]$$
$$- gf' + g'f + \frac{M}{1+m^2}(mf' - g) + rg = 0, \tag{16}$$

$$\theta''\left(1 + Rd(1 + (\theta_w - 1)\theta)^3\right) + \left(3(\theta_w - 1)\theta'^2(1 + \theta_w - 1)\theta)^2\right) + Prf\theta' + PrDu\phi'' = 0, \tag{17}$$

$$\phi'' + Scf\phi' + ScSr\theta'' = 0. \tag{18}$$

The boundary restrictions are transformed as:

$$f = \varrho, f' = 1 + \chi_1 f(0)'', g = 0, \theta' = -\chi_2(1 - \theta(0)), g = 0, \phi' = -\chi_3(1 - \phi(0)) \text{ at } \eta = 0,$$
$$f' \to 0, \theta \to 0, \phi \to 0, g \to 0 \text{ as } \eta \to \infty. \tag{19}$$

Here, the symbol $We$ represents the Weissenberg number, $\varrho$ shows the mass transfer parameter which describes suction $(\varrho > 0)$ and injection $(\varrho < 0)$. The symbol $Rd$ denotes the radiation parameter, whereas $Pr$, $Sc$, and $Du$ are respectively the Prandtl, Soret, and Dufour numbers. The symbols $\chi_2$, $\chi_3$ are the thermal and concentration profiles slip parameters, respectively. These parameters have the following definitions:

$$Sc = \frac{v}{D_m}, Pr = \frac{v}{a}, Rd = \frac{16\sigma * T_\infty^3}{3kk_e}, Du = \frac{D_m K_T(C_f - C_\infty)}{c_s c_p v(T_f - T_\infty)}, r = \frac{-v}{ak},$$

$$We = \frac{\lambda^2 l^2 a^3}{v}, \chi_1 = L\sqrt{\frac{a}{v}}, \chi_2 = \frac{h_1}{k}\sqrt{\frac{a}{v}}, \chi_3 = \frac{h_2}{D_m}\sqrt{\frac{a}{v}}, \varrho = -\frac{v_w}{\sqrt{av}}, M = \frac{\sigma B_0^2}{\rho a} \tag{20}$$

The basic physical quantities of engineering interest (Sherwood and Nusselt numbers, and the skin frictions (along $x$ and $z$ axis) are defined by [79]:

$$Sh(Re_x)^{-1/2} = -\phi(0)', \tag{21}$$

$$Nu(Re_x)^{-1/2} = -\left(1 + \frac{1}{Rd}\theta_{(w)}(0)^3\right)\theta(0)', \tag{22}$$

$$C_{fx}(Re_x)^{1/2} = \frac{f''(0)}{2}\left(2 + We(n-1)(f''(0))^2\right), \tag{23}$$

$$C_{fz}(Re_x)^{1/2} = \frac{g'(0)}{2}\left(2 + (n-1)We(g'(0))^2\right), \tag{24}$$

where $Re_x = \frac{xu_w}{\nu}$ represents the Reynolds number.

### 3. Results and Discussion

The homotopy analysis method (HAM) is an analytical procedure which is employed for solving the nonlinear coupled DEs. From its introduction in 1992 [80], HAM has been heavily used by investigators for solving the nonlinear coupled ODEs. The wide range of uses and applications of HAM are because of its convergence properties and initial guess wide range [71,81,82]. The procedure that HAM follows is based on the transformation $\hat{\Psi} : \hat{X} \times [0,1] \to \hat{Y}$, where $\hat{X}$ and $\hat{Y}$ are the topological spaces. The linear operators are defined as follow:

$$L_{\hat{f}}(\hat{f}) = \hat{f}''', \; L_{\hat{g}}(\hat{g}) = \hat{g}'', \; L_{\hat{\theta}}(\hat{\theta}) = \hat{\theta}'', \text{ and } L_{\hat{\phi}}(\hat{\phi}) = \hat{\phi}''. \tag{25}$$

We have employed HAM in this study for solving Equations (15)–(19). The achieved results are depicted through different graphs and the effects of related parameters over the Carreau fluid hydromagnetic behavior are investigated and explained in detail. Furthermore, the present study results are compared with the published work and the agreement ascertains the accuracy of HAM.

The dependence of $f'(\eta)$ (gradient in the velocity x-component) and $g(\eta)$ (velocity y-component) on augmenting magnetic parameter $M$ are respectively depicted in Figure 2a,b. The values of $M$ used in this Figure are $= 0.5, 1.0, 1.5, 2.0$. It is clear from Figure 2a, that at fixed $M$, $f'(\eta)$ declines with the rising $\eta$. The decline in $f'(\eta)$ is much faster at smaller $\eta$ values. Furthermore, the increasing magnetic parameter $M$ values result in a downfall in the $f'(\eta)$ profile. It is obvious from the Figure 2a that reduction in the $f'(\eta)$ profile is more visible in the range $\eta = 0.4$ to $\eta = 2.6$. The downfall in the $f'(\eta)$ profile may be associated with the augmenting Lorentz forces due to the enhancing $M$, which causes to reduce the non-uniformity in the fluid velocity. Figure 2b shows that the velocity $g(\eta)$ changes inversely with the rising $\eta$ at fixed $M$. The velocity field augments with the enhancing $M$. The enhancing behavior of $g(\eta)$ with uplifting $M$ is more dominant upto $\eta = 2.6$. Thus, the augmenting Lorentz forces due to rising $M$ accelerate the fluid flow.

The impact of the Hall parameter $(m)$ on the velocity gradient $f'(\eta)$ and velocity $g(\eta)$ is displayed respectively in Figure 3a,b. The different values of $m$ used in this computation are $0.5, 1.0, 1.5, 2.0$. It is evident from Figure 3a, that initially $f'(\eta)$ augments and then drops with the increasing $\eta$ values at fixed Hall parameter value. It is further observed that the enhancing $m$ results in the increase of the $f'(\eta)$ profile. The enhancing behavior of $f'(\eta)$ is more apparent from $\eta = 0.4$ to $\eta = 2.4$. The variation of the velocity $g(\eta)$ with enhancing $m$ is displayed in Figure 3b. It can be seen from this figure that at smaller $m$, the velocity drops with augmenting $\eta$. As the value of $m$ is increased to $m = 1.5$ and $m = 2.0$, the trend in the $g(\eta)$ profiles changes. Now initially the velocity increases, reaches to a maximum, and then declines with the increasing $\eta$. The variation in the velocity profiles is more dominant at smaller $\eta$ as can be seen from the figure. The increasing trend with the augmenting $m$ is due to the higher Hall potentials produced in the fluid which augment the fluid velocity as well as the gradient in the velocity.

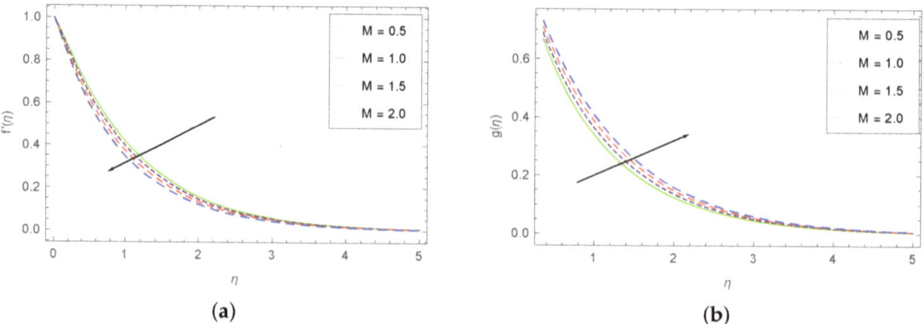

**Figure 2.** (**a**) $f'(\eta)$ variation with $M$, and (**b**) $g(\eta)$ dependence on $M$.

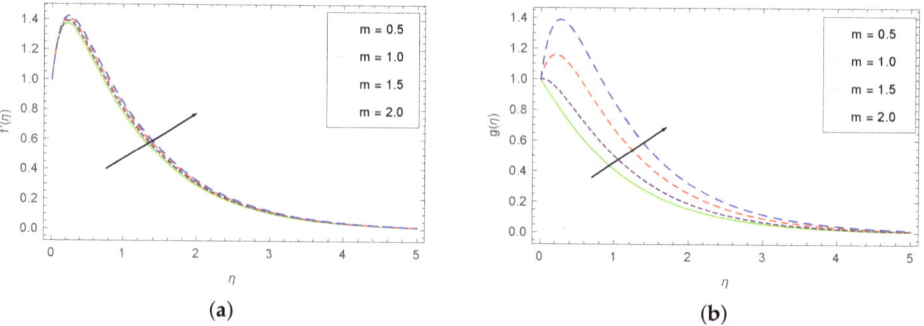

**Figure 3.** (**a**) $f'(\eta)$ dependence on $m$ and (**b**) $g(\eta)$ variation with $m$.

Figure 4a,b show the variations of $f'(\eta)$ and $g(\eta)$ with varying Weissenberg number ($We$). The values of $We$ used in the present computation are $0.30, 0.50, 0.70, 0.90$. From these two figures, it is observed that both $f'(\eta)$ and $g(\eta)$ display almost similar decreasing trend with the increasing $We$. Thus, it is clear that the enhancing viscous nature of the Carreau fluid associated with the rising $We$ constricts the fluid flow and hence reduces the fluid velocity.

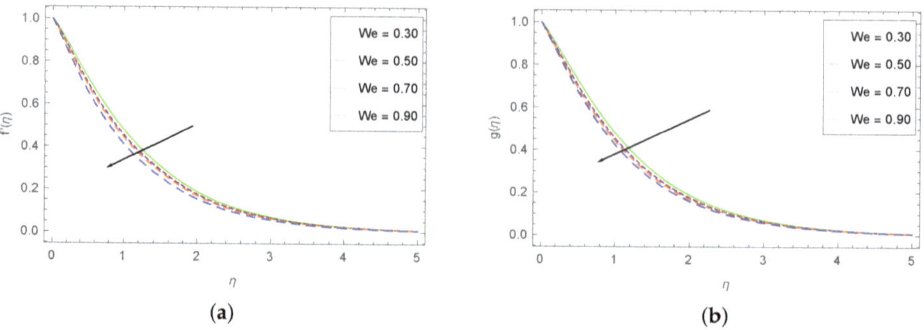

**Figure 4.** (**a**) $f'(\eta)$ dependence on $We$ and (**b**) $g(\eta)$ dependence on $We$.

The variation of $f'(\eta)$ with shrinking parameter ($\varrho$) and the index of power law ($n$) is depicted respectively in the Figure 5a,b. The values of $\varrho$ used are $4.0, 5.0, 6.0, 7.0$, while those of $n$ are $1.6, 2.2, 2.5, 2.8$. It is observed from Figure 5a that at fixed $\varrho$, $f'(\eta)$ first drops, reaches to minimum and then enhances with increasing $\eta$. Similarly, the $f'(\eta)$ profiles first drop and then augment with enhancing $\varrho$. Thus, due to suction ($\varrho > 0$) during the Carreau fluid flow, the $f'(\eta)$ profiles rise beyond $\eta = 1$. Beyond $\eta = 3.8$, all the curves for different $\varrho$ overlap with one another. Figure 5b shows the variation of $f'(\eta)$ with changing values of the power law index $n$. The Figure shows that, at fixed $n$, $f'(\eta)$ augments with higher $\eta$ values. The rate of enhancement in $f'(\eta)$ is larger for lower $\eta$ values in comparison with larger $\eta$. By enhancing the values of $n$, the $f'(\eta)$ profiles drop. The spacing between $f'(\eta)$ curves at different $n$ increases with rising $\eta$ as clear from the Figure.

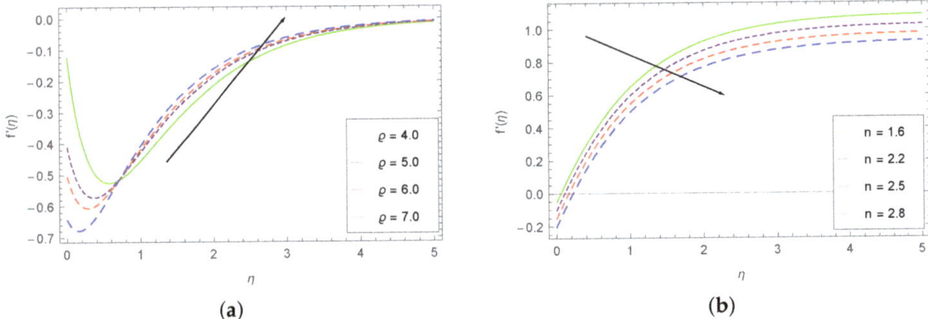

**Figure 5.** (**a**) $f'(\eta)$ dependence on $\varrho$ and (**b**) dependence of $f'(\eta)$ on $n$.

The variation in $f'(\eta)$ and $\phi(\eta)$ (fluid concentration) with the enhancing values of $\chi_1$ are displayed respectively in Figure 6a,b. The different values of $\chi_1$ used are $\chi_1 = 0.25, 0.50, 0.75, 0.95$. From Figure 6a, it is observed that at a given $\chi_1$, $f'(\eta)$ changes inversely with $\eta$. The rate of decline of $f'(\eta)$ is faster at lower $\eta$ in comparison with larger $\eta$ values. Furthermore, by augmenting the values of $\chi_1$, the $f'(\eta)$ profiles drop to smaller values. The spacing between the curves for different $\chi_1$ reduces with larger values of $\chi_1$. The different curves overlap beyond $\eta = 4.0$. The concentration field ($\phi(\eta)$) variation with changing $\chi_1$ is depicted in Figure 6b. It can be seen that at fixed $\chi_1$, the concentration field drops with enhancing $\eta$. The rate of decline of $\phi(\eta)$ is much larger at smaller $\eta$ values. An enhancement in the $\phi(\eta)$ profiles is observed with the rising $\chi_1$. The rate of enhancement of $\phi(\eta)$ is larger for the larger $\chi_1$. The $\phi(\eta)$ curves for different $\chi_1$ overlap with one another beyond $\eta = 3.6$.

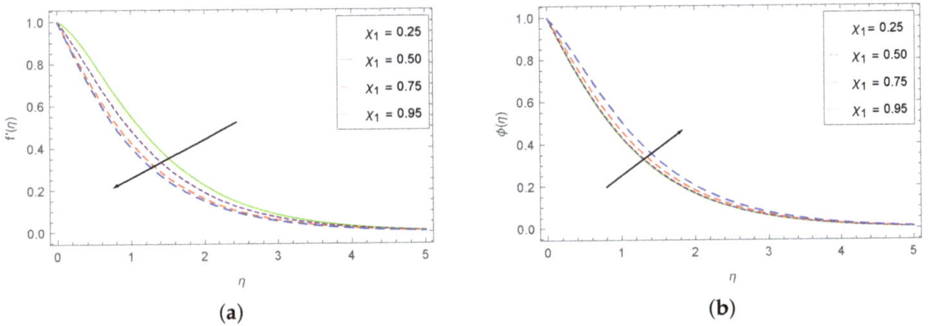

**Figure 6.** (**a**) $f'(\eta)$ variation with $\chi_1$ and (**b**) dependence of $\phi(\eta)$ on $\chi_1$.

The variation of the fluid temperature $\theta(\eta)$ with increasing $Pr$ (Prandtl number) and $Rd$ (radiation parameter) is displayed in Figure 7a,b. The values of $Pr$ are taken as $7.0, 10.0, 13.0, 16.0$, while those of $Rd$ are taken as $1.0, 2.0, 2.5, 3.0$. Figure 7a shows that the Carreau fluid temperature drops with the rising $\eta$ at fixed $Pr$. The rate of decrease of $\theta(\eta)$ is much faster at smaller $\eta$. As the $Pr$ values are increased, the temperature field profiles drop. The spacing between $\theta(\eta)$ curves is more prominent at the intermediate values of $\eta$. The drop in the fluid temperature with the enhancing Prandtl number is due to the smaller thermal diffusivity of the Carreua fluid, which causes a reduction in the temperature of the fluid. Figure 7b depicts the dependence of the temperature field on $Rd$. It can be observed that the fluid temperature augments with the rising $Rd$ values. The rate of enhancement in $\theta(\eta)$ with increasing $Rd$ is more drastic for smaller $\eta$ values. The $\theta(\eta)$ curves overlap beyond $\eta = 4.0$. The augmenting fluid temperature with the higher $Rd$ is due to the stronger heat source.

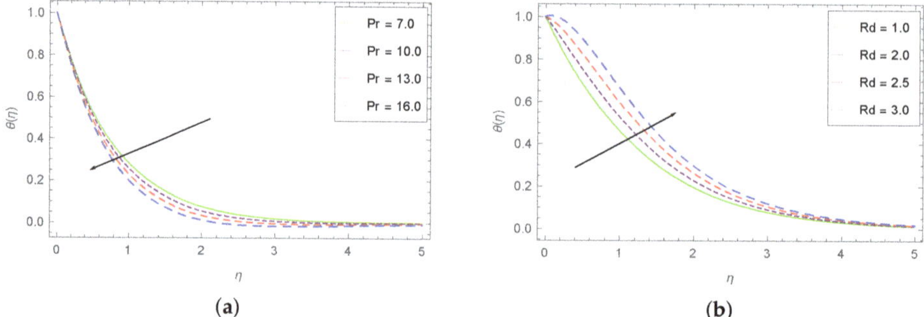


**Figure 7.** (a) $\theta(\eta)$ dependence on $Pr$ and (b) $\theta(\eta)$ dependence on $Rd$.

The influence of augmenting values of Weissenberg number ($We$) and Dufour number ($Du$) on $\theta(\eta)$ is displayed respectively in Figure 8a,b. The $We$ and $Du$ values are taken as $0.3, 0.5, 0.7, 0.9$ and $0.25, 0.45, 0.65, 0.95$, respectively. It is clear that $\theta(\eta)$ declines with the augmenting $\eta$ at constant $We$. The temperature field profiles of the fluid upsurge with the enhancing $We$ values. This means that the augmenting viscous nature of the Carreau fluid associated with the rising $We$ enhances the fluid temperature. The different $\theta(\eta)$ curves overlap beyond $\eta = 4.0$. Figure 8b displays the variation of the Carreau fluid temperature with the enhancing $Du$ (Dufour number). It is observed that the fluid temperature profiles rise with the enhancing $Du$ values. Thus, the accumulation of the fluid particles due to increasing $Du$ raises the fluid temperature.

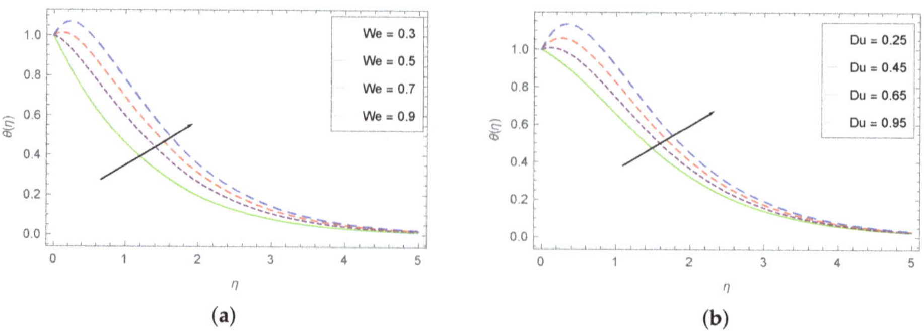


**Figure 8.** (a) $\theta(\eta)$ dependence on $We$ and (b) $\theta(\eta)$ dependence on $Du$.

Figure 9a depicts the fluid concentration $\phi(\eta)$ with varying Schmidt number $(Sc)$. From this Figure, we can observe that $\phi(\eta)$ changes inversely with rising $\eta$ at constant $Sc$. As $Sc$ changes from 0.10 to 0.40, 0.70, and 0.90, a decreasing behavior in the $\theta(\eta)$ profiles is seen. The different curves overlap beyond $\eta = 3.6$. The higher value of $Sc$ is analogous to smaller value of mass diffusivity, that causes the concentration of the Carreau fluid to drop as can be seen from the Figure. The dependence of $\phi(\eta)$ on the increasing $Sr$ (Soret number) is plotted in Figure 9b. The fluid concentration declines with augmenting $\eta$ at fixed $Sr$. An increase is observed in the fluid concentration with the enhancing $Sr$. As $Sr$ is related with the Carreau fluid temperature gradient, hence, higher $Sr$ denotes greater temperature difference, which causes an enhancement in the concentration.

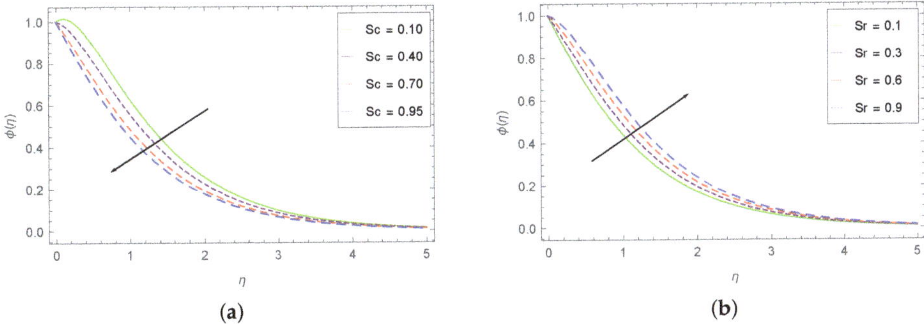

**Figure 9.** (a) $\phi(\eta)$ dependence on $Sc$ and (b) dependence of $\phi(\eta)$ on $Sr$.

### 4. Tables' Discussion

This section is devoted to comparing the results of this study to the work published and the computation of various quantities of engineering interest with the changing values of the associated parameters. The comparison is carried out in Tables 1–3. The quantities of engineering interest are computed in Tables 4 and 5. The comparison is made with the research work already carried out in ref. [75] as in the following:

The comparison of the skin friction values with varying $\varrho$ and $We$ is shown in Table 1. The $Rd, Sr, Du, M,$ and $Sc$ values are taken as 0. We see complete agreement between the results of our work and the already published work.

**Table 1.** Computation of $C_{fx}$ for changing $\varrho$ and $We$, for $Rd = Sr = M = Du = Sc = 0$.

| $\varrho$ | $We$ | Reference [75] | Present Results |
| --- | --- | --- | --- |
| 0.0 | 0.4 | 0.9913393 | 0.99133929 |
| 0.3 | 0.4 | 1.157602 | 1.1576019 |
| 0.6 | 0.4 | 1.348724 | 1.3487236 |
| 0.6 | 0.0 | 1.344032 | 1.3440315 |
| 0.6 | 0.2 | 1.348725 | 1.3487250 |
| 0.6 | 0.4 | 1.333041 | 1.3330407 |

Table 2 shows the comparison of Nusselt number $(Nu_x)$ for the varying $We, n, \varrho,$ and $Pr$, whereas $Rd, Sr, Du, M,$ and $Sc$ are taken as 0. The value of $\chi_1$ is taken as 0.2. Again both results are in tremendous agreement.

**Table 2.** Computation of $Nu_x$, when $Rd = Sr = M = Sc = Du = 0$.

| $We$ | $n$ | $\varrho$ | $Pr$ | $\chi_1$ | Reference [75] | Present Results |
|------|-----|-----------|------|----------|----------------|-----------------|
| 0.0 | 0.6 | 0.6 | 0.2 | 0.2 | 0.1668374 | 0.16683740 |
| 0.1 | 0.6 | 0.6 | 0.2 | 0.2 | 0.1666775 | 0.16667749 |
| 0.2 | 0.6 | 0.6 | 0.2 | 0.2 | 0.4741203 | 0.47412033 |
| 0.2 | 0.3 | 0.7 | 1.3 | 0.2 | 0.1666314 | 0.16663136 |
| 0.2 | 0.5 | 0.7 | 1.3 | 0.2 | 0.1666914 | 0.16669144 |
| 0.2 | 0.7 | 0.7 | 1.3 | 0.2 | 0.1667505 | 0.16675047 |

Table 3 compares the present and published results for $Sh_x$ (local Sherwood number) for changing $Sr$. We used $Du = We = 0$. The values of the parameters $n, Rd, Pr$, and $M$ are kept fixed in computing the values of $Sh_x$. The comparison proves that both computations are in agreement with each other.

**Table 3.** Computation of $Sh_x$, for $Du = We = 0$.

| $Sr$ | $n$ | $Rd$ | $Pr$ | $M$ | Reference [75] | Present Results |
|------|-----|------|------|-----|----------------|-----------------|
| 0.5 | 0.6 | 0.5 | 0.75 | 0.75 | 1.2110832 | 1.21108321 |
| 1.0 | 0.6 | 0.5 | 0.75 | 0.75 | 1.0581235 | 1.058124 |
| 1.5 | 0.6 | 0.5 | 0.75 | 0.75 | 0.905314 | 0.9053143 |
| 2.0 | 0.6 | 0.5 | 0.75 | 0.75 | 0.752515 | 0.7525148 |

The computation of $C_{fx}$ and $C_{fz}$ with the changing values of $M, m, \chi_1$, and $We$ for suction case is tabulated in Table 4. It is found that $C_{fx}$ first reduces and then increases with rising magnetic parameter ($M$). The skin friction $C_{fx}$ drops with the augmenting values of $m, \chi_1$ and $We$. The other skin friction component $C_{fz}$ augments (drops) with the rising values of $M, \chi_1$ and $We$ ($m$).

**Table 4.** Computation of $C_{fx}$ and $C_{fz}$ in the suction case.

| $M$ | $m$ | $\chi_1$ | $We$ | $Re_x^{\frac{1}{2}} C_{fx}$ | $Re_x^{\frac{1}{2}} C_{fz}$ |
|-----|-----|----------|------|------------------------------|------------------------------|
| 0.5 | 0.5 | 0.25 | 0.30 | $-1.27682794$ | 0.02280230 |
| 1 | — | — | — | $-1.47275352$ | 0.03841483 |
| 1.5 | — | — | — | $-1.52006594$ | 0.04188107 |
| 2 | — | — | — | $-1.24036061$ | 0.09561891 |
| — | 1 | — | — | $-1.17768683$ | 0.12810091 |
| — | 1.5 | — | — | $-1.30688120$ | 0.02219887 |
| — | 2 | — | — | $-1.33663381$ | 0.02161783 |
| — | — | 0.50 | — | $-1.29858315$ | 0.02300640 |
| — | — | 0.75 | — | $-1.31813402$ | 0.02318486 |
| — | — | 0.95 | — | $-1.32674352$ | 0.02452684 |
| — | — | — | 0.50 | $-1.38932465$ | 0.02918536 |
| — | — | — | 0.70 | $-1.40374526$ | 0.03326478 |
| — | — | — | 0.90 | $-1.40374537$ | 0.03326589 |

The variation of $Nu_x$ and $Sh_x$ with the augmenting $M, n, Sr, Du, Rd$, and $\chi_1$ in the blowing case is shown in Table 5. It is clear that both $Nu_x$ and $Sh_x$ drop and then enhance with enhancing $M$. The Nusselt number first increases and then decreases, whereas the Sherwood number enhances with the higher $n$. Furthermore the Nusselt number enhances with the augmenting values of all the remaining parameters, that is $Sr, Du, Rd$ and $\chi_1$. The Sherwood number rises (declines) with the increasing values of $Du$ and $Rd$ ($Sr$ and $\chi_1$).

**Table 5.** $Nu_x$, and $Sh_x$ in the case of blowing.

| $M$ | $n$ | $Sr$ | $Du$ | $Rd$ | $\chi_1$ | $Nu/Re_x^{\frac{-1}{2}}$ | $Sh/Re_x^{\frac{-1}{2}}$ |
|---|---|---|---|---|---|---|---|
| 0.5 | 1.6 | 0.1 | 0.25 | 1 | 0.25 | 0.114028 | 0.139181 |
| 1 | – | – | – | – | – | 0.106677 | 0.134409 |
| 1.5 | – | – | – | – | – | 0.100334 | 0.130193 |
| 2 | 1 | – | – | – | – | 0.106198 | 0.134071 |
| – | 2.2 | – | – | – | – | 0.109861 | 0.136627 |
| – | 2.5 | – | – | – | – | 0.113931 | 0.139391 |
| – | 2.8 | 0.2 | – | – | – | 0.113041 | 0.142802 |
| – | – | 0.4 | – | – | – | 0.108654 | 0.137044 |
| – | – | 0.6 | – | – | – | 0.104832 | 0.131925 |
| – | – | – | 0.2 | – | – | 0.125738 | 0.130361 |
| – | – | – | 0.4 | – | – | 0.112658 | 0.133111 |
| – | – | – | 0.6 | – | – | 0.101036 | 0.135657 |
| – | – | – | – | 0.5 | – | 0.111641 | 0.133395 |
| – | – | – | – | 1.0 | – | 0.106677 | 0.134407 |
| – | – | – | – | 1.5 | – | 0.102281 | 0.135322 |
| – | – | – | – | – | 0.1 | 0.110879 | 0.137607 |
| – | – | – | – | – | 0.3 | 0.102933 | 0.131509 |
| – | – | – | – | – | 0.5 | 0.096511 | 0.126388 |

## 5. Conclusions

This section concludes the main findings of the present research work. The investigation of mass and thermal energy transfer of the 3D Carreau fluid moving through a permeable and stretching (shrinking) sheet is undertaken by considering the effects of thermal radiations, cross diffusion, and Hall current. Suitable similarity relations are employed in order to transform the set of coupled PDEs to a system of coupled ODEs. The set of coupled ODEs is then solved through the well-known standard analytical technique of HAM. The influence of the relevant physical variables on the hydromagnetic behavior of the Carreau fluid 3D flow is examined through various plots. The variations of the coefficients of skin friction, local Nusselt, and Sherwood numbers with the changing parameter values are shown through various tables. The important findings of this work are outlined as below:

- The gradient in velocity $f'(\eta)$ reduces with the augmenting $M$, $We$, and $\chi_1$, whereas it enhances with the Hall parameter $(m)$ and the shrinking parameter $(\varrho)$.
- The velocity component $g(\eta)$ enhances with augmenting $M$, $m$, while it declines with Wesinberg number $(We)$.
- The Carreau fluid temperature enhances with the rising $Rd$, $We$, and $Du$, while it reduces with the increasing $Pr$ (Prandtl number) values.
- The concentration of the fluid augments with enhancing $\chi_1$ while it reduces with the augmenting $Sc$ values.
- The skin friction coefficients $(C_{fx})$ drops with the enhancing $m$, $\chi_1$, and $We$. The other component $(C_{fz})$ enhances with the rising $M$, $\chi_1$, and $We$.
- The local Nusselt number depreciates with the enhancing $Sr$, $Du$, $Rd$, and $\chi_1$, while the Sherwood number increases (depreciates) with the rising $n$, $Du$, and $Rd$ ($Sr$ and $\chi_1$).
- The published and obtained results show an agreement which validates the employed analytical procedure accuracy.

**Author Contributions:** Conceptualization, A.U., I., and M.M.S.; software, S.K., A.U., and M.A.; writing—original draft preparation, I., M.M.S., and T.A.; writing—review and editing, A.K., T.A., S.K., and W.K.M.; formal analysis, W.K.M., I., and M.A.; validation, I., S.K., and A.U.; methodology, A.U., T.A., M.A., and W.K.M.; investigation, W.K.M., A.K., and S.K.; resources, M.M.S., A.K., and S.K.; project administration, M.M.S., M.A., and W.K.M.; funding acquisition, A.K. and T.A. All authors have read and agreed to the published version of the manuscript.

**Funding:** This research received no external funding.

**Institutional Review Board Statement:** Not applicable.

**Informed Consent Statement:** Not applicable.

**Data Availability Statement:** The data used to support the findings of this study are available from the corresponding author upon request.

**Acknowledgments:** Aziz Khan and Thabet Abdeljawad acknowledge the support provided by Prince Sultan University for funding this work through research group Nonlinear Analysis Methods in Applied Mathematics (NAMAM)group number RG-DES-2017–01-17.

**Conflicts of Interest:** The authors declare no conflict of interest.

## Abbreviations

The below mentioned parameters and abbreviations with their possible dimensions are used in this article:

| | |
|---|---|
| $\sigma$ | Electrical conductivity $\frac{S}{m}$ |
| $B_0$ | magnetic field strength T |
| $Re_x$ | Local Reynolds number |
| $C_{fx}$ | Local Skin friction |
| $m$ | Hall parameter $\frac{m^3}{C}$ |
| $\varrho$ | Suction/ injection parameter |
| $J_w$ | Mass flux $\left(\frac{kg}{s\,m^2}\right)$ |
| $f$ | Dimensionless velocity |
| $\phi$ | Dimensionless concentration |
| $\theta$ | Dimensionless temperature |
| $\infty$ | Condition at infinity |
| $x, y,$ and $z$ | Coordinates (m) |
| $0$ | Reference condition |
| $\eta$ | Similarity variable |
| $Sc$ | Schmidt number |
| $\gamma$ | Thermal relaxation parameter |
| $Du$ | Dufour number |
| $u_w$ | Stretching velocity $\left(\frac{m}{s}\right)$ |
| $Pr$ | Prandtl number |
| $T$ | Fluid temperature (K) |
| $\rho$ | Density $\left(\frac{kg}{m^3}\right)$ |
| $v$ | Kinematic viscosity $\frac{m^2}{s}$ |
| $\mu$ | Dynamic viscosity mPa |
| $t$ | Time (s) |
| $C_p$ | Specific heat $\left(\frac{J}{kg\,K}\right)$ |
| $n$ | Power law index |
| $k_T$ | Thermal diffusion ratio |
| $L_1$ | Velocity slip factor |
| $h_2$ | Convective mass transfer coefficient |
| $h_1$ | Convective heat transfer coefficient |
| $\tau$ | Extra stress tensor |
| $M$ | Magnetic field interaction parameter |
| $\Pi$ | Strain rate tensor |
| $\sigma^*$ | Stefan Boltzmann constant |
| $\chi_1$ | Velocity slip parameter |
| $\chi_2$ | Thermal profile slip parameter |
| $\chi_3$ | Concentration profile slip parameter |
| $We$ | Weissenberg number |

## References


1. Moradikazerouni, A.; Afrand, M.; Alsarraf, J.; Mahian, O.; Wongwises, S.; Tran, M.D. Comparison of the effect of five different entrance channel shapes of a micro-channel heat sink in forced convection with application to cooling a supercomputer circuit board. *Appl. Therm. Eng.* **2019**, *150*, 1078–1089. [CrossRef]
2. Moradikazerouni, A.; Afrand, M.; Alsarraf, J.; Wongwises, S.; Asadi, A.; Nguyen, T.K. Investigation of a computer CPU heat sink under laminar forced convection using a structural stability method. *Int. J. Heat Mass Transf.* **2019**, *134*, 1218–1226. [CrossRef]
3. Crane, L.J. Flow past a stretching plate. *Z. Angew. Math. Phys. ZAMP* **1970**, *21*, 645–647. [CrossRef]
4. Eid, M.R.; Mahny, K.L. Flow and heat transfer in a porous medium saturated with a Sisko nanofluid over a nonlinearly stretching sheet with heat generation/absorption. *Heat Transf. Asian Res.* **2018**, *47*, 54–71. [CrossRef]
5. Vajravelu, K. Viscous flow over a nonlinearly stretching sheet. *Appl. Math. Comput.* **2001**, *124*, 281–288. [CrossRef]
6. Liu, I.C.; Wang, H.H.; Peng, Y.F. Flow and heat transfer for three-dimensional flow over an exponentially stretching surface. *Chem. Eng. Commun.* **2013**, *200*, 253–268. [CrossRef]
7. Hayat, T.; Asad, S.; Mustafa, M.; Alsaedi, A. Boundary layer flow of Carreau fluid over a convectively heated stretching sheet. *Appl. Math. Comput.* **2014**, *246*, 12–22. [CrossRef]
8. Schlichting, H.; Gersten, K. *Boundary-Layer Theory*; Springer: Berlin/Heidelberg, Germany, 2016.
9. Dash, G.; Tripathy, R.; Rashidi, M.; Mishra, S. Numerical approach to boundary layer stagnation-point flow past a stretching/shrinking sheet. *J. Mol. Liq.* **2016**, *221*, 860–866. [CrossRef]
10. Ajlouni, A.W.M.; Al-Rabai'ah, H.A. Fractional-calculus diffusion equation. *Nonlinear Biomed. Phys.* **2010**, *4*, 3. [CrossRef] [PubMed]
11. Ishak, A.; Nazar, R.; Pop, I. Hydromagnetic flow and heat transfer adjacent to a stretching vertical sheet. *Heat Mass Transf.* **2008**, *44*, 921. [CrossRef]
12. Xu, H.; Liao, S.J.; Pop, I. Series solutions of unsteady three-dimensional MHD flow and heat transfer in the boundary layer over an impulsively stretching plate. *Eur. J. Mech. B/Fluids* **2007**, *26*, 15–27. [CrossRef]
13. Ishak, A.; Jafar, K.; Nazar, R.; Pop, I. MHD stagnation point flow towards a stretching sheet. *Phys. A Stat. Mech. Its Appl.* **2009**, *388*, 3377–3383. [CrossRef]
14. Naganthran, K.; Nazar, R.; Pop, I. A study on non-Newtonian transport phenomena in a mixed convection stagnation point flow with numerical simulation and stability analysis. *Eur. Phys. J. Plus* **2019**, *134*, 1–14. [CrossRef]
15. Naganthran, K.; Nazar, R.; Pop, I. Stability analysis of impinging oblique stagnation-point flow over a permeable shrinking surface in a viscoelastic fluid. *Int. J. Mech. Sci.* **2017**, *131*, 663–671. [CrossRef]
16. Vajravelu, K.; Hadjinicolaou, A. Convective heat transfer in an electrically conducting fluid at a stretching surface with uniform free stream. *Int. J. Eng. Sci.* **1997**, *35*, 1237–1244. [CrossRef]
17. Pop, I.; Na, T.Y. A note on MHD flow over a stretching permeable surface. *Mech. Res. Commun.* **1998**, *25*, 263–269. [CrossRef]
18. Freidoonimehr, N.; Rashidi, M.M.; Mahmud, S. Unsteady MHD free convective flow past a permeable stretching vertical surface in a nano-fluid. *Int. J. Therm. Sci.* **2015**, *87*, 136–145. [CrossRef]
19. Ahmad, H.; Seadawy, A.R.; Khan, T.A.; Thounthong, P. Analytic approximate solutions for some nonlinear Parabolic dynamical wave equations. *J. Taibah Univ. Sci.* **2020**, *14*, 346–358. [CrossRef]
20. Jamaludin, A.; Naganthran, K.; Nazar, R.; Pop, I. Thermal radiation and MHD effects in the mixed convection flow of Fe3O4–water ferrofluid towards a nonlinearly moving surface. *Processes* **2020**, *8*, 95. [CrossRef]
21. Lund, L.A.; Omar, Z.; Khan, I.; Baleanu, D.; Nisar, K.S. Dual similarity solutions of MHD stagnation point flow of Casson fluid with effect of thermal radiation and viscous dissipation: Stability analysis. *Sci. Rep.* **2020**, *10*, 1–13. [CrossRef] [PubMed]
22. Reddy, M.G. Influence of magnetohydrodynamic and thermal radiation boundary layer flow of a nanofluid past a stretching sheet. *J. Sci. Res.* **2014**, *6*, 257–272. [CrossRef]
23. Gnaneswara Reddy, M. Thermal radiation and chemical reaction effects on MHD mixed convective boundary layer slip flow in a porous medium with heat source and Ohmic heating. *EPJP* **2014**, *129*, 41. [CrossRef]
24. AboEldahab, E.M. Radiation effect on heat transfer in an electrically conducting fluid at a stretching surface with a uniform free stream. *J. Phys. D Appl. Phys.* **2000**, *33*, 3180. [CrossRef]
25. Abo-Eldahab, E.M.; El Gendy, M.S. Radiation effect on convective heat transfer in an electrically conducting fluid at a stretching surface with variable viscosity and uniform free stream. *Phys. Scr.* **2000**, *62*, 321. [CrossRef]
26. Reddy, M.G. Effects of Thermophoresis, Viscous Dissipation and Joule Heating on Steady MHD Flow over an Inclined Radiative Isothermal Permeable Surface with Variable Thermal Conductivity. *J. Appl. Fluid Mech.* **2014**, *7*, 51–61.
27. Anwar, T.; Kumam, P.; Khan, I.; Thounthong, P. Generalized Unsteady MHD Natural Convective Flow of Jeffery Model with ramped wall velocity and Newtonian heating; A Caputo-Fabrizio Approach. *Chin. J. Phys.* **2020**, *68*, 849–865. [CrossRef]
28. Fayz-Al-Asad, M.; Alam, M.N.; Ahmad, H.; Sarker, M.; Alsulami, M.; Gepreel, K.A. Impact of a closed space rectangular heat source on natural convective flow through triangular cavity. *Results Phys.* **2021**, *23*, 104011. [CrossRef]
29. Anwar, T.; Kumam, P.; Baleanu, D.; Khan, I.; Thounthong, P. Radiative heat transfer enhancement in MHD porous channel flow of an Oldroyd-B fluid under generalized boundary conditions. *Phys. Scr.* **2020**, *95*, 115211. [CrossRef]
30. Farooq, A.; Kamran, M.; Bashir, Y.; Ahmad, H.; Shahzad, A.; Chu, Y.M. On the flow of MHD generalized maxwell fluid via porous rectangular duct. *Open Phys.* **2020**, *18*, 989–1002. [CrossRef]

31. Ma, Y.; Shahsavar, A.; Moradi, I.; Rostami, S.; Moradikazerouni, A.; Yarmand, H.; Zulkifli, N.W.B.M. Using finite volume method for simulating the natural convective heat transfer of nano-fluid flow inside an inclined enclosure with conductive walls in the presence of a constant temperature heat source. *Phys. A Stat. Mech. Its Appl.* **2019**, 123035. [CrossRef]
32. Tian, Z.; Rostami, S.; Taherialekouhi, R.; Karimipour, A.; Moradikazerouni, A.; Yarmand, H.; Zulkifli, N.W.B.M. Prediction of rheological behavior of a new hybrid nanofluid consists of copper oxide and multi wall carbon nanotubes suspended in a mixture of water and ethylene glycol using curve-fitting on experimental data. *Phys. A Stat. Mech. Its Appl.* **2020**, *549*, 124101. [CrossRef]
33. Shah, Z.; Islam, S.; Gul, T.; Bonyah, E.; Khan, M.A. The electrical MHD and hall current impact on micropolar nanofluid flow between rotating parallel plates. *Results Phys.* **2018**, *9*, 1201–1214. [CrossRef]
34. Shah, Z.; Islam, S.; Ayaz, H.; Khan, S. Radiative heat and mass transfer analysis of micropolar nanofluid flow of Casson fluid between two rotating parallel plates with effects of Hall current. *J. Heat Transf.* **2019**, *141*, 022401. [CrossRef]
35. Khan, A.; Shah, Z.; Islam, S.; Khan, S.; Khan, W.; Khan, A.Z. Darcy–Forchheimer flow of micropolar nanofluid between two plates in the rotating frame with non-uniform heat generation/absorption. *Adv. Mech. Eng.* **2018**, *10*, 1687814018808850. [CrossRef]
36. Hayat, T.; Rashid, M.; Imtiaz, M.; Alsaedi, A. Magnetohydrodynamic (MHD) flow of Cu-water nanofluid due to a rotating disk with partial slip. *AIP Adv.* **2015**, *5*, 067169. [CrossRef]
37. Vo, D.D.; Alsarraf, J.; Moradikazerouni, A.; Afrand, M.; Salehipour, H.; Qi, C. Numerical investigation of $\gamma$-AlOOH nano-fluid convection performance in a wavy channel considering various shapes of nanoadditives. *Powder Technol.* **2019**, *345*, 649–657. [CrossRef]
38. Alsarraf, J.; Moradikazerouni, A.; Shahsavar, A.; Afrand, M.; Salehipour, H.; Tran, M.D. Hydrothermal analysis of turbulent boehmite alumina nanofluid flow with different nanoparticle shapes in a minichannel heat exchanger using two-phase mixture model. *Phys. A Stat. Mech. Its Appl.* **2019**, *520*, 275–288. [CrossRef]
39. Abouelregal, A.E.; Ahmad, H. Thermodynamic modeling of viscoelastic thin rotating microbeam based on non-Fourier heat conduction. *Appl. Math. Model.* **2021**, *91*, 973–988. [CrossRef]
40. Asadi, A.; Aberoumand, S.; Moradikazerouni, A.; Pourfattah, F.; Żyła, G.; Estellé, P.; Mahian, O.; Wongwises, S.; Nguyen, H.M.; Arabkoohsar, A. Recent advances in preparation methods and thermophysical properties of oil-based nanofluids: A state-of-the-art review. *Powder Technol.* **2019**, *352*, 209–226. [CrossRef]
41. Abouelregal, A.E.; Moustapha, M.V.; Nofal, T.A.; Rashid, S.; Ahmad, H. Generalized thermoelasticity based on higher-order memory-dependent derivative with time delay. *Results Phys.* **2021**, *20*, 103705. [CrossRef]
42. Shi, E.; Zang, X.; Jiang, C.; Mohammadpourfard, M. Entropy generation analysis for thermomagnetic convection of paramagnetic fluid inside a porous enclosure in the presence of magnetic quadrupole field. *J. Therm. Anal. Calorim.* **2020**, *139*, 2005–2022. [CrossRef]
43. Soleiman, A.; Abouelregal, A.E.; Ahmad, H.; Thounthong, P. Generalized thermoviscoelastic model with memory dependent derivatives and multi-phase delay for an excited spherical cavity. *Phys. Scr.* **2020**, *95*, 115708. [CrossRef]
44. Alam, M.K.; Memon, K.; Siddiqui, A.; Shah, S.; Farooq, M.; Ayaz, M.; Nofal, T.A.; Ahmad, H. Modeling and analysis of high shear viscoelastic Ellis thin liquid film phenomena. *Phys. Scr.* **2021**, *96*, 055201. [CrossRef]
45. Abouelregal, A.E.; Ahmad, H. Response of thermoviscoelastic microbeams affected by the heating of laser pulse under thermal and magnetic fields. *Phys. Scr.* **2020**, *95*, 125501. [CrossRef]
46. Nisar, K.S.; Khan, U.; Zaib, A.; Khan, I.; Morsy, A. A novel study of radiative flow involving micropolar nanoliquid from a shrinking/stretching curved surface including blood gold nanoparticles. *Eur. Phys. J. Plus* **2020**, *135*, 1–19. [CrossRef]
47. Carreau, P.J. Rheological equations from molecular network theories. *Trans. Soc. Rheol.* **1972**, *16*, 99–127. [CrossRef]
48. Ali, N.; Hayat, T. Peristaltic motion of a Carreau fluid in an asymmetric channel. *Appl. Math. Comput.* **2007**, *193*, 535–552. [CrossRef]
49. Goodarzi, M.; Javid, S.; Sajadifar, A.; Nojoomizadeh, M.; Motaharipour, S.H.; Bach, Q.V.; Karimipour, A. Slip velocity and temperature jump of a non-Newtonian nanofluid, aqueous solution of carboxy-methyl cellulose/aluminum oxide nanoparticles, through a microtube. *Int. J. Numer. Methods Heat Fluid Flow* **2019**, *29*, 1606–1628. [CrossRef]
50. Maleki, H.; Alsarraf, J.; Moghanizadeh, A.; Hajabdollahi, H.; Safaei, M.R. Heat transfer and nanofluid flow over a porous plate with radiation and slip boundary conditions. *J. Cent. South Univ.* **2019**, *26*, 1099–1115. [CrossRef]
51. Hayat, T.; Saleem, N.; Ali, N. Effect of induced magnetic field on peristaltic transport of a Carreau fluid. *Commun. Nonlinear Sci. Numer. Simul.* **2010**, *15*, 2407–2423. [CrossRef]
52. Tshehla, M. The flow of Carreau fluid down an incline with a free surface. *Int. J. Phys. Sci* **2011**, *6*, 3896–3910.
53. Ellahi, R.; Riaz, A.; Nadeem, S.; Ali, M. Peristaltic flow of Carreau fluid in a rectangular duct through a porous medium. *Math. Probl. Eng.* **2012**, *2012*. [CrossRef]
54. Machireddy, G.R.; Polarapu, P.; Bandari, S. Effects of magnetic field and Ohmic heating on viscous flow of a nanofluid towards a nonlinear permeable stretching sheet. *J. Nanofluids* **2016**, *5*, 459–470. [CrossRef]
55. Jiaqiang, E.; Jin, Y.; Deng, Y.; Zuo, W.; Zhao, X.; Han, D.; Peng, Q.; Zhang, Z. Wetting models and working mechanisms of typical surfaces existing in nature and their application on superhydrophobic surfaces: A review. *Adv. Mater. Interfaces* **2018**, *5*, 1701052.
56. Tassaddiq, A.; Khan, I.; Nisar, K.S.; Singh, J. MHD flow of a generalized Casson fluid with Newtonian heating: A fractional model with Mittag–Leffler memory. *Alex. Eng. J.* **2020**, *59*, 3049–3059. [CrossRef]
57. Khan, Z.A.; Haq, S.U.; Khan, T.S.; Khan, I.; Nisar, K.S. Fractional Brinkman type fluid in channel under the effect of MHD with Caputo-Fabrizio fractional derivative. *Alex. Eng. J.* **2020**, *59*, 2901–2910. [CrossRef]

58. Krisna, P.M.; Sandeep, N.; Sugunamma, V. Effects of radiation and chemical reaction on MHD convective flow over a permeable stretching surface with suction and heat generation. *Walailak J. Sci. Technol. (WJST)* **2015**, *12*, 831–847.

59. Sandeep, N.; Sulochana, C.; Sugunamma, V. Radiation and magnetic field effects on unsteady mixed convection flow over a vertical stretching/shrinking surface with suction/injection. *Ind. Eng. Lett.* **2015**, *5*, 127–136.

60. Jonnadula, M.; Polarapu, P.; Reddy, G. Influence of thermal radiation and chemical reaction on MHD flow, heat and mass transfer over a stretching surface. *Procedia Eng.* **2015**, *127*, 1315–1322. [CrossRef]

61. Maleki, H.; Safaei, M.R.; Togun, H.; Dahari, M. Heat transfer and fluid flow of pseudo-plastic nanofluid over a moving permeable plate with viscous dissipation and heat absorption/generation. *J. Therm. Anal. Calorim.* **2019**, *135*, 1643–1654. [CrossRef]

62. Gheynani, A.R.; Akbari, O.A.; Zarringhalam, M.; Shabani, G.A.S.; Alnaqi, A.A.; Goodarzi, M.; Toghraie, D. Investigating the effect of nanoparticles diameter on turbulent flow and heat transfer properties of non-Newtonian carboxymethyl cellulose/CuO fluid in a microtube. *Int. J. Numer. Methods Heat Fluid Flow* **2019**, *29*, 1699–1723. [CrossRef]

63. Maleki, H.; Safaei, M.R.; Alrashed, A.A.; Kasaeian, A. Flow and heat transfer in non-Newtonian nanofluids over porous surfaces. *J. Therm. Anal. Calorim.* **2019**, *135*, 1655–1666. [CrossRef]

64. Nazari, S.; Ellahi, R.; Sarafraz, M.; Safaei, M.R.; Asgari, A.; Akbari, O.A. Numerical study on mixed convection of a non-Newtonian nanofluid with porous media in a two lid-driven square cavity. *J. Therm. Anal. Calorim.* **2020**, *140*, 1121–1145. [CrossRef]

65. Biswal, S.; Sahoo, P. Hall effect on oscillatory hydromagnetic free convective flow of a visco-elastic fluid past an infinite vertical porous flat plate with mass transfer. *Proc. Natl. Acad. Sci. India Sect. A* **1999**, *1*, 45–58.

66. Changal, R.M.; Ananda, R.N.; Vijaya, K.V.S. Hall-current effects on unsteady MHD flow between stretching sheet and an oscillating porous upper parallel plate with constant suction. *Therm. Sci.* **2011**, *15*, 527–536. [CrossRef]

67. Datta, N.; Jana, R.N. Oscillatory magnetohydrodynamic flow past a flat plate with Hall effects. *J. Phys. Soc. Jpn.* **1976**, *40*, 1469–1474. [CrossRef]

68. Aboeldahab, E.M.; Elbarbary, E.M. Hall current effect on magnetohydrodynamic free-convection flow past a semi-infinite vertical plate with mass transfer. *Int. J. Eng. Sci.* **2001**, *39*, 1641–1652. [CrossRef]

69. Khan, I.; Memon, A.A.; Memon, M.A.; Bhatti, K.; Shaikh, G.M.; Baleanu, D.; Alhussain, Z.A. Finite Element Least Square Technique for Newtonian Fluid Flow through a Semicircular Cylinder of Recirculating Region via COMSOL Multiphysics. *J. Math.* **2020**, *2020*. [CrossRef]

70. Rajput, U.; Kanaujia, N. Chemical reaction in MHD flow past a vertical plate with mass diffusion and constant wall temperature with hall current. *Int. J. Eng. Sci. Technol.* **2016**, *8*, 28–38. [CrossRef]

71. Shah, Z.; Dawar, A.; Khan, I.; Islam, S.; Ching, D.L.C.; Khan, A.Z. Cattaneo-Christov model for electrical magnetite micropoler Casson ferrofluid over a stretching/shrinking sheet using effective thermal conductivity model. *Case Stud. Therm. Eng.* **2019**, *13*, 100352. [CrossRef]

72. Ullah, A.; Alzahrani, E.O.; Shah, Z.; Ayaz, M.; Islam, S. Nanofluids thin film flow of Reiner-Philippoff fluid over an unstable stretching surface with Brownian motion and thermophoresis effects. *Coatings* **2019**, *9*, 21. [CrossRef]

73. Alharbi, S.O.; Dawar, A.; Shah, Z.; Khan, W.; Idrees, M.; Islam, S.; Khan, I. Entropy generation in MHD eyring–powell fluid flow over an unsteady oscillatory porous stretching surface under the impact of thermal radiation and heat source/sink. *Appl. Sci.* **2018**, *8*, 2588. [CrossRef]

74. Hayat, T.; Ahmed, B.; Abbasi, F.; Alsaedi, A. Numerical investigation for peristaltic flow of Carreau–Yasuda magneto-nanofluid with modified Darcy and radiation. *J. Therm. Anal. Calorim.* **2019**, *137*, 1359–1367. [CrossRef]

75. Ullah, A.; Hafeez, A.; Mashwani, W.K.; Kumam, W.; Kumam, P.; Ayaz, M. Non-Linear Thermal Radiations and Mass Transfer Analysis on the Processes of Magnetite Carreau Fluid Flowing Past a Permeable Stretching/Shrinking Surface under Cross Diffusion and Hall Effect. *Coatings* **2020**, *10*, 523. [CrossRef]

76. Carreau, P.; Kee, D.D.; Daroux, M. An analysis of the viscous behaviour of polymeric solutions. *Can. J. Chem. Eng.* **1979**, *57*, 135–140. [CrossRef]

77. Ghadikolaei, S.; Hosseinzadeh, K.; Yassari, M.; Sadeghi, H.; Ganji, D. Analytical and numerical solution of non-Newtonian second-grade fluid flow on a stretching sheet. *Therm. Sci. Eng. Prog.* **2018**, *5*, 309–316. [CrossRef]

78. Vo, D.D.; Shah, Z.; Sheikholeslami, M.; Shafee, A.; Nguyen, T.K. Numerical investigation of MHD nanomaterial convective migration and heat transfer within a sinusoidal porous cavity. *Phys. Scr.* **2019**, *94*, 115225. [CrossRef]

79. Olajuwon, I.B. Convection heat and mass transfer in a hydromagnetic Carreau fluid past a vertical porous plate in presence of thermal radiation and thermal diffusion. *Therm. Sci.* **2011**, *15*, 241–252. [CrossRef]

80. Liao, S.J. The Proposed Homotopy Analysis Technique for the Solution Of Nonlinear problems. Ph.D. Thesis, Shanghai Jiao Tong University Shanghai, Shanghai, China, 1992.

81. Khan, N.S.; Zuhra, S.; Shah, Z.; Bonyah, E.; Khan, W.; Islam, S. Slip flow of Eyring-Powell nanoliquid film containing graphene nanoparticles. *AIP Adv.* **2018**, *8*, 115302. [CrossRef]

82. Sohail, M.; Naz, R.; Shah, Z.; Kumam, P.; Thounthong, P. Exploration of temperature dependent thermophysical characteristics of yield exhibiting non-Newtonian fluid flow under gyrotactic microorganisms. *AIP Adv.* **2019**, *9*, 125016. [CrossRef]

*Review*

# Fluid Films as Models for Understanding the Impact of Inhaled Particles in Lung Surfactant Layers

**Eduardo Guzmán** [1,2]

[1]  Departamento de Química Física, Facultad de Ciencias Químicas, Universidad Complutense de Madrid, Ciudad Universitaria s/n, 28040 Madrid, Spain; eduardogs@quim.ucm.es; Tel.: +34-91-394-4107

[2]  Instituto Pluridisciplinar, Universidad Complutense de Madrid, Paseo Juan XXIII 1, 28040 Madrid, Spain

**Abstract:** Pollution is currently a public health problem associated with different cardiovascular and respiratory diseases. These are commonly originated as a result of the pollutant transport to the alveolar cavity after their inhalation. Once pollutants enter the alveolar cavity, they are deposited on the lung surfactant (LS) film, altering their mechanical performance which increases the respiratory work and can induce a premature alveolar collapse. Furthermore, the interactions of pollutants with LS can induce the formation of an LS corona decorating the pollutant surface, favoring their penetration into the bloodstream and distribution along different organs. Therefore, it is necessary to understand the most fundamental aspects of the interaction of particulate pollutants with LS to mitigate their effects, and design therapeutic strategies. However, the use of animal models is often invasive, and requires a careful examination of different bioethics aspects. This makes it necessary to design in vitro models mimicking some physico-chemical aspects with relevance for LS performance, which can be done by exploiting the tools provided by the science and technology of interfaces to shed light on the most fundamental physico-chemical bases governing the interaction between LS and particulate matter. This review provides an updated perspective of the use of fluid films of LS models for shedding light on the potential impact of particulate matter in the performance of LS film. It should be noted that even though the used model systems cannot account for some physiological aspects, it is expected that the information contained in this review can contribute on the understanding of the potential toxicological effects of air pollution.

**Keywords:** dynamics; fluid interfaces; inhalation; lung surfactant; nanoparticles; pollutants; rheology

**Citation:** Guzmán, E. Fluid Films as Models for Understanding the Impact of Inhaled Particles in Lung Surfactant Layers. *Coatings* **2022**, *12*, 277. https://doi.org/10.3390/coatings12020277

Academic Editor: George Kokkoris

Received: 27 December 2021
Accepted: 16 February 2022
Published: 19 February 2022

**Publisher's Note:** MDPI stays neutral with regard to jurisdictional claims in published maps and institutional affiliations.



## 1. Introduction

In recent years, the injection of pollutants into the atmosphere as a result of the combustion processes of fossil fuels in industries, power plants, heating systems and vehicles, and the residues of the production processes of nanotechnological industries has undergone a spectacular growth. This opens many questions related to the potential harmful effects of such compounds for human health [1] as is evidenced from the recent statistics of the World Health Organization (WHO) [2], which include long-term exposure to air pollutants as one of the most important sources of cardiovascular diseases, and mortality (around one third of the deaths by strokes, lung cancer, or cardiac diseases are associated with air pollution) [3,4]. Therefore, air pollution should currently be included as a very important public health issue in modern s·ociety [5,6], making it necessary to evaluate the potential impact of pollutants in normal physiological functions, paying special attention to the role of inhaled pollutants [7–9].

Pollutant inhalation is associated with the emergence of pathogenic states and respiratory diseases, e.g., acute respiratory distress syndrome (ARDS), asthma, chronic obstructive pulmonary diseases, fibrosis, or lung cancer [10,11]. Therefore, it is critical to perform a careful examination of the effect of particles in respiratory function [12].

Among pollutants, fine particulate matter (PM2.5, particles with diameter $\leq$2.5 µm) is counted as one of the most important sources of respiratory diseases [13–15]. This type of particulate matter can be inhaled and transported along the respiratory tract to the alveoli, where they interact with the lung surfactant (LS) film [16,17], modifying the LS chemical composition and lateral organization upon incorporation, which in turn alter LS functionality. Furthermore, the interaction between LS and particles can induce different pathways which favor particle translocation towards other tissues and organs [18,19]. This may be the origin of several adverse health effects [20,21].

LS is a mixture containing mainly lipids and proteins which overlays the inner wall of alveoli in mammals (i.e., the alveolar lining) [7,22], contributing to two very important physiological aspects: (i) regulation of the mechanical properties of the alveoli during breathing, preventing alveolar collapse during exhalation [23–25], and (ii) protection against inhaled matter, mainly pathogens (e.g., SARS-CoV-2) [7,26–31]. Therefore, any dysfunction on the LS mechanical properties or modification of its composition may be harmful for health [32–38]. However, to date there is an important lack of knowledge about the true impact of inhaled particles in the respiratory cycle, even though it is clear that the deposition of particles on LS alters the breathing mechanics, the use of particles, even inhaled ones, in therapy and diagnosis is undergoing continuous growth [39–43]. Therefore, it is necessary to deepen the current understanding of the interactions of different types of particulate pollutants with LS for predicting their potential toxicity [44–46]. However, it is not always easy to perform a direct evaluation of the interaction of pollutants with LS under physiological relevant conditions, and most in vivo tests are invasive, including exposure to particles followed by bronchoalveolar lavage or even the sacrifice of small animals [15,47,48]. This makes it necessary to seek non-invasive in vitro approaches providing a first evaluation of the impact of inhaled particles in the physico-chemical properties of LS films [49–52].

The use of models based on fluid films of different lipids or mixtures of lipids and proteins at water/vapor interfaces, or their deposits onto solid surfaces as Langmuir–Blodgett or Langmuir–Schaefer films, is widespread for the evaluation of the impact of specific chemicals on the activity of different biological systems under physiological relevant conditions [49,53–64]. Therefore, the use of interfacial science may help in the evaluation of the effect of inhaled pollutants or particles in LS films [58,59,65–68] by exploiting the importance of different physical and chemical aspects of the liquid/vapor interface formed between the alveolar lining and the gas contained in the alveolar cavity in LS physiology [69,70]. It should be stressed that studies using model systems are in most of the cases an oversimplification, allowing mimicking of only some specific physico-chemical aspects related to the complex biophysical function of LS film. Furthermore, the most common practice relies on the use of models including a limited number of components, i.e., minimal systems, which many times makes it difficult to extrapolate the results obtained by using model systems to the real in vivo situation. On the other side, model systems do not include, in most of the cases, some specific hydrodynamics and aerodynamics aspects guiding the interaction between LS and inhaled particles under true in vivo conditions.

It should be noted that the limitations associated with the use of model systems based on fluid interfaces for evaluating the potential toxicity of inhaled particles upon their interaction with LS cannot hide the many possibilities offered by these types of model systems for preliminary assessment of the harmful effects of inhaled pollutants in the normal respiratory function [59,65,67]. This review tries to provide an updated physico-chemical perspective about the current understanding on the potential toxicity of inhaled materials upon interaction with LS films, which requires the introduction of some general points related to the LS performance, including its chemical composition and the role of such composition in LS interfacial properties. Furthermore, considering that the aim of this work is focused on the incorporation of inhaled particles into LS films, some general aspects of such process will be also discussed. On the other side, considering that the review tries to provide insights into the utility of fluid films as models for studying the impact of particles on LS performance, it is necessary to introduce the most common experimental tools used for exploring the interaction of LS with particulate matter, and the chemical composition of such models together with the influence of the physico-chemical characteristics of the particles in the model properties. The last part of the review will highlight some relevant information about the interaction of particles with LS extracted from the exploitation of the tools offered for the interfacial science.

## 2. Lung Surfactant from Chemistry to a Functional Layer

LS is a very complex mixture containing lipids (around 90% of the total weight) and different proteins (comprising about the 10% of the total weight) [24,71], with the specific proteolipidic composition of LS playing a key role and being essential for the control of its structure, properties, and function. This means that the specific composition of LS determines the formation of a surfactant film at the surface of the liquid film covering the alveolar epithelium, contributing to the surface tension reduction and facilitating the respiratory mechanics [32,72]. Furthermore, the formation of the LS film acts as the main biological barrier protecting against the entrance of inhaled particles into the mammal bloodstream. Figure 1 presents a scheme summarizing the average composition of mammal LS.

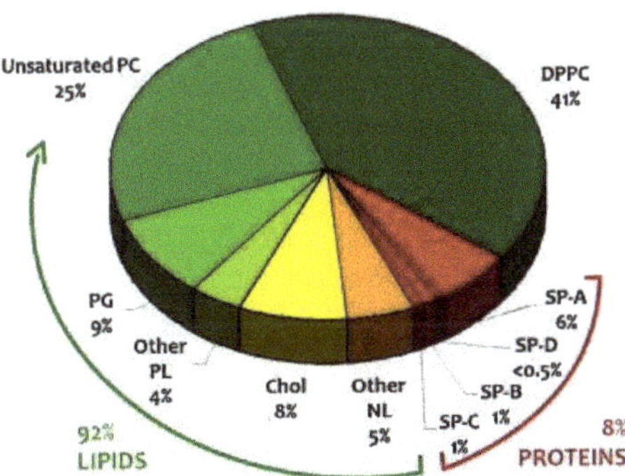


**Figure 1.** Average composition, as percentage in relation to the total weight, of mammal LS. PL: phospholipids; PC: phosphatidylcholines; PG: phosphatidylglycerol; Chol: cholesterol; NL: neutral lipids; SP-A/B/C/D: surfactant protein A/B/C/D. Reprinted from Parra and Pérez-Gil [24], with permission from Elsevier, Copyright 2015.

The lipid fraction includes three major components, phospholipids, cholesterol, and neutral lipids, corresponding to around 79%, 8%, and 5% of the total weight of LS, respectively. Among phospholipids, zwitterionic phosphatidylcholines (PC) emerge as the most important contribution to the average composition of LS (in the range of 60–70% of the total weight of LS), with anionic species, such phosphatidylglycerols (PG) and phosphatidylinositols (PI) together with other minor components, e.g., phosphatidylethanolamines (PE) and sphingomyelin, appearing in smaller concentrations [73,74]. A more detailed analysis of the average phospholipid composition of mammal LS shows that 1,2-dipalmitoyl-sn-glycero-3-phosphocholine (DPPC) is the main component of the lipid fraction, and in turn, of the LS (in the range 40–50% of the total LS weight) [75]. It should be noted that the exact content of DPPC in LS is inversely proportional to the rate of surface area change, i.e., the respiratory rate, occurring during breathing [71]. Other phosphatidylcholines that appear commonly in mammal LS are 1-palmitoyl-2-oleyl-sn-glycerol-3-phosphatidylcholine, 1-palmitoyl-2-palmitoleyl-sn-glycero-3-phosphatidylcholine, and 1-palmitoyl-2-myristoil-sn-glycero-3-phosphatidylcholine. It should be noted that the specific proportion between DPPC and other phosphatidylcholines is correlated to the specific respiratory rate and the alveolar development [73,76–79]. Beyond phosphatidylcholines, anionic phospholipids, e.g., phosphatidylglycerols and some phosphatidylinositols, account for 10% of the total weight of the LS, with an absence of myristoyl components and a low concentration of dipalmitoyl among these types of compounds [73,80]. The presence of anionic PG and PI presents a very important role in the interactions between the lipid fraction and the cationic surfactant proteins (SP-B and SP-C) [22].

The presence of cholesterol is also a very important feature in LS, allowing the modulation of lipid packing in LS structures, and hence any change of the cholesterol concentration may dramatically impact the lateral organization and functionality of LS [81]. Furthermore, the presence of neutral lipids in LS, e.g., cholesterol esters, triglycerides, diglycerides, and free fatty acids, is very important for surfactant properties, and its content has been modulated during evolution for optimizing the activity of the surfactant [82].

The protein fraction is commonly formed by four proteins, the so-called surfactant proteins (SP), SP-A, SP-B, SP-C, and SP-D, which are named following the order of their first detection. Furthermore, some additional proteins, e.g., SP-G and SP-H, can appear in specific mammal species [83]. Among the four most common proteins, there are two oligomeric fibrous ones with hydrophilic characters (SP-A and SP-D), presenting a very important role in the innate defense of the lung. Furthermore, they perform a certain role in maintaining the alveolar surfactant homeostasis [84]. On the other side, SP-B and SP-C are hydrophobic proteins, playing a very important role in the control of the interfacial properties of LS, and hence in its functionality [75,85].

SP-B is a positively charged polypeptide with a molecular weight of about 8.7 kDa, adopting $\alpha$-helical structure [86]. This protein performs a very important role on the LS adsorption at the liquid/gas interface, and in the dynamics exchange between the interfacial LS layer and the subphase enriched in compounds with low surface activity [86,87]. Furthermore, SP-B plays a very important role in the storage of LS modulating the assembly of lipids into the lamellar bodies, and their deficiencies are lethal [88]. On the other side, SP-C is a small hydrophobic protein (molecular weight around 3.7 kDa) with helical structure, and their deficiency is not lethal but can induce chronic respiratory disorders [72,89]. SP-C is associated with the promotion of vesicle fragmentation helping in the membrane curvature [90,91].

*Lung Surfactant Interfacial Activity: A Key Aspect for the Respiratory Function*

The interfacial activity of LS is closely correlated to the phospholipid ability for self-organizing at the interface but also to the LS composition, temperature, and lateral pressure [72,92]. Thus, below the melting temperature (temperature for the order-disorder transition), the lateral mobility of phospholipids is hindered due to the strong interfacial packing, whereas once the melting temperature is passed, the fluidity of phospholipid membranes is enhanced. The emergence of the transition between ordered and disordered phases is correlated to the specific molecular characteristics of the considered phospholipid. In particular, for the main component of the LS, i.e., the 1,2-Dipalmitoyl-sn-glycero-3-phosphatidylcholine (DPPC), the transition occurs around 41 °C, whereas the transition temperature can drop down to values below 0 °C for membranes of phospholipid having unsaturated acyl chains. Furthermore, the presence of cholesterol is essential for modulating the lipid packing. On the other side, when interfacial films, i.e., monolayers, are considered, the modulation of the phase behavior is commonly associated with surface pressure changes $\Pi = \gamma_0 - \gamma$, with $\gamma_0$ and $\gamma$ being the surface tension for a pristine water/vapor (around 72 mN/m at 37 °C), and the surface tension of the monolayer, respectively. Therefore, if the interfacial density of phospholipids at the interface is low, it can be assumed that the monolayer behaves as a gas-like system, with the increase of the interfacial density pushing the monolayer through phases with higher lateral order, where the lipid mobility starts to be hindered, e.g., liquid expanded phase ($L_E$) or liquid condensed phase ($L_C$), to reach a 2D solid-like conformation for high values of $\Pi$ [93].

Moreover, lipid membranes can undergo 3D assembly processes in aqueous medium to form a broad range of structure with lamellar and non-lamellar ordering. It should be noted that the specific molecular characteristics of lipids, and in particular their hydrophilic-lipophilic balance (HLB), are essential in the modulation of the LS adsorption at the liquid/gas interface and reorganization during breathing [72]. Figure 2 show a sketch of different lipid structures that can emerge in LS under physiological conditions.

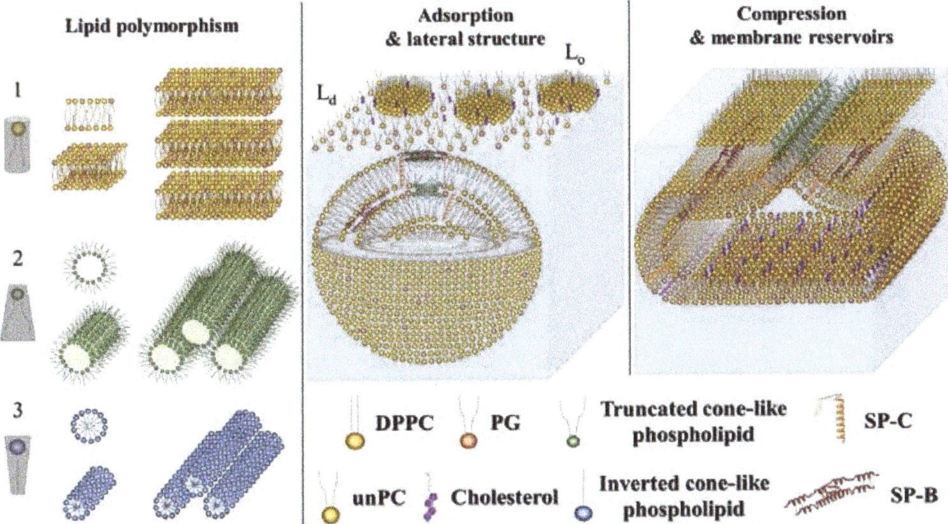


**Figure 2.** Different structures emerging in LS. Reprinted from Castillo-Sánchez et al. [72], with permission under Open access CC BY 4.0 license, https://creativecommons.org/licenses/by/4.0/ (accessed on 20 December 2021).

The lipid polymorphism in LS surfactant plays a very important role for ensuring the correct respiratory function, which requires the fulfillment of three conditions: (i) efficient interfacial adsorption in a small time-scale; (ii) surface tension reduction down to very low values under compression without any weakening of the mechanical stability, and (iii) efficient respreading during expansion, ensuring the stabilization of the surface of the alveoli under recurrent compression-expansion cycles [72].

*De novo* secreted LS undergoes a continuous adsorption to the gas/liquid interface for guarantying the normal respiratory function. This entails that lipids are incorporated into the interface following a process governed by LS composition, and lipid concentration and structure, with the transition bilayer-to-monolayer being essential for the formation of functional interfacial films (see Figure 2). This requires the formation of non-lamellar structures at specific locations with the contribution of anionic lipids and SP-B and SP-C proteins [92,94–97]. In particular, SP-B protein presents a very important role in the first contact of LS with the interface, and also promotes membrane aggregation, fusion, and permeabilization, which may be related to the rapid lipid exchange between the interfacial film and the fluid phase underneath, commonly enriched in the less surface-active components, forming the so-called reservoirs [86,87,98–102].

The adsorption and spreading of the LS at the gas/liquid interface leads to the formation of an interfacial film which leads to a strong decrease of the surface tension during compression (at exhalation), minimizing the respiratory work [22]. Experimental evidences shown that the absorption of LS at water/vapor interface leads to an immediate drop of the surface tension from 70 mN/m to a value around 25 mN/m, and this latter value can be furtherly reduced to an almost negligible one upon a change of the interfacial area in the range

10–15%, which is equivalent to the final steps of the exhalation. This is possible because the composition of the interfacial film can be remodeled (squeezing-out process) to reduce the surface tension during exhalation [103], which requires excluding those molecules from the interface which cannot attain high values of surface pressure (unsaturated phospholipids, cholesterol, and surfactant proteins), a film enriched in DPPC remaining at the interface. This can lead to the formation of rather solid structures under high lateral pressures as those occurring in the alveolar surface during exhalation. This increases the lung compliance, and stabilizes the alveolar volume, which in turn contributes towards prevention of the premature alveolar collapse during exhalation [23–25]. Simultaneously to the compositional remodeling, the interfacial film is bound through protein-mediated interactions to reservoirs containing excluded lipids, proteins, and freshly secreted LS [92]. The association of the reservoirs with the interface plays a central role in the new remodeling of the interfacial composition occurring during inspiration by supplementing new molecules to the interface [102,104–109]. Furthermore, these types of structures contribute to the mechanical stability as a result of the high cohesion induced by protein-protein and protein-lipid interactions [87,104]. Therefore, disaturated lipids drive the surface tension decrease during exhalation, with surface-active proteins (SP-B and SP-C) acting as helper systems for ensuring such minimization of the surface tension.

The lateral organization of the LS interfacial film is reminiscent of a coexistence between two liquid phases, where domains enriched in DPPC appear surrounded by a disorder lipid phase. These domains grow during exhalation leading to a solid structure at maximum compression [92], which is reverted during inspiration as result of the quick replenishment of the gas/liquid interface by the adsorption of lipids from the reservoirs. This leads to the increase of the surface tension leading to the formation disordered phase. Therefore, the continuous remodeling of the LS films modulated by the surface proteins SP-B and SP-C connecting the interfacial film with the reservoirs ensures effortless breathing [32,72,110]. Figure 3 present a sketch showing the different phenomena occurring during the compression-expansion cycles of alveolar cavity, indicating the remodeling processes and their effect on the surface tension.

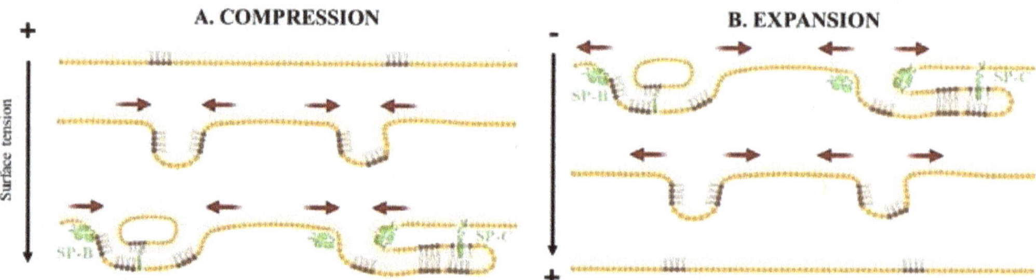


**Figure 3.** Sketch of the different remodeling processes occurring during compression-expansion cycles of the alveolar cavity. Reprinted from Da Silva et al. [32], with permission from Elsevier, Copyright 2021.

A last important process associated with the interfacial activity of the LS is related to the recycling of lipids and proteins excluded from the interfacial layer as result of the activity of alveolar macrophages and type II pneumocytes [111]. In particular, the SP-C protein presents a very central role in the recycling process contribution to the formation of small vesicles that can be easily taken up [90]. On the other side, SP-B stimulated the secretion of lamellar bodies in type II pneumocyte [112], whereas SP-A contributes to the mechanism regulating the inhibition of LS secretion when their concentration is enough in the alveoli [113].

### 3. Impact of Inhaled Particle Deposition in Lung Surfactant Film

The transport of inhaled particles along the respiratory tract involves several processes: Brownian diffusion, gravitational sedimentation, inertial impaction, and interception [114]. However, the specific role of each individual process is far from clear due to the high complexity of the breathing dynamics. On the other side, inhaled aerosols commonly present a high polydispersity, which limits their penetration into deep lung regions, and only a small fraction of particles with sizes below 4 μm can go through the respiratory airways to reach the alveolar cavity. This can be understood considering the tortuous way, including 23 tubular bifurcations, encountered for the particle from their inhalation through the nose to their final destination inside the alveoli [7].

It is commonly accepted that the biophysical effects of inhaled substances start when they penetrate into the alveoli, leading to a partial inhibition of the LS function, and in turn in the lung function [32]. It should be stressed that the high surface area of the alveoli (up to 100 m$^2$) leads to relatively small concentrations of deposited particles on the alveolar fluid. However, the high surface-to-volume (or surface-to-mass) ratio of deposited particles leads to a situation in which inhaled particles induce stronger effects than those that are expected considering only the deposited mass, i.e., the acquired dose [42,55,57,65,115,116]. Therefore, the evaluation of the safety of inhaled particles must consider the role of several parameters, including the dimensions of the particulate material and its concentration, the particle molecular structure, and other physico-chemical properties (hydrophobicity and charge) [69,114,117–125].

It should be stressed that the deposition of particles on the LS layer is commonly associated with several events that are triggered after the interaction of the particles with the LS. Such interaction may occur directly with specific components of the LS film at the interface, or indirectly through the competition of the inhaled particles with the LS components for the space available at the interface [63,126–129]. Thus, the interaction of particles with the LS film may induce an inhibition of the LS functionality. In particular, particles can critically impact normal respiratory function [66], with the alteration of the Marangoni flows occurring in the LS upon particle deposition playing a critical role [130,131]. This is particularly important because Marangoni flows are essential in the control of the compositional remodeling of the LS film during breathing. Therefore, if the Young–Laplace law is assumed as an oversimplified representation of the alveolar mechanics during breathing [132], it may be expected that an alveolar compression under high surface tension conditions may induce an alveolar collapse. However, the ability of LS to decrease the surface tension during exhalation, and the consequent reduction of the respiratory work, takes the system far from collapse conditions. This is, in part, a result of the high content of LS in DPPC, and other disaturated phospholipids, which lead to the formation, at room temperature, of highly condensed phase with thigh packing at very low surface tensions. However, the incorporation of particles into the LS film may influence the mechanical aspects associated with the normal physiological respiratory function in different ways [32,65,66].

The deposition of particles on the LS film may inhibit the formation of condensed phases during alveolar compression, which in turn limits its ability to reduce the surface tension, and drives a premature alveolar collapse, e.g., hydrophobic particles may remain trapped within the hydrophobic regions of the LS film, modifying the intricate balance of the interactions emerging between the components and hindering the formation of tightly packed phases [133,134]. This can lead to a situation in which the surface tension is not low enough during compression, and it is impossible to overcome the forces pulling the alveolar walls together [135]. It should be noted that collapsed alveoli can be reopened during inspiration. However, alveolar collapse leads to a reduction of the tidal volume [32]. On the other side, particle incorporation modifies the remodeling dynamics of LS film, modifying the area of the hysteresis loop corresponding to the compression-expansion cycles. This is the result of the modification of the boundary conditions controlling the mass transport between the interfacial film and the adjacent fluid due to two possible effects: (i) incomplete clearance of inhaled particles from LS film or (ii) modification of the fluid phase composition as a result of the expulsion of particles from the interface during the compression [130,131,136–138]. The latter effect adds an additional mechanism in which particles can influence the functionality of LS as a result of the adsorption of LS compounds onto the particle surface, contributing to the formation of an LS corona onto particle surface which influence their fate. This modifies the particle wettability and the LS composition, which alters the ability of LS to reduce the surface tension and the remodeling process [139]. It is commonly accepted that SP-A binds to hydrophilic particles, facilitating the subsequent adsorption of the rest of the LS components, which allows particle clearance from the lung [140]. The situation changes when hydrophobic particles are considered, with the binding of hydrophobic components starting the formation of the corona [141]. According to the above discussion, it may be expected that particles penetrating into the alveolar cavity can be embedded within the LS membrane and undergo different processes in a highly dynamic environment. Thus, particles will be subjected to the same compression-expansion processes than the LS film, which can induce their release from the LS towards the alveolar spaces, making it possible for particles to interact and be internalized by the cellular barrier [142,143]. Therefore, it may be expected that particle inhalation can present short-term and long-term harmful effects in the normal physiology of the respiratory process.

## 4. Surface Science Approaches for Evaluating the Interaction of Particles with the Lung Surfactant Film: Tools and Models

### 4.1. Experimental Tools

The use of colloidal and interfacial approaches allows overcoming some of the main limitations associated with the in vivo evaluation of the interaction between inhaled particles and LS, providing important insights on the most fundamental bases governing the interaction of particles with LS, avoiding the use of invasive methodologies [58,59,66].

The use of Langmuir film balances for studying spread films of lipids or lipid-protein mixtures at liquid/vapor interfaces are probably the most exploited approaches for deepening on the knowledge of LS mechanics, and its modification as results of the incorporation of particles or different types of chemicals, providing very insightful information on the quasi-static mechanical properties of LS films [58,144]. It is true that the studies based on the use of Langmuir film balances do not provide true biophysical information. However, they help in the understanding of several aspects with relevance for the incorporation of particles into LS films in terms of the modification of the surface pressure of pristine LS layers as a result of the incorporation of particles, i.e., the changes in the surface pressure vs. molecular area plot for the interfacial film, the so-called interfacial isotherm [108,117,118,128,129,145].

The use of Langmuir film balances also allows the study of the rheological response of interfacial layers under compression-expansion cycles (oscillatory barrier experiments) which are somehow reminiscent of the deformations of the alveoli during breathing. During these experiments, the surface pressure is monitored by using a surface balance fitted with a contact probe, normally a Wilhelmy plate, whereas the area available for the film is modified by the coupled motion of two barriers arranged parallel in opposite extremes of the trough [146,147]. These experiments require surface pressure changes occurring in comparable time-scales to the area changes. Despite the many possibilities offered for the use of Langmuir film balance, it should be stressed that the probed frequencies and deformation amplitudes are far from that what occur during normal breathing, and in turn they only give semi-quantitative information that can be exploited for evaluating the incorporation of particles into LS films.

Langmuir film balances can also be exploited for transferring the interfacial film from the surface of the liquid to a solid substrate by the Langmuir–Blodgett or Langmuir–Schaefer methods, which makes an ex situ study possible of the interaction of the LS film with particles using microscopy techniques, commonly atomic force microscopy (AFM). These types of studies provide insights on how particles modify the interfacial morphology of LS films [148–150]. Figure 4 shows sketches of the Langmuir film balance and the two alternatives for depositing films on solid surfaces.

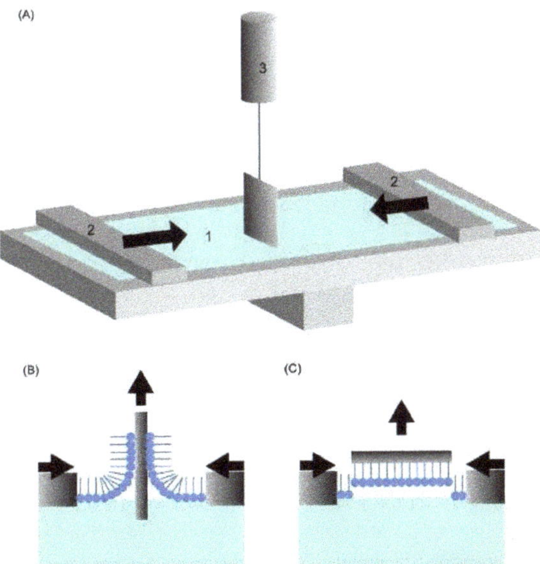

**Figure 4.** (**A**) Scheme of a Langmuir film balance fitted with a contact probe (Wilhelmy plate) and two symmetric barriers. The numbers indicate: 1. Fluid interface, 2. Symmetric barriers and 3. Force balance fitted with a Wilhelmy plate. (**B**) Scheme of the vertical transference of a film from a fluid interface to a solid substrate following the Langmuir–Blodgett method. (**C**) Scheme of the horizontal transference of a film from a fluid interface to a solid substrate following the Langmuir–Schaefer method. Reprinted from Wang et al. [67], with permission from Elsevier, Copyright 2020.

The highly dynamic character of the alveoli during breathing requires deepening of the mechanical characterization of LS layers under dilational deformations mimicking the expiration-inspiration cycles [116,151]. This information is difficult to obtain by using Langmuir film balances as was stated above (further details can be found in the recent review by Ravera et al. [68]). However, the use of techniques based on oscillating drops or bubbles, e.g., pulsating bubble surfactometer or captive bubble tensiometer [152–156], help in the evaluation of the potential impact of the particle incorporation in the mechanics of LS films under realistic simulations of the physiological situation. Thus, it is possible to evaluate the remodeling process and stability of LS films upon their exposure to inhaled particles.

The use of the captive bubble tensiometer relies on the creation of an air bubble suspended in a liquid phase commonly containing the LS which adsorbs to the liquid/vapor interface. Once the LS film is ready at the interface, particles are deposited on the interface by injection in its vicinity through the liquid phase. In some cases, the particles can be premixed with the LS before the formation of the bubble, which provides a less realistic representation of the inhalation process. The use of a captive bubble tensiometer makes it possible to evaluate the fast adsorption of the LS film to the water/vapor interface, and the remodeling processes, as well as the surface activity of LS under changes of the interfacial and its stability [157,158]. Another alternative for evaluating the impact of particles in the mechanics of LS films is the use of a pulsating bubble surfactometer. This technique is based on the creation of an air bubble suspended on a capillary tube inside a chamber filled with LS which undergoes compression-expansion cycles by the action of a piston pulsator [152,157].

Conventional oscillating bubbles have also been used for exploring the effect of particles in LS [159,160]. This approach relies on the oscillation of the bubble surface at a frequency close to that corresponding to the breathing rate (around 15 min$^{-1}$), allowing the evaluation of different relevant properties for the breathing cycle, including the minimum surface tension reached under compression and the area of the hysteresis loop associated with the compression-expansion of the area available, and how these parameters are influenced as a result of the incorporation of particles [68]. Oscillating drops in pendant drop tensiometers also allow obtaining information about the mechanical performance of LS layers under sinusoidal changes of the interfacial area [159,161,162]. Thus, the analysis of the dependence of the dilational viscoelasticity of the layer on the deformation frequency helps in understanding how particles modify the kinetics associated with the exchange of material between the interfacial film and the adjacent fluid layer. This becomes very important because any modification of the remodeling kinetics may be either a signature of inhibition, degradation of the LS activity, or in combination [137,163,164]. Therefore, the combination of experimental results and suitable theoretical models may help in the evaluation of the particle impact in the interfacial relaxation [165]. This is in agreement with the work by Kondej and Sosnowski [166], where the changes in surface viscosity and elasticity obtained against dilational deformations of the interface are proposed as a signature of the inhibition of the LS activity due to the particle incorporation. As an alternative, the analysis of the evolution of the linearity of the rheological response in terms of the total harmonics distortion (THD) is also a very promising tool for evaluating the inhibition of the LS film mechanics [165].

The use of constrained drop surfactometer (see Figure 5) has recently emerged as a very promising tool for studying LS films under conditions which are very close to the true physiological one [70].

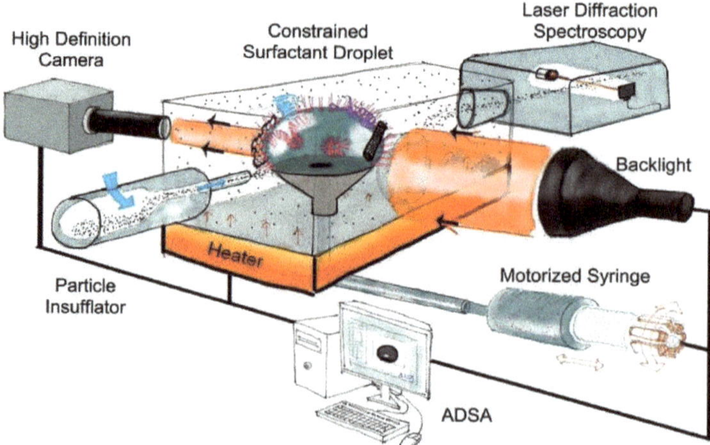


**Figure 5.** Sketch of the typical configuration of a constrained drop surfactometer. Reprinted from Valle et al. [70], with permission from ACS, Copyright 2015.

Constrained drop surfactometer consists of a single sessile droplet constrained on a knife-sharp edge pedestal where the LS is accommodated. This configuration avoids surfactant leakage even at the lowest values of surface tension [167], allowing real time evaluation of the volume, surface area, and surface tension of the droplets using a modified axisymmetric drop analysis (closed-loop axisymmetric drop analysis). These act as feedback control systems enabling the direct measurement and modification of the above-mentioned parameters. The main novelty of the use of constrained drop surfactometer in relation to other traditional configurations relies on the possibility to perform precise harmonic oscillatory deformations of the interface with predefined amplitudes and frequencies (up to 0.5 Hz), which is useful for determining the surface dilational modulus before and during the exposure of the LS film to particulate matter [167].

The above discussion shows clearly that the evaluation of the effect of particles on the interfacial properties of LS films is a very interesting tool for deepening knowledge on the effect of pollutants in the dynamic response of the air/liquid interface in the alveoli during breathing [168]. This is possible by evaluating aspects, such as the minimum and maximum surface tension values reached upon compression and expansion, respectively, which provides important insights on the surfactant functionality. However, in recent years, many researches have turned their interests towards the understanding of the relationship existing between the changes in the interfacial packing of LS layers and the worsening of their mechanical properties [22,169]. This is important because as was stated above, the LS present a very complex organization at the interface and in the adjacent liquid phase [170], and the use of techniques such as Brewster angle microscopy (BAM), AFM, surface force apparatus (SFA), ellipsometry, infrared reflection absorption spectroscopy (IRRAS), or epifluorescence microscopy, and their combinations with tensiometric techniques (mainly Langmuir film balances) are very powerful tools for deepening knowledge on the impact of particles on the organization of LS films in the micrometric and submicrometric scale [171–174]. Furthermore, there are more sophisticated tools, including neutron reflectivity or synchrotron grazing angle X-ray diffractions which may also be useful in the evaluation of the impact of particulate matter on LS interfacial organization [108,134,175–182].

*4.2. Chemistry of Lung Surfactant Models*

The methodologies used for evaluating the interaction between pollutants and LS components rely on the use of different models, ranging from simple lipid monolayers to complex surfactant extracts obtained from bronchoalveolar lavage fluid. Therefore, it is necessary to know the main differences of the physico-chemical properties of the chosen model with respect to the true LS to draw conclusions with biophysical relevance. Therefore, any LS model should mimic some of the physical properties related to the physiological function of LS: (i) ability to decrease the surface tension of the water/vapor interface down to a quasi-null value (<2 mN/m) upon reduction of the interfacial area, (ii) effective compositional remodeling during compression at very low surface tension values, and (iii) fast re-adsorption and respreading during expansion of the area of the interface [7,58,66,67].

DPPC fulfills the first requirement, which has pushed the use of DPPC monolayers as minimal models for understanding some physico-chemical aspects related to the impact of particles in LS films. Furthermore, the interfacial behavior of this lipid has been extensively studied by many researchers [59,66,175,183–187]. However, the use of DPPC as a model for understanding the performance of LS is rather limited because its inefficiency in the reservoir formation and its slow respreading at the interface during expansion. This has driven the research on the use of more complex models, including DPPC in combination with other lipids or fatty acids, e.g., palmitic acid, cholesterol or DOPC (1,2-dioleoyl-sn-glycero-3-phosphocholine), among others [117,136,165,173,188–191], which provide a simplified picture of the complex behavior of LS under physiological conditions. The use of these models can provide a deeper understanding of the behavior of LS films because some studies have evidenced the important role of specific lipids, e.g., 1-palmitoyl-2-oleoyl-sn-glycero-3-phosphocholine (POPC) or 1-palmitoyl-2-oleoyl-sn-glycero-3-(phospho-rac-(1-glycerol)) (POPG), on the modulation of the phase behavior of LS film [173,189,190,192]. On the other side, it has been reported that the inclusion of DOPC in LS models provides a suitable environment for gaining information of the remodeling process, contributing to the squeezing-out of lipids and reservoir formation upon compression, and to the respreading of the LS upon expansion [193].

It should be noted that the biophysical interpretation of the results obtained from studies using LS models composed by single lipids or their mixtures is very difficult, especially because the absence of the surface-active proteins makes it difficult to control the exchange of molecules between the liquid/vapor interface and the adjacent aqueous subphase during the compression-expansion cycles of the interfacial layers, and hence the control over the formation of reservoirs and the re-adsorption/respreading of material at the fluid interface becomes very poor, which are very critical processes for the normal physiological function of the LS related to the surface properties and mechanical stability of the interfacial layer [22,104,106,194]. Therefore, the use of most realistic models, based on natural extracts or laboratory mixtures, having a composition similar to that what is found in mammal LS is required for a biophysical quantification of the potential effects of particles in LS films. This has been solved by using different commercial LS formulations (clinical LS) as models [195]. These types of surfactants are commonly used in surfactant replacement therapy (SRT), and for treating neonatal respiratory distress syndrome (NRDS) [196,197], having gained interest as palliative treatment for reducing the impact of SARS-CoV-2 in respiratory physiology [30,198]. Table 1 summarized some of the currently used clinical lung surfactants and their origins.

**Table 1.** Summary of the most extended clinical lung surfactants, together with their type, origin, and producer.

| Commercial Name | Origin | Producer |
| --- | --- | --- |
| Infasurf | lavage of calf lung fluid | ONY Biotech Inc., Amherst, NY, USA |
| Curosurf | lavage of porcine lung fluid | Chiesi Farmaceutici S.p.A, Parma, Italy |
| Survanta | lavage of bovine lung fluid | AbbVie Inc., North Chicago, IL, USA |
| BLES | lavage of bovine lung fluid | BLES Biochemicals Inc., London, ON, Canada |
| Alveofact | lavage of bovine lung fluid | Lyomark Pharma, Oberhaching, Germany |
| Venticute | synthetic | Byk Gulden Pharmaceuticals, Konstanz, Germany |
| Surfaxin | synthetic | Discovery Laboratory Inc., Warrington, PA, USA |
| Exosurf | synthetic | GlaxoSmithKline, Brentford, UK |

The importance of the use of realistic LS models is easily understood considering the differences on the effect of particles in the behavior of interfacial films of DPPC and commercial LS formulation [199]. In particular, DPPC films containing particles do not easily undergo any remodeling process, emerging as a fast destabilization of the interfacial film when the surface tension reaches a low value. This pushes the monolayer to a premature collapse [133,200], which drives the formation of bilayer folds and other 3D structures. These structures are not easily re-adsorbed into the interfacial film upon expansion. However, the presence of other lipids and the surface-active proteins SP-B and SP-C results in a minimization of the destabilization of the interfacial film. Thus, whereas from the studies of the interaction of particles with DPPC films a clear inhibition of the LS activity may be expected, this can be, at least in part, ruled out from similar studies in which commercial LS formulations are used as model [199,201–203]. Similar conclusions can be extracted from the analysis of the minimum surface tension that can be reached by LS model in presence of particles. Thus, the incorporation of particles into interfacial films of commercial LS formulations leads to a smaller increase of the minimum surface tension reached upon compression than when particles are incorporated in DPPC layers. This can be understood considering that the presence of surface-active proteins favors the remodeling process of the interfacial layer, and in particular the reservoir formation and the re-adsorption/respreading kinetics of the squeezed out material during the expansion of the interfacial area [115,199].

The Tanaka mixture containing only lipids (DPPC, POPG, and palmitic acid in weight ratio 68:22:9) emerges as a very simple and robust model for study involving the interfacial properties of LS [204]. This is possible because the addition of POPG to DPPC monolayers plays a crucial role in the modulation of the interfacial compressibility of the interfacial film by reducing the rigidity of pristine DPPC films, which is essential for the LS packing and its ability to reduce the interfacial tension. Furthermore, the presence of POPG has a very important contribution in the formation and stabilization of reservoirs. On the other side, palmitic acid is essential for the fluidization of films at high values of surface tension, making the molecular rearrangement in the LS films easy. Furthermore, palmitic acid contributes to the interfacial packing, improving the film rigidity at low surface tension values, avoiding a premature collapse.

Despite that many studies dealing with LS models consider the use of single monolayer, the LS complexity can be modelled by using other types of colloidal structures, including bilayers or multilayers [49]. This presents interest for gaining additional insights into the LS behavior and how this is modified by particle incorporation, especially considering the important role of bulk structures, e.g., reservoirs, in the exchange of material between the interface and the adjacent fluid phase during the compression-expansion cycles [205].

From the above discussion, it is clear that there are many possibilities for exploring LS films, especially from a mechanical point of view. However, the used of models with biophysical relevance should consider two main aspects: (i) the temperature used should be close to the physiological one (37 °C) and (ii) the specific hydrodynamic conditions of the experiments which govern the Marangoni flows within the interfacial film, and between the interface and the adjacent fluid phase [65,66,107,109,130,131].

## 5. Evaluating the Interaction of Lung Surfactant and Particles: Methodological Approaches

The investigation of the interaction of particles with LS films requires, in most cases, to put into contact a film preformed at the water/vapor interface and the particles, which can be commonly done by aerosolization or deposition. In particular cases, the studies are performed by premixing the lung surfactant and the particles before the preparation of the interfacial films [32]. The use of these different approaches can decisively impact the biophysical interpretation of the experimental results [206]. Furthermore, the existence of different approaches for incorporating particles into LS layers makes it difficult to compare the different impacts found for similar particles in different studies. Thus, the incorporation of hydrophilic particles into LS layers is commonly explored by spreading the LS model film onto a subphase containing a dispersion of particles [119,188], or by the injection of the particles into the subphase once the LS layer is preformed at the water/vapor interface [116,207]. These methodologies are far from the real situation that is found during particle inhalation. However, they emerge as very useful alternatives for evaluating the direct interaction of particles with LS layers [67].

The situation becomes even more complex when the interaction of hydrophobic particles with LS layers is considered. In this case, there are multiple possibilities for incorporating particles into LS films. The first approach used for the incorporation of hydrophobic into LS films relies on the co-spreading at the pristine water/vapor interface of a mixed dispersion containing the LS model and the particles [42,208]. Thus, it is possible to tune, at will, the exposure dose in relation to the LS concentration, ensuring that the exposure of the surfactant to the particles is high enough. However, the obtained films are probably very far from what are expected once inhaled particles interact with LS layer, the formation of particles decorated with a surfactant corona emerging even before the formation of the LS layers at the water/vapor interface. A more realistic approach considers the deposition of particles at the interface from the airside once the surface layer is formed at the water/vapor interface [118,133], which leads to a more realistic picture of what happens in the LS film upon pollutant inhalation. However, organic solvent is used for the spreading of either the LS layer, the particles, or both [206,209], which may modify the LS-particle interactions, or the lateral packing and homogeneity of the film, and the difficulties associated with the mimicking of the specific mass transport conditions occurring during in vivo breathing are two main drawbacks of the studies using the approach discussed above [67,70,130,131].

Aerosolization of the particles onto a preformed LS film by using a dry power insufflator can provide a more realistic representation of the inhalation process, even though the use of organic solvents continues being necessary for spreading the LS layer [199]. Figure 6 presents sketches of three commonly used approaches for the study of particles with LS models, as well as examples of the differences in the surface pressure-area per molecule isotherms obtained using different preparation procedures.

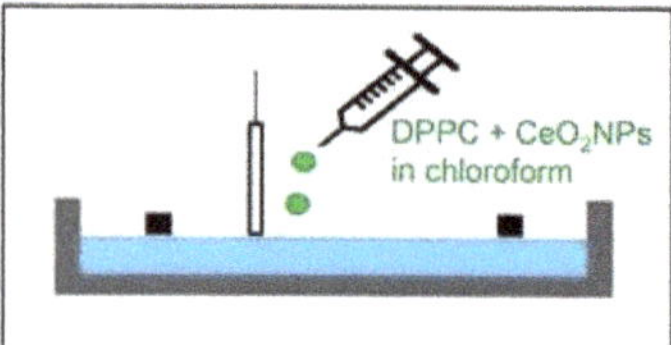


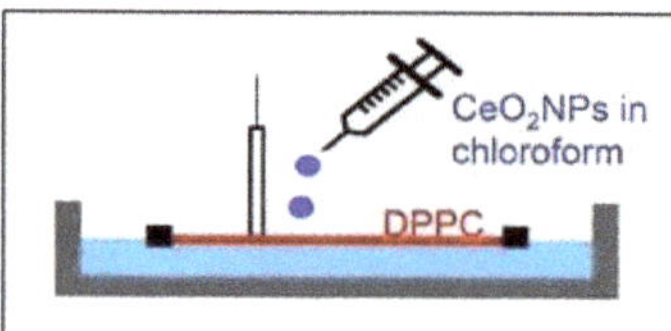


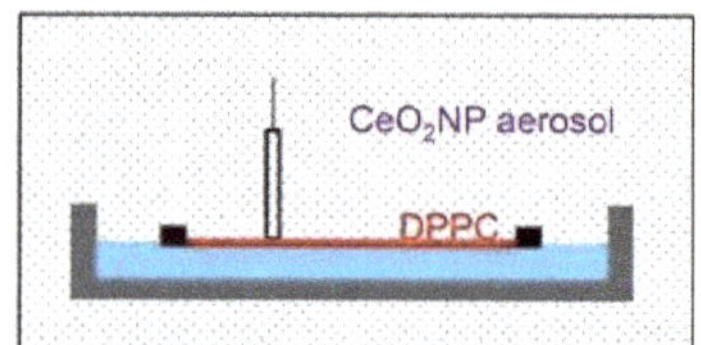


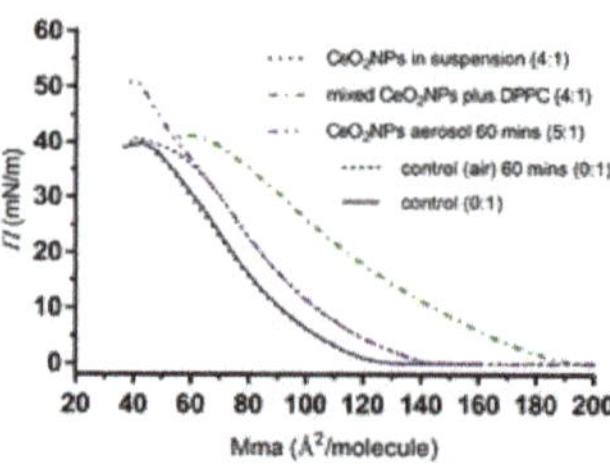


**Figure 6.** Sketches representing the three different approaches used for incorporating particles into LS models layers based on DPPC, and the differences in the surface pressure-area per molecule isotherm obtained depending on the procedure used for preparing the mixed layers. Adapted from Miguel-Diez et al. [206], with permission from Elsevier, Copyright 2019.

The above discussion considers the study of the interaction of particles with spread monolayers of LS models obtained by mixing different raw materials in an organic solvent. However, a very different situation emerges when commercial LS formulations are used as models. These are commonly supplied as saline aqueous dispersions of LS components as vesicles and micelles of lipids associated with LS proteins. Therefore, the use of these types of formulations require the preparation of interfacial films by the direct adsorption of the LS components from the bulk dispersion to the aqueous solution/vapor interface. This requires depositing the particles on the interfacial film by injection into the aqueous subphase after the formation of the interfacial layer (hydrophilic particles) or by spreading through the airside onto the formed layer (hydrophobic particles). On the other side, hydrophilic particles can also be premixed with the LS dispersion before forming the interfacial film. The latter leads to the co-adsorption of LS molecules and particles at the liquid/vapor interface [210]. It should be stressed that commercial LS formulation can also be extracted in an organic solvent, and then applied by spreading at a pristine/water vapor interface [211].

It is true that most of the studies using LS models for exploring the impact of particles in the respiratory mechanics present a situation that appears very far from what happens when the pollutants are inhaled. However, these studies are very useful for gaining an important understanding of the most fundamental physico-chemical bases of the interaction of inhaled pollutants and LS films, and their potential harmful effects [59,65,66].

## 6. How Do the Particle Physico-Chemical Properties Affect Their Interaction with LS Layers?

Once particles are deposited on the LS layer, their specific physico-chemical properties define the particle-LS interactions, and their consequences. Furthermore, it is very important to control the particle dose that interacts with the LS film because in most cases the harmful effects associated with inhaled particles present a strong dose dependence [70,193,212–214]. It should be noted that in general, particles' affects to the respiratory function are at two different levels: (i) LS biological function and (ii) LS metabolism [7]. Therefore, a deep understanding of the impact of particles on the physico-chemical properties of LS layers requires examination of the specific characteristics of the probed particles [65,128]. Figure 7 summarized the potential impact of particles with different physico-chemical properties in the interfacial properties of layers formed by LS models.

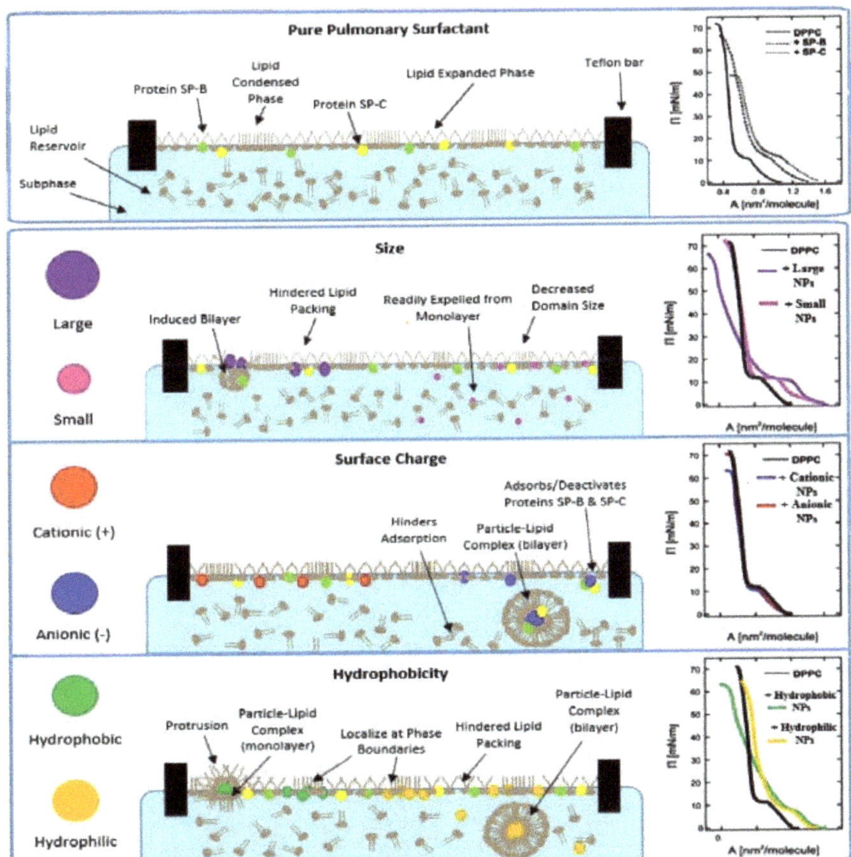


**Figure 7.** Schematic representation of the effect of different types of particles in the interfacial organization of LS layers, and the surface pressure-area per molecule isotherms. Reprinted from Arick et al. [65], with permission from Elsevier, Copyright 2015.

### 6.1. Interaction of Particles and Lung Surfactant Films: A Matter of Size

Particle size is a very important parameter controlling the interaction of particles with cells, affecting particle uptake. Furthermore, particle size affects their cytotoxicity and the induction of inflammatory response [215,216]. Therefore, it may be expected that size can also influence the interaction between inhaled pollutants and the LS film.

Dwidedi et al. [57] studied the interaction of hydrophobic poly(organosiloxane) particles with two different sizes (12 and 136 nm) with two LS models, and found that the smallest particles do not induce any significant modification of the phase behavior of DPPC layers, even though the morphology of the $L_c$ domains undergoes a slight change. However, when the incorporation of the biggest particles was considered, the lifting-off of the surface-pressure isotherm was shifted to lower values of the compression degree, and the DPPC film undergoes a strong modification of its phase behavior. The situation changes when 1,2-dipalmitoyl-sn-glycero-3-phosphoglycerol (DPPG) and the SP-C were used together with DPPC as model LS. In this case, it was again found that the smaller the particle size the smaller the modification of the phase behavior. However, the incorporation of big particles (diameter 136 nm) induces a strong modification of the interfacial phase behavior of the LS model. These modifications were found to be stronger at the highest surface pressures (above 35–40 mN/m), which is critical for the normal physiological function of the LS film, and in particular for the remodeling of the composition of the interfacial, inhibiting the incorporation of vesicles from the adjacent fluid phase to the interface. Furthermore, particles with a diameter of 136 nm lead to a fluidization of the interfacial film, i.e., reduce the elasticity of the LS film, independently of the model used. This may be explained considering that particle incorporation leads to a disruption of the lateral packing of the LS film, modifying the cohesion between the molecules in the monolayer. This results from the complexity of the interaction balance governing the interfacial packing, which includes different contributions, e.g., steric hindrance, excluded area effects, and other type of interactions [200]. This leads to a strong influence of the particle size and their concentration in the performance of LS film [118,119,133,179].

The effect of the particle size in the biophysical properties of LS films is strongly dependent on the specific chemistry of the probed particles. This is clear considering the existence of some studies in which the effect of particles is rather independent of their size, and emerges only dependent on the particle nature [173,178,217], whereas in other cases, size emerges as the main aspect governing the impact of particles with LS films.

Orsi et al. [178] reported that the modification of the phase behavior, lateral packing of the molecules at the interface and interfacial dynamics of DPPC films upon the incorporation of particles with size in the range 9–60 nm emerged rather independent of the specific particle size. Furthermore, the incorporation of particles into the DPPC films hindered the formation of condensed phase, leading to an interfacial organization that was reminiscent of what is expected for Pickering emulsions (2D Pickering emulsion-like structure) in which particles decorated with DPPC molecules are distributed around ordered domains of DPPC. This organization results in a reduction of the line tension of the domains in relation to that which is found for pure DPPC, and consequently their growth is hindered, i.e., the formation of domains with smaller size is found [179] in agreement with the molecular dynamic simulations by Curtis et al. [217]. Figure 8 presents a schematic representation of the structure emerging from the incorporation of hydrophilic particles into DPPC films.

Contrary to what was reported for the incorporation of hydrophilic silicon dioxide particles, Ku et al. [218] found that the incorporation of gelatin particles into DPPC monolayers leads to a modification of the interfacial behavior of the lipid in such a way that the resultant emerges strongly dependent on the specific dimensions of the incorporated particles. Furthermore, it was found that the largest particles (in this case with an average diameter around 236 nm) present the strongest interaction with the LS model layer altering both the surface pressure-area per molecule and the surface potential-area per molecule isotherms, and the reduction of particle dimensions weakens such interaction. Particle incorporation into DPPC pushes the phase behavior to more expanded states, making the rearrangement of lipid molecules at the fluid interface difficult.

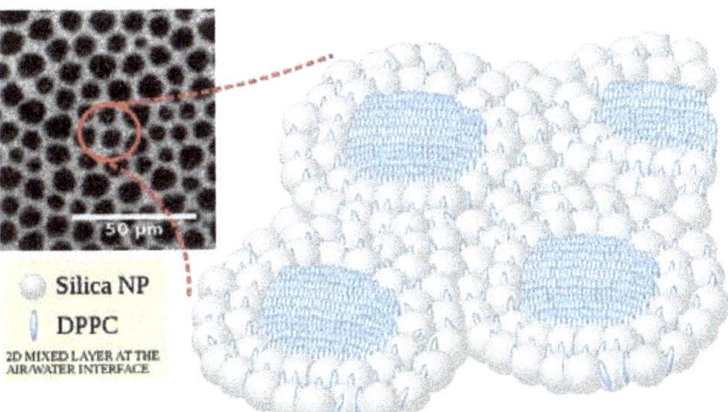



**Figure 8.** Sketch showing the interfacial organization of interfacial films of DPPC upon the incorporation of hydrophilic silicon dioxide and experimental image of such organization obtained by epifluorescence microscopy. Reprinted from Orsi et al. [178], with permission under Open access CC BY 4.0 license, https://creativecommons.org/licenses/by/4.0/ (2015) (accessed on 20 December 2021).

The above discussion evidences clearly that the influence of the size on the impact of particles in the physico-chemical properties of LS model is far from clear, and the current framework is controversial. An additional contribution to this controversy emerges from the study by Kodama et al. [219] in which the interaction of particles with different dimensions with a commercial LS formulation (Survanta) was explored. They found that only very small particles (average diameter about 20 nm) drive a significant modification of the interfacial phase behavior of the LS model. The multiple scenarios found for the interaction of particles and LS models can be understood considering that the effect of particles is dependent on a complex interplay of different factors, including the types of particles and LS model as well as the specific interfacial behavior of the model used.

It should be noted that together with particle size, there are physical parameters related to the particles, e.g., total surface area and specific surface area, chemical nature, and surface charge, that influence their impact on the behavior of LS layers. This is better understood considering that the alteration of LS behavior as a result of the particle incorporation emerges from two different directions: (i) particle aggregation in the LS layers and (ii) specific particle-LS interactions [219]. This perspective agrees with the results found when the interaction of particles in vivo is analyzed. The association of particles with LS films is strongly correlated to the specific nature of the interactions occurring within the system, and the ability of particles to be coated for an LS corona [220]. On the other side, the aggregation plays a very important role in the deposition of particles along the respiratory tract, as well as on the interaction with LS and clearance mechanism. It should be noted that together with the size, there are many other physico-chemical parameters of the particles modulating their interactions with LS [7,66].

### 6.2. Role of Particle Surface Charge and Wettability on Their Interactions with Lung Surfactant Films

The ability of particles to interact with biological structures is strongly dependent on the particle surface charge and their hydrophilic-lipophilic balance, i.e., wettability. This is understood considering that both parameters present a very critical role in how particles are incorporated within biological interfaces and interact with biomolecules, affecting metabolic pathways of the LS and their biophysical function [65,143,221]. In particular, it has been found that charged polystyrene particles lead to a stronger inflammatory response in lung than neutral ones [222], and hence it may be expected that electrostatic interactions present a prominent role in the modification of the respiratory physiology [200,212,223]. This can be ascribed to the higher strength of electrostatic interactions in relation to other forces [224].

The incorporation of negatively charged hydrophilic silicon dioxide particles into DPPC monolayers can occur through their electrostatic interaction with the ammonium terminal group of the lipid, allowing a modification of the dipolar moment of the DPPC molecules at the interface which in turn modifies its ability for reorienting at the water/vapor interface and the molecular packing. This hinders the formation of ordered phases, and reduces the rigidity of the film. On the other side, hydrophilic silicon particles undergo an effective clearance from the interface at the highest values of the surface pressure, suggesting the formation of a lipid corona on the particle surface [136]. Similar conclusions were found by Farnoud and Fiegel [55] when analyzing the interaction of negatively charged polystyrene carboxylate with DPPC films. Furthermore, they also reported that the incorporation of charged particles into lipid monolayers affects the size of the compression-expansion hysteresis loop, which plays a critical role in the exchange of material between the interface and the adjacent subphase, affecting to the compositional remodeling of LS layers, and in turn, to its normal physiological function.

The inhibitory character of the LS function associated with its interaction with charge particles was also evidenced by molecular dynamic simulations [225]. It is worth mentioning that the specific nature of the particle charge does not present a significant effect for the interaction of monolayers formed by zwitterionic lipids such as DPPC. However, it may be expected that the effect of the cationic and anionic particles on true LS becomes more complex due to the presence of lipids and proteins with different positively and negatively charged moieties.

Negatively charged polylactide particles induce a strong inhibitory effect of the behavior of bovine LS extract (Curosurf) as a result of the association between the particles and the SP-B protein. Furthermore, the interaction of particles with SP-C protein also has a very important influence in the inhibitory effect of the negatively charged polylactide particles. However, the interaction of positively charged particles with the LS model results in a reduced inhibitory effect which confirms the importance of the charge nature on their inhibitory character [19,103,226]. This agrees with the different scenario found upon the interaction of positively and negatively charged aluminum oxide, silicon dioxide, and latex nanoparticles, with LS layers. Thus, negatively charged particles present a negligible impact on the LS function, whereas positive charged ones lead to the formation of aggregates with LS vesicles. This interaction, which occurs with an LS having an average negative charge, is enhanced as the charge density of the particles is increased [208]. The different impact of charged particles on the interfacial properties of lipid mixtures and LS formulations containing proteins was independently demonstrated by Behyan et al. [189] in their study about the interaction of positively and negatively charged silicon dioxide particles with different LS models, a DPPC/POPG mixture, and an extract from calf lung (Infasurf). They found that whereas the behavior of Infasurf was only modified by the cationic particles, that of the natural extract was influenced by both, positively and negatively charged particles.

The particle wettability also emerges as a very important parameter influencing the particle interaction with LS films [200]. Thus, the incorporation of hydrophobic particles into DPPC layers modifies lateral packing of molecules at the interface and the cohesion interactions between the lipids at the interface, which results in the inhibition of the ability of DPPC for reducing the surface tension of the interface, and the reduction of the film rigidity. Furthermore, hydrophobic particles do not undergo an effective clearance upon the compression of the interface [118,133]. On the contrary, hydrophilic particles undergo an effective clearance from the interface upon compression [200]. The above picture is in agreement with the work by Zhang et al. [214] on the interactions of hydrophobic gold particles and DPPC monolayers, in which the particle hydrophobicity was found to be critical for ensuring the retention of the particles within the alveolar lining film for long periods of time. Furthermore, the interactions of particles with the lipids result in a decrease of the layer rigidity in accordance with the finding by Guzmán et al. [118,133]. Similar results in relation to the retention of hydrophobic and hydrophilic particles were found when the interaction of particles with different hydrophilicity and natural LS models was analyzed. The enhanced retention of the hydrophobic particles is associated with the emergence of strong van der Waals interactions between particles and the hydrophobic tails of the lipid molecules [57,227], which makes the modification of the monolayer organization easy as result of the particle trapping [228].

In general, it was found that the incorporation of hydrophobic particles, e.g., hydrophobic montmorillonite, silicon dioxide, carbon black, or graphene oxide, into DPPC films shifts the surface pressure-area per molecule isotherm to more expanded states as a result of an excluded area effect induced by particles [118,129,133,229]. On the other side, hydrophilic particles, including halloysite or bentonice, induce the opposite effect, pushing DPPC monolayers towards more compressed states and reducing the average distance between lipid molecules [229]. The above situation is very different to what emerges when commercial LS formulations are used as models. In these cases, the isotherm appears shifted to more compressed states with independence of the wetting properties of the particles. This leads to a strong inhibition of the LS performance, which is characterized by a worsening of the ability of LS to reduce the surface tension, and compositional remodeling [19,227]. However, the influence of hydrophilic and hydrophobic particles follows very different pathways. Thus, hydrophilic particles undergo a direct penetration into LS layers, whereas hydrophobic particles require being wrapped by LS components to be incorporated in LS films [228,230] in agreement with the finding by Hu et al. [19]. They combined experiments and simulations, and found that hydrophilic particles undergo a fast translocation through the LS films during compression, whereas hydrophobic particles lead to the formation of structural protrusion in the LS film, and hence the inhibition induced by hydrophilic particles emerge faster than those induced for the hydrophobic one.

*6.3. Impact of Particle Shape on the Interactions with LS Layers*

Particle shape, and in particular shape anisotropy, emerges as a very important parameter controlling the physico-chemical properties of colloidal particles, and their ability for self-organizing at the fluid interface [231], and hence it may be expected that it plays a very important role in the modification of the LS performance as a result of the particle incorporation [232]. This is clear by comparing the modifications of the phase behavior of DPPC layers upon the incorporation of three different types of particles, two surface inactive anisotropic clays (plate-like bentonite and halloysite nanotubes) and spherical silicon dioxide particles. The results found that the increase of the particle anisotropy facilitates the particle clearance upon compression [229]. However, there are no systematic experimental studies evaluating the impact of the particle anisotropy on LS layers. However, some simulations (molecular dynamics) evidenced that the length-to-diameter aspect ratio of the particles plays a very important role on the control of their penetration and disturbance of LS function [233].

The influence of the particle anisotropy on the modification of LS layers was further explored by Kondej and Sosnowski [161]. They studied how carbon particles with different geometry (nanotubes and nanohorns) modify the performance of LS layers and found that the increase of the surface area of the particles induces a stronger frustration in the LS behavior. It should be stressed that most of the effect on LS behavior associated with particle anisotropy can be ascribed to the influence of the capillary forces [234].

*6.4. Does the Particle Chemistry Matter in Their Interactions with Lung Surfactant Films?*

The impact of the particle chemical nature on LS performance is difficult to systematize [59,65], and only some general aspects will be included about the impact of particle chemistry in LS layers. Silicon dioxide particles alter the LS performance in such a way that it is strongly dependent on the number of SiOH groups on the particle surface. Thus, the increase of the surface density of silanol groups leads to the emergence of silicosis and other lung diseases associated with the inhalation of silicon dioxide particles [235]. The importance of the chemistry of the particles in the modification of LS properties is also found by comparing the effect of carbon black and fumed silicon dioxide in the interfacial properties of DPPC layers [128]. Despite the physical characteristic of the particles possibly being very similar, the specific chemical nature of the particles governs the balance of interactions occurring within the interface, and that consequently modifies the degree of interfacial disruption and the aggregation of the particles at the interface.

Figure 9 presents a summary of the impact of different physico-chemical properties of the particles in the LS function.

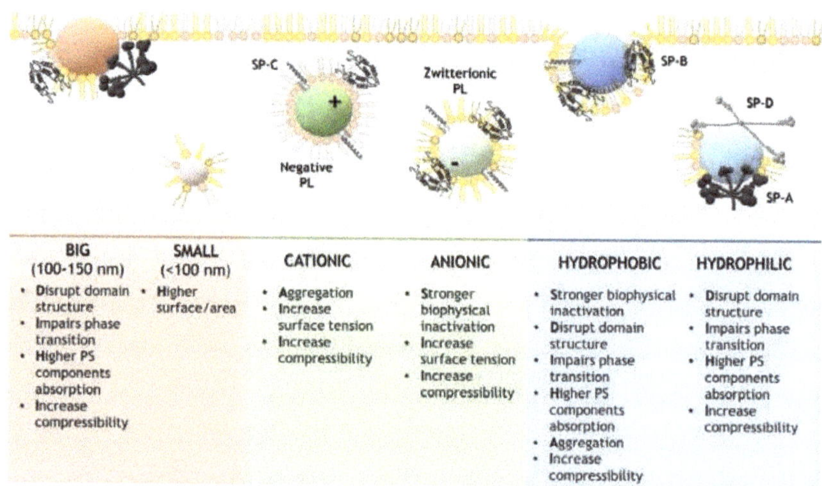


| BIG (100-150 nm) | SMALL (<100 nm) | CATIONIC | ANIONIC | HYDROPHOBIC | HYDROPHILIC |
|---|---|---|---|---|---|
| • Disrupt domain structure<br>• Impairs phase transition<br>• Higher PS components absorption<br>• Increase compressibility | • Higher surface/area | • Aggregation<br>• Increase surface tension<br>• Increase compressibility | • Stronger biophysical inactivation<br>• Increase surface tension<br>• Increase compressibility | • Stronger biophysical inactivation<br>• Disrupt domain structure<br>• Impairs phase transition<br>• Higher PS components absorption<br>• Aggregation<br>• Increase compressibility | • Disrupt domain structure<br>• Impairs phase transition<br>• Higher PS components absorption<br>• Increase compressibility |

**Figure 9.** Impact of different physico-chemical properties of the particles in the LS performance. General perspective of the main effects and interactions of particles upon contact with LS. Reprinted from García-Mouton et al. [7], with permission from Elsevier, Copyright 2019.

## 7. Some Experimental Results of the Interaction of Particles with Interfacial Lung Surfactant Models

The understanding of the potential harmful effects associated with the incorporation of particles in LS layers requires a deeper analysis of the impact of particles in the tensiometric properties and dynamic response of the interfacial films, as well as their effect on the lateral organization and structure. Furthermore, it is also very important to evaluate the distribution of particles between phases with different orders because it can impact on the translocation of particles across the pulmonary fluid and the clearance processes. This section will review the effect of different types of particles on the behavior of LS models, paying attention mainly to realistic systems including lipids and the surface-active proteins.

The incorporation of particles, mainly those of hydrophobic nature, into LS films is commonly associated with a partial inhibition of the LS function. This was evidenced in the study by Valle et al. [227] where the interaction of polymer particles with LS films was evaluated. These particles lead to a contraction of the area available for LS molecules, i.e., shift the isotherm to more compressed states, which significantly alter the ability for surface tension reduction (see Figure 10a). The modification of the tensiometric properties as a result of the incorporation of particles is also reflected in changes in the lateral organization of the molecules at the interface as evidenced by AFM micrographs of Langmuir-Blodgett films at different surface pressures (see Figure 10b). These images evidence that particle incorporation makes the lateral packing of the molecules difficult, hindering the nucleation and growth of domains containing ordered phases, which disturbs the monolayer-to-multilayer transition. On the other side, the analysis of the AFM images also evidences a higher tendency of particles to aggregate into LS layers as their hydrophobicity increases, which may be a signature of the important role of hydrophobicity in the control of the particle retention and translocation across the pulmonary fluid. Last but not least, the incorporation of particles increases the area of the hysteresis loop of the compression-expansion cycles of the LS layers, which is expected to present a critical impact on the normal function of LS (see Figure 10c). Recently, Beck-Broichsitter et al. [236] evidenced that the inhibition of the LS activity upon particle deposition is strongly correlated to their ability to sequester the surface-active proteins. Therefore, the shielding of the particles to avoid the formation of an LS corona reduces their harmful effects and allows their exploitation as carriers for inhalable drugs.

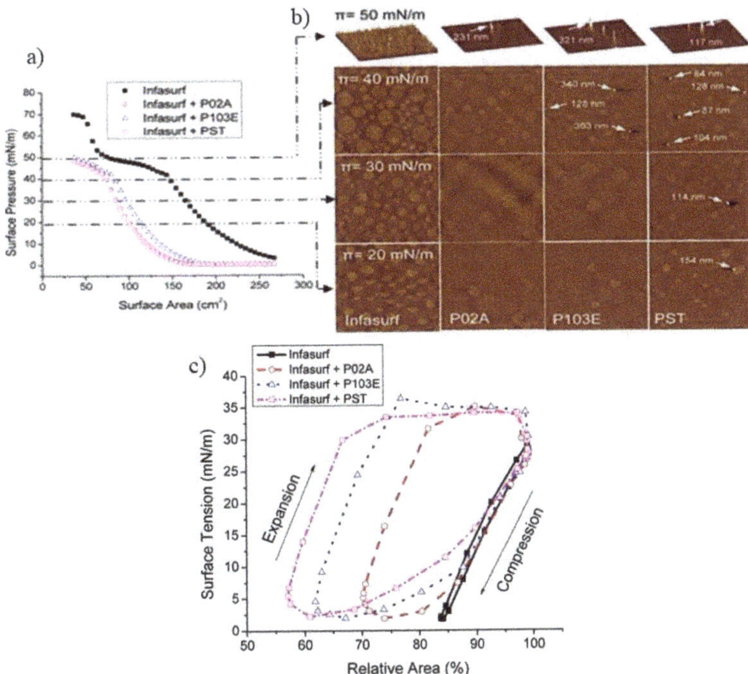


**Figure 10.** (**a**) Modification of the compression isotherms of an LS model for the compression isotherm of an LS model. (**b**) AFM micrograph showing the lateral structure of Langmuir–Blodgett films at three different surface pressures of the LS model and LS models upon the incorporation of three different types of hydrophobic polymer particles. All 2D images are shown at a resolution of 50 μm × 50 μm and have a z-range of 5 nm. The high-pressure images are shown in 3D to capture the topographic contrast between the particles and multilayer structures. The presence of NPs, denoted by white arrows, increases with increasing hydrophobicity. (**c**) Compression–expansion cycles for the pure LS model and LS model upon the incorporation of three different types of hydrophobic polymer particles. Dynamic cycling was conducted under physiological temperature (37 °C) and cycling speed (3 s/cycle). Reprinted from Valle et al. [227], with permission from American Chemical Society, Copyright 2014.

The deposition of hydrophilic hydroxyapatite particles leads to similar effects than those discussed above for hydrophobic polymer particles [69]. In addition, there is a strong time dependent behavior in the inhibitory character of the LS function as a result of the inhalation of hydroxyapatite particles, reaching the maximum inhibition after 7 h of exposure. It should be noted that even though hydroxyapatite particles are effectively cleared from LS films, they modify the lateral packing of the LS molecules at the fluid interface by hindering the formation of condensed phase, altering the remodeling process of the film and the reservoir formation. LS films at physiologically relevant conditions present a morphology characterized by the presence of a homogeneously distributed fluid multilayer in which insertions of DPPC domains appear. This structure changes upon the deposition of particles on the LS films, with the appearance of crystalline folds along the direction of lateral compression, which leads to the inhibition of the LS performance.

It should be noted that the interaction of particles with LS models containing proteins emerges very different in most of the cases to what happens for model mixtures in absence of surface-active proteins. The role of these proteins in the formation of the LS corona on the surface of the particles is essential for the inhibitory role of particles [236]. The important role of the proteins is clear considering that in most cases the interaction of hydrophilic particles, e.g., silicon dioxide, $Fe_3O_4$, or titanium dioxide, with models based only on lipids, takes the surface pressure-area per molecule to more expanded states. This shifting is enhanced with the increase of the particle concentration, which may be explained considering that the interaction occurs through electrostatic particles, with the particles being retained as granular domains within the monolayer [117,119,188,223].

The importance of the model composition is also supported by the results obtained by Tatur and Badia [115] using a multi-technique approach (surface tension measurements, ellipsometry, Brewster angle microscopy, and AFM of Langmuir-Schaeffer films). They compared the effect of alkylated gold particles on the interfacial behavior of DPPC and discovered that the particles do not alter the tensiometric properties of Survanta films, but they lead to a strong modification to those of the DPPC layers, changing both the phase behavior and the lateral packing of the molecules. Thus, gold nanoparticles hinder the nucleation and growth of condensed phase domains, leading to the change of the shape of the domains from a multilobe geometry for pristine DPPC monolayers to a circular one in the presence of particles. This is similar to what was reported for the incorporation of silico dioxide particles into DPPC films [179]. The different effect of gold particles on DPPC and Survanta films can be understood by considering that their accumulation occurs within the disorder region, and this leads to the incorporation of particles within the phase containing the surface-active proteins. This minimizes their effect on the condensation of ordered phases when Survanta films are considered. The above results suggest that alkylated gold particles present a rather limited inhibition of the LS function under the experimental conditions. However, a true biophysical interpretation of the results should consider its effect under conditions mimicking the physiological one. This becomes very important because Hossain et al. [237] reported the inhibition of LS function as a result of the interaction with gold nanoparticles. These make the reduction of the surface tension of the water/vapor interface difficult, reducing the lateral packing of LS molecules at the interface and dragging LS molecules into the adjacent subphase

Tobacco smoke results in similar inhibition of the LS performance as particles, as was evidenced for Survanta and Curosurf films under physiological relevant conditions [238]. Thus, the response of LS layers after exposure to tobacco smoke to consecutive compression-expansion cycle evidences a clear reduction of the efficiency of the compositional remodeling process, which agrees with the reduced formation of reservoir evidenced from the analysis of the Langmuir–Blodgett deposits. Furthermore, the tobacco smoke reduces the ability of the LS to reduce the surface tension under compression conditions, increasing the respiratory work. The comparison of the effect of tobacco smoke with that associated with the vapor ejected by electronic cigarettes evidenced unexpected results [172]. On one side, both alter the lateral packing of LS film, whereas on the other side only tobacco smoke modifies the tensiometric properties of LS films, resulting in a premature collapse of the layer. These differences can be understood only by considering the different chemical nature of the vapor obtained from each type of cigarette. The influence of tobacco smoke can be correlated to the effect of other types of hydrophobic carbonaceous particles (nanotubes or nanohorns), which alter the viscoelastic response of LS layers (Survanta) under dynamic conditions, enhancing the monolayer rigidity. However, the effect of hydrophilic carbonaceous particles is rather limited which may be explained considering their effective clearance [161].

The inhibition of the LS function is not only associated with particle inhalation, with the exposure to other chemicals (trimethoxyoctylsilane, methyl 3-oxo-2-pentylcyclopentaneacetate, and diisopentyl ether) also being associated with the emergence of acute inhalation toxicity as was evidenced by combining interfacial science techniques and biophysical methodologies. It should be noted that the understanding of the potential inhalation toxicity associated with these types of chemicals is very important because LS is the main target of impregnation spray products [239]. The interaction of LS layers with the above products leads to a partial inactivation of the LS function, resulting in a fluidization of the condensed phases due to the insertion of the chemical between LS molecules which weakens the cohesion in the layer and reduces the stability of the LS function. There are many other chemicals, including benzalkonium chloride and cetylpyridinium chloride, that present a strong inhibitory characteristic of the LS function. These molecules alter the lateral organization of the molecules at the water/vapor interface, affecting the ability of LS to reduce the surface tension [240].

## 8. Beyond Experiments: Exploiting Computational Tools for Elucidating the Interaction of Lung Surfactant with Inhaled Particles

The use of computational tools for the evaluation of the interaction of particles with LS is rare due to the complexity of these types of systems, which require a strong computational effort. However, some studies have evidenced that molecular dynamic simulations can help the understanding of the dynamics of the composition remodeling process, and how this is modified by the incorporation of particles [86].

Hu et al. [19] demonstrated, by using coarse-grained molecular dynamics, the differences in the translocation pathway of hydrophilic and hydrophobic particles, which agrees with experimental results. Thus, hydrophilic particles undergo an easy translocation through the LS films upon compression, whereas hydrophobic particles were retained within the LS layer. This different behavior was found to be the result of the different abilities of particles for the binding of the surface-active material, with the binding of SP-B being essential for the clearance of hydrophilic particles.

The interaction of hydrophobic carbonaceous particles with LS films has also been explored by using molecular dynamics, which has shown the direct interaction of such particles with the surfactant proteins, altering the interfacial organization of the LS film. Furthermore, the simulations evidence that asymmetric particles are placed at the interface under compression in a direction perpendicular to the monolayer [241,242], which agrees with the experimental picture found for the interaction of graphene nanosheets and LS model films [129,243]. On the other side, molecular dynamic simulations have provided additional evidences on the dose-dependent inhibition of the LS performance upon the deposition of carbonaceous particles [244], which agrees with the experimental findings [63,118,129].

Molecular dynamic simulations were also used for confirming that particles alter the exchange of material between the interface and the adjacent fluid subphase, which plays a critical role in the LS functioning [164]. This agrees with the molecular dynamic simulations by Li et al. [245] which evidenced that the transport of particles to the LS film can disrupt the ultrastructure and fluidity of the interface, resulting in a premature alveolar collapse.

## 9. What Can We Learn Using Model Lung Surfactant Films?

The effect of particle incorporation into LS films is the emergence of specific dependency on the specific nature of the considered particles, and the use LS model. However, most of the studies present some general features that are relevant for extracting biophysical relevant information. In particular, the incorporation of particles into LS films worsens the mechanical properties of the surfactant-laden interface by hindering the reorientation and lateral packing of the surfactant molecules at the water/vapor interface. This alters the compositional remodeling of the interfacial layer. Furthermore, the use of model systems evidences that the interaction of LS with particles results many times in the formation of a surfactant corona on the particles surface, which leads to a compositional modification of the LS, and can facilitate the translocation of the particles beyond the alveoli, resulting in their transport along the bloodstream to different organs and tissues. Thus, the interaction of particles with LS can alter and inhibit the LS functionality through different mechanisms.

Moreover, the formation of the surfactant corona can also modify the dynamic of material exchange between the interfacial layer and the adjacent fluid subphase, which alters the particle clearance during breathing. However, it is necessary to obtain a further understanding on how specific components of LS may influence the biological alterations induced by particles, and in particular the influence of the adsorption of surfactant proteins on the particle fate, and the physiological function of LS.

The use of LS models allows inferring that the inhalation, and subsequent deposition of particles on LS leads to an important alteration of the normal physiological respiratory function. However, it is necessary to deepen knowledge of the specific effects induced by particles with different properties (charge, wettability, chemical nature, morphology, or size) for obtaining a suitable evaluation of the potential harmful effects associated with inhaled pollutants. This is a very important for unraveling the possible elimination pathways of the particles after their deposition on the LS film. Furthermore, the understanding of the specific effects associated with particle nature is very important in the modulation of the in vivo interaction of LS and pollutants.

In summary, the use of fluid films as LS models presents an important role in the safety evaluation of particulate pollutants. However, there are many open questions that require refining of some specific physico-chemical and mechanical aspects of the currently available models.

## 10. Conclusions

This review analyzes some of the most relevant colloidal and interfacial approaches exploited for the in vitro evaluation of the potential toxicity of inhaled pollutants. It is true that the results obtained by using these types of tools are often times difficult to extrapolate to a true biophysical situation, they are useful for a fundamental understanding of some of the physico-chemical aspects associated with the harmful effects of particulate matter in the LS function. In particular, the use of model based on monolayers at fluid interfaces can help to understand how the deposition of particle leads to either inhibition, inactivation, or both, of the LS function due to the alteration of the packing/organization of the molecules at the interface. This hinders the formation of ordered phases, and inhibits the compositional remodeling of LS layers at low values of the surface tensions, which reduces the LS ability for reaching surface tension values close to zero, increasing the mechanical work associated with breathing. However, the use of colloidal and interfacial models makes it difficult to mimic the entire respiratory cycle and its modification by particles, thus they provide very limited access to aerodynamic and hydrodynamic aspects of the problems which many times are very important aspects for understanding the true effect of inhaled materials in respiratory physiology. Despite the clear limitations associated with the use of model systems, it is clear that these types of systems provide a general perspective of the potential risks and hazards associated with particle inhalation. However, a complete picture of these aspects can only be obtained by combining physico-chemical models with complementary biophysical and medical studies.

**Funding:** This work was funded by MICINN (Spain) under grant PID2019-106557GB-C21 and by the E.U. in the framework of the European Innovative Training Network-Marie Sklodowska-Curie Action NanoPaInt (grant agreement 955612).

**Institutional Review Board Statement:** Not applicable.

**Informed Consent Statement:** Not applicable.

**Data Availability Statement:** Not applicable.

**Conflicts of Interest:** The author declares no conflict of interest. The funders had no role in the design of the study; in the collection, analyses, or interpretation of data; in the writing of the manuscript, or in the decision to publish the results.

# References


1. Kishor, R.; Purchase, D.; Saratale, G.D.; Saratale, R.G.; Ferreira, L.F.R.; Bilal, M.; Chandra, R.; Bharagava, R.N. Ecotoxicological and health concerns of persistent coloring pollutants of textile industry wastewater and treatment approaches for environmental safety. *J. Environ. Chem. Eng.* **2021**, *9*, 105012. [CrossRef]
2. WHO. How Air Pollution Is Destroying Our Health. Available online: https://www.who.int/airpollution/news-and-events/how-air-pollution-is-destroying-our-health (accessed on 16 August 2021).
3. Miller, K.A.; Siscovick, D.S.; Sheppard, L.; Shepherd, K.; Sullivan, J.H.; Anderson, G.L.; Kaufman, J.D. Long-Term Exposure to Air Pollution and Incidence of Cardiovascular Events in Women. *N. Engl. J. Med.* **2007**, *356*, 447–458. [CrossRef] [PubMed]
4. Brook, R.D.; Franklin, B.; Cascio, W.; Hong, Y.; Howard, G.; Lipsett, M.; Luepker, R.; Mittleman, M.; Samet, J.; Smith, S.C.; et al. Air Pollution and Cardiovascular Disease. *Circulation* **2004**, *109*, 2655–2671. [CrossRef]
5. Pavese, G.; Alados-Arboledas, L.; Cao, J.; Satheesh, S.K. Carbonaceous Particles in the Atmosphere: Experimental and Modelling Issues. *Adv. Meteorol.* **2014**, *2014*, 529850. [CrossRef]
6. Aili, A.; Xu, H.; Kasim, T.; Abulikemu, A. Origin and Transport Pathway of Dust Storm and Its Contribution to Particulate Air Pollution in Northeast Edge of Taklimakan Desert, China. *Atmosphere* **2021**, *12*, 113. [CrossRef]
7. Garcia-Mouton, C.; Hidalgo, A.; Cruz, A.; Pérez-Gil, J. The Lord of the Lungs: The essential role of pulmonary surfactant upon inhalation of nanoparticles. *Eur. J. Pharm. Biopharm.* **2019**, *144*, 230–243. [CrossRef]
8. Oberdörster, G.; Maynard, A.; Donaldson, K.; Castranova, V.; Fitzpatrick, J.; Ausman, K.; Carter, J.; Karn, B.; Kreyling, W.; Lai, D.; et al. Principles for characterizing the potential human health effects from exposure to nanomaterials: Elements of a screening strategy. *Part. Fibre Toxicol.* **2005**, *2*, 8. [CrossRef]
9. Chio, C.-P.; Liao, C.-M. Assessment of atmospheric ultrafine carbon particle-induced human health risk based on surface area dosimetry. *Atmos. Environ.* **2008**, *42*, 8575–8584. [CrossRef]
10. Fan, L.; Liu, S. Respirable nano-particulate generations and their pathogenesis in mining workplaces: A review. *Int. J. Coal Sci. Technol.* **2021**, *8*, 179–198. [CrossRef]
11. Serra, D.S.; Araujo, R.S.; Oliveira, M.L.M.; Cavalcante, F.S.A.; Leal-Cardoso, J.H. Lung injury caused by occupational exposure to particles from the industrial combustion of cashew nut shells: A mice model. *Arch. Environ. Occup. Health* **2021**, *76*, 1–11. [CrossRef]
12. Ghorani-Azam, A.; Riahi-Zanjani, B.; Balali-Mood, M. Effects of air pollution on human health and practical measures for prevention in Iran. *J. Res. Med. Sci.* **2016**, *21*, 65. [CrossRef] [PubMed]
13. Peters, A.; Dockery, D.W.; Muller, J.E.; Mittleman, M.A. Increased Particulate Air Pollution and the Triggering of Myocardial Infarction. *Circulation* **2001**, *103*, 2810–2815. [CrossRef]
14. Liu, S.-T.; Liao, C.-Y.; Kuo, C.-Y.; Kuo, H.-W. The Effects of PM2.5 from Asian Dust Storms on Emergency Room Visits for Cardiovascular and Respiratory Diseases. *Int. J. Environ. Res. Public Health* **2017**, *14*, 428. [CrossRef] [PubMed]
15. Kendall, M.; Guntern, J.; Lockyer, N.P.; Jones, F.H.; Hutton, B.M.; Lippmann, M.; Tetley, T.D. Urban PM2.5 surface chemistry and interactions with bronchoalveolar lavage fluid. *Inhal. Toxicol.* **2004**, *16* (Suppl. S1), 115–129. [CrossRef] [PubMed]
16. Manojkumar, N.; Srimuruganandam, B. Investigation of on-road fine particulate matter exposure concentration and its inhalation dosage levels in an urban area. *Build. Environ.* **2021**, *198*, 107914. [CrossRef]
17. Borghi, F.; Spinazzè, A.; Mandaglio, S.; Fanti, G.; Campagnolo, D.; Rovelli, S.; Keller, M.; Cattaneo, A.; Cavallo, D.M. Estimation of the Inhaled Dose of Pollutants in Different Micro-Environments: A Systematic Review of the Literature. *Toxics* **2021**, *9*, 140. [CrossRef]
18. Cao, Y.; Zhao, Q.; Geng, Y.; Li, Y.; Huang, J.; Tian, S.; Ning, P. Interfacial interaction between benzo[a]pyrene and pulmonary surfactant: Adverse effects on lung health. *Environ. Pollut.* **2021**, *287*, 117669. [CrossRef]
19. Hu, G.; Jiao, B.; Shi, X.; Valle, R.P.; Fan, Q.; Zuo, Y.Y. Physicochemical Properties of Nanoparticles Regulate Translocation across Pulmonary Surfactant Monolayer and Formation of Lipoprotein Corona. *ACS Nano* **2013**, *7*, 10525–10533. [CrossRef]
20. Muthusamy, S.; Peng, C.; Ng, J.C. Effects of binary mixtures of benzo[a]pyrene, arsenic, cadmium, and lead on oxidative stress and toxicity in HepG2 cells. *Chemosphere* **2016**, *165*, 41–51. [CrossRef]
21. Widziewicz, K.; Rogula-Kozłowska, W.; Loska, K.; Kociszewska, K.; Majewski, G. Health Risk Impacts of Exposure to Airborne Metals and Benzo(a)Pyrene during Episodes of High PM10 Concentrations in Poland. *Biomed. Environ. Sci.* **2018**, *31*, 23–36. [CrossRef]
22. Lopez-Rodriguez, E.; Pérez-Gil, J. Structure-function relationships in pulmonary surfactant membranes: From biophysics to therapy. *Biochim. Biophys. Acta* **2014**, *1838*, 1568–1585. [CrossRef] [PubMed]
23. Sosnowski, T.; Podgórski, A. Assessment of the Pulmonary Toxicity of Inhaled Gases and Particles with Physicochemical Methods. *Int. J. Occup. Saf. Ergon.* **1999**, *5*, 431–447. [CrossRef] [PubMed]
24. Parra, E.; Pérez-Gil, J. Composition, structure and mechanical properties define performance of pulmonary surfactant membranes and films. *Chem. Phys. Lipids* **2015**, *185*, 153–175. [CrossRef] [PubMed]

25. De Souza Carvalho, C.; Daum, N.; Lehr, C.M. Carrier interactions with the biological barriers of the lung: Advanced in vitro models and challenges for pulmonary drug delivery. *Adv. Drug Deliv. Rev.* **2014**, *75*, 129–140. [CrossRef] [PubMed]
26. Kuroki, Y.; Takahashi, M.; Nishitani, C. Pulmonary collectins in innate immunity of the lung. *Cell Microbiol.* **2007**, *9*, 1871–1879. [CrossRef] [PubMed]
27. Griese, M. Pulmonary surfactant in health and human lung diseases: State of the art. *Eur. Respir. J.* **1999**, *13*, 1455–1476. [CrossRef] [PubMed]
28. Sosnowski, T.R. Inhaled aerosols: Their role in COVID-19 transmission, including biophysical interactions in the lungs. *Curr. Opin. Colloid Interface Sci.* **2021**, *54*, 101451. [CrossRef]
29. Rezaei, M.; Netz, R.R. Airborne virus transmission via respiratory droplets: Effects of droplet evaporation and sedimentation. *Curr. Opin. Colloid Interface Sci.* **2021**, *55*, 101471. [CrossRef]
30. Veldhuizen, R.A.W.; Zuo, Y.Y.; Petersen, N.O.; Lewis, J.F.; Possmayer, F. The COVID-19 pandemic: A target for surfactant therapy? *Exp. Rev. Respir. Med.* **2021**, *15*, 597–608. [CrossRef]
31. Zuo, Y.Y.; Uspal, W.E.; Wei, T. Airborne Transmission of COVID-19: Aerosol Dispersion, Lung Deposition, and Virus-Receptor Interactions. *ACS Nano* **2020**, *14*, 16502–16524. [CrossRef]
32. Da Silva, E.; Vogel, U.; Hougaard, K.S.; Pérez-Gil, J.; Zuo, Y.Y.; Sørli, J.B. An adverse outcome pathway for lung surfactant function inhibition leading to decreased lung function. *Curr. Res. Toxicol.* **2021**, *2*, 225–236. [CrossRef] [PubMed]
33. Lucassen, J. Dynamic dilational properties of composite surfaces. *Colloids Surf.* **1992**, *65*, 139–149. [CrossRef]
34. Ling, X.; Mayer, A.; Yang, X.; Bournival, G.; Ata, S. Motion of Particles in a Monolayer Induced by Coalescing of a Bubble with a Planar Air-Water Interface. *Langmuir* **2021**, *37*, 3648–3661. [CrossRef] [PubMed]
35. Noskov, B.A.; Bykov, A.G. Dilational rheology of monolayers of nano- and micropaticles at the liquid-fluid interfaces. *Curr. Opin. Colloid Interface Sci.* **2018**, *37*, 1–12. [CrossRef]
36. Maestro, A. Tailoring the interfacial assembly of colloidal particles by engineering the mechanical properties of the interface. *Curr. Opin. Colloid Interface Sci.* **2019**, *39*, 232–250. [CrossRef]
37. Guzmán, E.; Abelenda-Núñez, I.; Maestro, A.; Ortega, F.; Santamaria, A.; Rubio, R.G. Particle-laden fluid/fluid interfaces: Physico-chemical foundations. *J. Phys. Cond. Matter* **2021**, *33*, 333001. [CrossRef]
38. Maestro, A.; Santini, E.; Guzmán, E. Physico-chemical foundations of particle-laden fluid interfaces. *Eur. Phys. J. E* **2018**, *41*, 97. [CrossRef]
39. Sohail, M.; Guo, W.; Li, Z.; Xu, H.; Zhao, F.; Chen, D.; Fu, F. Nanocarrier-based Drug Delivery System for Cancer Therapeutics: A Review of the Last Decade. *Curr. Med. Chem.* **2021**, *28*, 3753–3772. [CrossRef]
40. Radivojev, S.; Luschin-Ebengreuth, G.; Pinto, J.T.; Laggner, P.; Cavecchi, A.; Cesari, N.; Cella, M.; Melli, F.; Paudel, A.; Fröhlich, E. Impact of simulated lung fluid components on the solubility of inhaled drugs and predicted in vivo performance. *Int. J. Pharm.* **2021**, *606*, 120893. [CrossRef]
41. Kirkpatrick, C.J.; Bonfield, W. NanoBioInterface: A multidisciplinary challenge. *J. R. Soc. Interface* **2010**, *7*, S1–S4. [CrossRef]
42. Harishchandra, R.K.; Saleem, M.; Galla, H.-J. Nanoparticle interaction with model lung surfactant monolayers. *J. R. Soc. Interface* **2010**, *7*, S15–S26. [CrossRef] [PubMed]
43. Chaudhury, A.; Debnath, K.; Bu, W.; Jana, N.R.; Basu, J.K. Penetration and preferential binding of charged nanoparticles to mixed lipid monolayers: Interplay of lipid packing and charge density. *Soft Matter* **2021**, *17*, 1963–1974. [CrossRef] [PubMed]
44. Iannelli, R.; Bianchi, V.; Macci, C.; Peruzzi, E.; Chiellini, C.; Petroni, G.; Masciandaro, G. Assessment of pollution impact on biological activity and structure of seabed bacterial communities in the Port of Livorno (Italy). *Sci. Total Environ.* **2012**, *426*, 56–64. [CrossRef]
45. Wan, F.; Nylander, T.; Foged, C.; Yang, M.; Baldursdottir, S.G.; Nielsen, H.M. Qualitative and quantitative analysis of the biophysical interaction of inhaled nanoparticles with pulmonary surfactant by using quartz crystal microbalance with dissipation monitoring. *J. Colloid Interface Sci.* **2019**, *545*, 162–171. [CrossRef] [PubMed]
46. Halappanavar, S.; Rahman, L.; Nikota, J.; Poulsen, S.S.; Ding, Y.; Jackson, P.; Wallin, H.; Schmid, O.; Vogel, U.; Williams, A. Ranking of nanomaterial potency to induce pathway perturbations associated with lung responses. *NanoImpact* **2019**, *14*, 100158. [CrossRef]
47. Kadoya, C.; Ogami, A.; Morimoto, Y.; Myojo, T.; Oyabu, T.; Nishi, K.; Yamamoto, M.; Todoroki, M.; Tanaka, I. Analysis of Bronchoalveolar Lavage Fluid Adhering to Lung Surfactant-Experiment on Intratracheal Instillation of Nickel Oxide with Different Diameters. *Ind. Health* **2012**, *50*, 31–36. [CrossRef]
48. Kendall, M. Fine airborne urban particles (PM2.5) sequester lung surfactant and amino acids from human lung lavage. *Am. J. Physiol.-Lung Cell. Mol. Physiol.* **2007**, *293*, L1053–L1058. [CrossRef]
49. Clifton, L.A.; Campbell, R.A.; Sebastiani, F.; Campos-Terán, J.; Gonzalez-Martinez, J.F.; Björklund, S.; Sotres, J.; Cárdenas, M. Design and use of model membranes to study biomolecular interactions using complementary surface-sensitive techniques. *Adv. Colloid Interface Sci.* **2020**, *277*, 102118. [CrossRef]
50. Cedervall, T.; Lynch, I.; Lindman, S.; Berggård, T.; Thulin, E.; Nilsson, H.; Dawson, K.A.; Linse, S. Understanding the nanoparticle-protein corona using methods to quantify exchange rates and affinities of proteins for nanoparticles. *Proc. Natl. Acad. Sci. USA* **2007**, *104*, 2050–2055. [CrossRef]
51. Lundqvist, M.; Stigler, J.; Cedervall, T.; Berggård, T.; Flanagan, M.B.; Lynch, I.; Elia, G.; Dawson, K. The Evolution of the Protein Corona around Nanoparticles: A Test Study. *ACS Nano* **2011**, *5*, 7503–7509. [CrossRef]

52. Casals, E.; Pfaller, T.; Duschl, A.; Oostingh, G.J.; Puntes, V. Time Evolution of the Nanoparticle Protein Corona. *ACS Nano* **2010**, *4*, 3623–3632. [CrossRef] [PubMed]
53. Stefaniu, C.; Brezesinski, G.; Möhwald, H. Langmuir monolayers as models to study processes at membrane surfaces. *Adv. Colloid Interface Sci.* **2014**, *208*, 197–213. [CrossRef] [PubMed]
54. Podgórski, A.; Sosnowski, T.R.; Gradoń, L. Deactivation of the Pulmonary Surfactant Dynamics by Toxic Aerosols and Gases. *J. Aerosol Med.* **2001**, *14*, 455–466. [CrossRef] [PubMed]
55. Farnoud, A.M.; Fiegel, J. Low concentrations of negatively charged sub-micron particles alter the microstructure of DPPC at the air–water interface. *Colloids Surf. A* **2012**, *415*, 320–327. [CrossRef]
56. Sosnowski, T.R.; Koliński, M.; Gradoń, L. Alteration of Surface Properties of Dipalmitoyl Phosphatidylcholine by Benzo[a]pyrene: A Model of Pulmonary Effects of Diesel Exhaust Inhalation. *J. Biomed. Nanotechnol.* **2012**, *8*, 818–825. [CrossRef] [PubMed]
57. Dwivedi, M.V.; Harishchandra, R.K.; Koshkina, O.; Maskos, M.; Galla, H.-J. Size Influences the Effect of Hydrophobic Nanoparticles on Lung Surfactant Model Systems. *Biophys. J.* **2014**, *106*, 289–298. [CrossRef] [PubMed]
58. Sosnowski, T.R.; Kubski, P.; Wojciechowski, K. New experimental model of pulmonary surfactant for biophysical studies. *Colloids Surf. A* **2017**, *519*, 27–33. [CrossRef]
59. Guzmán, E.; Santini, E. Lung surfactant-particles at fluid interfaces for toxicity assessments. *Curr. Opin. Colloid Interface Sci.* **2019**, *39*, 24–39. [CrossRef]
60. Ariga, K. Don't Forget Langmuir-Blodgett Films 2020: Interfacial Nanoarchitectonics with Molecules, Materials, and Living Objects. *Langmuir* **2020**, *36*, 7158–7180. [CrossRef]
61. Bertsch, P.; Bergfreund, J.; Windhab, E.J.; Fischer, P. Physiological fluid interfaces: Functional microenvironments, drug delivery targets, and first line of defense. *Acta Biomater.* **2021**, *130*, 32–53. [CrossRef]
62. Rubio, R.G.; Guzmán, E.; Ortega, F.; Liggieri, L. Monolayers of Cholesterol and Cholesteryl Stearate at the Water/Vapor Interface: A Physico-Chemical Study of Components of the Meibum Layer. *Colloids Interfaces* **2021**, *5*, 30. [CrossRef]
63. Guzmán, E.; Santini, E.; Ferrari, M.; Liggieri, L.; Ravera, F. Evaluation of the impact of carbonaceous particles in the mechanical performance of lipid Langmuir monolayers. *Colloids Surf. A* **2022**, *634*, 127974. [CrossRef]
64. Guzmán, E.; Santini, E.; Ferrari, M.; Liggieri, L.; Ravera, F. Evaluating the impact of hydrophobic silicon dioxide in the interfacial properties of lung surfactant films. *Environ. Sci. Technol.* **2022**, *56*. [CrossRef] [PubMed]
65. Arick, D.Q.; Choi, Y.H.; Kim, H.C.; Won, Y.-Y. Effects of nanoparticles on the mechanical functioning of the lung. *Adv. Colloid Interface Sci.* **2015**, *225*, 218–228. [CrossRef]
66. Sosnowski, T.R. Particles on the lung surface—Physicochemical and hydrodynamic effects. *Curr. Opin. Colloid Interface Sci.* **2018**, *36*, 1–9. [CrossRef]
67. Wang, F.; Liu, J.; Zeng, H. Interactions of particulate matter and pulmonary surfactant: Implications for human health. *Adv. Colloid Interface Sci.* **2020**, *284*, 102244. [CrossRef]
68. Ravera, F.; Miller, R.; Zuo, Y.Y.; Noskov, B.A.; Bykov, A.G.; Kovalchuk, V.I.; Loglio, G.; Javadi, A.; Liggieri, L. Methods and models to investigate the physicochemical functionality of pulmonary surfactant. *Curr. Opin. Colloid Interface Sci.* **2021**, *55*, 101467. [CrossRef]
69. Fan, Q.; Wang, Y.E.; Zhao, X.; Loo, J.S.C.; Zuo, Y.Y. Adverse Biophysical Effects of Hydroxyapatite Nanoparticles on Natural Pulmonary Surfactant. *ACS Nano* **2011**, *5*, 6410–6416. [CrossRef]
70. Valle, R.P.; Wu, T.; Zuo, Y.Y. Biophysical Influence of Airborne Carbon Nanomaterials on Natural Pulmonary Surfactant. *ACS Nano* **2015**, *9*, 5413–5421. [CrossRef]
71. Bernhard, W. Lung surfactant: Function and composition in the context of development and respiratory physiology. *Ann. Anat.* **2016**, *208*, 146–150. [CrossRef]
72. Castillo-Sánchez, J.C.; Cruz, A.; Pérez-Gil, J. Structural hallmarks of lung surfactant: Lipid-protein interactions, membrane structure and future challenges. *Arch. Biochem. Biophys.* **2021**, *703*, 108850. [CrossRef] [PubMed]
73. Lang, C.J.; Postle, A.D.; Orgeig, S.; Possmayer, F.; Bernhard, W.; Panda, A.K.; Jürgens, K.D.; Milsom, W.K.; Nag, K.; Daniels, C.B. Dipalmitoylphosphatidylcholine is not the major surfactant phospholipid species in all mammals. *Am. J. Physiol. Regul. Integr. Comp. Physiol.* **2005**, *289*, R1426–R1439. [CrossRef] [PubMed]
74. Goerke, J. Pulmonary surfactant: Functions and molecular composition. *Biochim. Biophys. Acta* **1998**, *1408*, 79–89. [CrossRef]
75. Bernhard, W.; Mottaghian, J.; Gebert, A.; Rau, G.A.; von der Hardt, H.; Poets, C.F. Commercial versus Native Surfactants. *Am. J. Respir. Crit. Care Med.* **2000**, *162*, 1524–1533. [CrossRef] [PubMed]
76. Pynn, C.J.; Henderson, N.G.; Clark, H.; Koster, G.; Bernhard, W.; Postle, A.D. Specificity and rate of human and mouse liver and plasma phosphatidylcholine synthesis analyzed in vivo. *J. Lipid Res.* **2011**, *52*, 399–407. [CrossRef]
77. Bernhard, W.; Gebert, A.; Vieten, G.; Rau, G.A.; Hohlfeld, J.M.; Postle, A.D.; Freihorst, J. Pulmonary surfactant in birds: Coping with surface tension in a tubular lung. *Am. J. Physiol. Regul. Integr. Comp. Physiol.* **2001**, *281*, R327–R337. [CrossRef]
78. Bernhard, W.; Hoffmann, S.; Dombrowsky, H.; Rau, G.A.; Kamlage, A.; Kappler, M.; Haitsma, J.J.; Freihorst, J.; von der Hardt, H.; Poets, C.F. Phosphatidylcholine Molecular Species in Lung Surfactant. *Am. J. Respir. Cell Mol. Biol.* **2001**, *25*, 725–731. [CrossRef]
79. Bernhard, W.; Postle, A.D.; Rau, G.A.; Freihorst, J. Pulmonary and gastric surfactants. A comparison of the effect of surface requirements on function and phospholipid composition. *Comp. Biochem. Physiol. A Mol. Integr. Physiol.* **2001**, *129*, 173–182. [CrossRef]

80. Rau, G.A.; Vieten, G.; Haitsma, J.J.; Freihorst, J.; Poets, C.; Ure, B.M.; Bernhard, W. Surfactant in newborn compared with adolescent pigs: Adaptation to neonatal respiration. *Am. J. Respir. Cell Mol. Biol.* **2004**, *30*, 694–701. [CrossRef]
81. Bernardino de la Serna, J.; Perez-Gil, J.; Simonsen, A.C.; Bagatolli, L.A. Cholesterol rules: Direct observation of the coexistence of two fluid phases in native pulmonary surfactant membranes at physiological temperatures. *J. Biol. Chem.* **2004**, *279*, 40715–40722. [CrossRef]
82. Orgeig, S.; Daniels, C.B.; Johnston, S.D.; Sullivan, L.C. The pattern of surfactant cholesterol during vertebrate evolution and development: Does ontogeny recapitulate phylogeny? *Reprod. Fertil. Dev.* **2003**, *15*, 55–73. [CrossRef] [PubMed]
83. Schröder, H.; Sollfrank, L.; Paulsen, F.; Bräuer, L.; Schicht, M. Recombinant expression of surfactant protein H (SFTA3) in Escherichia coli. *Ann. Anat.* **2016**, *208*, 129–134. [CrossRef] [PubMed]
84. Haagsman, H.P.; Diemel, R.V. Surfactant-associated proteins: Functions and structural variation. *Comp. Biochem. Physiol. A Mol. Integr. Physiol.* **2001**, *129*, 91–108. [CrossRef]
85. Zasadzinski, J.A.; Ding, J.; Warriner, H.E.; Bringezu, F.; Waring, A.J. The physics and physiology of lung surfactants. *Curr. Opin. Colloid Interface Sci.* **2001**, *6*, 506–513. [CrossRef]
86. Liekkinen, J.; Enkavi, G.; Javanainen, M.; Olmeda, B.; Pérez-Gil, J.; Vattulainen, I. Pulmonary Surfactant Lipid Reorganization Induced by the Adsorption of the Oligomeric Surfactant Protein B Complex. *J. Mol. Biol.* **2020**, *432*, 3251–3268. [CrossRef] [PubMed]
87. Martinez-Calle, M.; Prieto, M.; Olmeda, B.; Fedorov, A.; Loura, L.M.S.; Pérez-Gil, J. Pulmonary surfactant protein SP-B nanorings induce the multilamellar organization of surfactant complexes. *Biochim. Biophys. Acta* **2020**, *1862*, 183216. [CrossRef] [PubMed]
88. Edwards, V.; Cutz, E.; Viero, S.; Moore, A.M.; Nogee, L. Ultrastructure of Lamellar Bodies in Congenital Surfactant Deficiency. *Ultrastruct. Pathol.* **2005**, *29*, 503–509. [CrossRef]
89. Lukovic, D.; Cruz, A.; Gonzalez-Horta, A.; Almlen, A.; Curstedt, T.; Mingarro, I.; Pérez-Gil, J. Interfacial Behavior of Recombinant Forms of Human Pulmonary Surfactant Protein SP-C. *Langmuir* **2012**, *28*, 7811–7825. [CrossRef]
90. Roldan, N.; Nyholm, T.K.M.; Slotte, J.P.; Pérez-Gil, J.; García-Álvarez, B. Effect of Lung Surfactant Protein SP-C and SP-C-Promoted Membrane Fragmentation on Cholesterol Dynamics. *Biophys. J.* **2016**, *111*, 1703–1713. [CrossRef]
91. Roldan, N.; Pérez-Gil, J.; Morrow, M.R.; García-Álvarez, B. Divide & Conquer: Surfactant Protein SP-C and Cholesterol Modulate Phase Segregation in Lung Surfactant. *Biophys. J.* **2017**, *113*, 847–859. [CrossRef]
92. Pérez-Gil, J. Structure of pulmonary surfactant membranes and films: The role of proteins and lipid–protein interactions. *Biochim. Biophys. Acta* **2008**, *1778*, 1676–1695. [CrossRef] [PubMed]
93. Kaganer, V.M.; Möhwald, H.; Dutta, P. Structure and phase transitions in Langmuir monolayers. *Rev. Mod. Phys.* **1999**, *71*, 779–819. [CrossRef]
94. Perkins, W.R.; Dause, R.B.; Parente, R.A.; Minchey, S.R.; Neuman, K.C.; Gruner, S.M.; Taraschi, T.F.; Janoff, A.S. Role of Lipid Polymorphism in Pulmonary Surfactant. *Science* **1996**, *273*, 330–332. [CrossRef]
95. Chavarha, M.; Khoojinian, H.; Schulwitz, L.E.; Biswas, S.C.; Rananavare, S.B.; Hall, S.B. Hydrophobic Surfactant Proteins Induce a Phosphatidylethanolamine to Form Cubic Phases. *Biophys. J.* **2010**, *98*, 1549–1557. [CrossRef]
96. Chavarha, M.; Loney, R.W.; Kumar, K.; Rananavare, S.B.; Hall, S.B. Differential Effects of the Hydrophobic Surfactant Proteins on the Formation of Inverse Bicontinuous Cubic Phases. *Langmuir* **2012**, *28*, 16596–16604. [CrossRef]
97. Chavarha, M.; Loney, R.W.; Rananavare, S.B.; Hall, S.B. An Anionic Phospholipid Enables the Hydrophobic Surfactant Proteins to Alter Spontaneous Curvature. *Biophys. J.* **2013**, *104*, 594–603. [CrossRef]
98. Hobi, N.; Giolai, M.; Olmeda, B.; Miklavc, P.; Felder, E.; Walther, P.; Dietl, P.; Frick, M.; Pérez-Gil, J.; Haller, T. A small key unlocks a heavy door: The essential function of the small hydrophobic proteins SP-B and SP-C to trigger adsorption of pulmonary surfactant lamellar bodies. *Biochim. Biophys. Acta* **2016**, *1863*, 2124–2134. [CrossRef]
99. Olmeda, B.; García-Álvarez, B.; Gómez, M.J.; Martínez-Calle, M.; Cruz, A.; Pérez-Gil, J. A model for the structure and mechanism of action of pulmonary surfactant protein B. *FASEB J.* **2015**, *29*, 4236–4247. [CrossRef]
100. Parra, E.; Moleiro, L.H.; López-Montero, I.; Cruz, A.; Monroy, F.; Pérez-Gil, J. A combined action of pulmonary surfactant proteins SP-B and SP-C modulates permeability and dynamics of phospholipid membranes. *Biochem. J.* **2011**, *438*, 555–564. [CrossRef]
101. Possmayer, F.; Keating, N.; Zuo, Y.; Petersen, N.; Veldhuizen, R. Pulmonary Surfactant Reduces Surface Tension to Low Values near Zero through a Modified Squeeze-Out Mechanism. *Biophys. J.* **2012**, *102*, 414a. [CrossRef]
102. Keating, E.; Zuo, Y.Y.; Tadayyon, S.M.; Petersen, N.O.; Possmayer, F.; Veldhuizen, R.A. A modified squeeze-out mechanism for generating high surface pressures with pulmonary surfactant. *Biochim. Biophys. Acta* **2012**, *1818*, 1225–1234. [CrossRef] [PubMed]
103. Pérez-Gil, J.; Keough, K.M. Interfacial properties of surfactant proteins. *Biochim. Biophys. Acta* **1998**, *1408*, 203–217. [CrossRef]
104. Bernardino de la Serna, J.; Vargas, R.; Picardi, V.; Cruz, A.; Arranz, R.; Valpuesta, J.M.; Mateu, L.; Pérez-Gil, J. Segregated ordered lipid phases and protein-promoted membrane cohesivity are required for pulmonary surfactant films to stabilize and protect the respiratory surface. *Faraday Discuss.* **2013**, *161*, 535–548. [CrossRef] [PubMed]
105. Akanno, A.; Guzmán, E.; Fernández-Peña, L.; Llamas, S.; Ortega, F.; Rubio, R.G. Equilibration of a Polycation–Anionic Surfactant Mixture at the Water/Vapor Interface. *Langmuir* **2018**, *34*, 7455–7464. [CrossRef] [PubMed]
106. Schürch, D.; Ospina, O.L.; Cruz, A.; Pérez-Gil, J. Combined and independent action of proteins SP-B and SP-C in the surface behavior and mechanical stability of pulmonary surfactant films. *Biophys. J.* **2010**, *99*, 3290–3299. [CrossRef]
107. Campbell, R.A.; Tummino, A.; Noskov, B.A.; Varga, I. Polyelectrolyte/surfactant films spread from neutral aggregates. *Soft Matter* **2016**, *12*, 5304–5312. [CrossRef]

108. Sheridan, A.J.; Slater, J.M.; Arnold, T.; Campbell, R.A.; Thompson, K.C. Changes to DPPC Domain Structure in the Presence of Carbon Nanoparticles. *Langmuir* **2017**, *33*, 10374–10384. [CrossRef]

109. Tummino, A.; Toscano, J.; Sebastiani, F.; Noskov, B.A.; Varga, I.; Campbell, R.A. Effects of Aggregate Charge and Subphase Ionic Strength on the Properties of Spread Polyelectrolyte/Surfactant Films at the Air/Water Interface under Static and Dynamic Conditions. *Langmuir* **2018**, *34*, 2312–2323. [CrossRef]

110. Cañadas, O.; Olmeda, B.; Alonso, A.; Pérez-Gil, J. Lipid–Protein and Protein–Protein Interactions in the Pulmonary Surfactant System and Their Role in Lung Homeostasis. *Int. J. Mol. Sci.* **2020**, *21*, 3708. [CrossRef]

111. Olmeda, B.; Martínez-Calle, M.; Pérez-Gil, J. Pulmonary surfactant metabolism in the alveolar airspace: Biogenesis, extracellular conversions, recycling. *Ann. Anat.* **2017**, *209*, 78–92. [CrossRef]

112. Martínez-Calle, M.; Olmeda, B.; Dietl, P.; Frick, M.; Pérez-Gil, J. Pulmonary surfactant protein SP-B promotes exocytosis of lamellar bodies in alveolar type II cells. *FASEB J.* **2018**, *32*, 4600–4611. [CrossRef] [PubMed]

113. White, M.K.; Strayer, D.S. Surfactant Protein a Regulates Pulmonary Surfactant Secretion via Activation of Phosphatidylinositol 3-Kinase in Type II Alveolar Cells. *Exp. Cell Res.* **2000**, *255*, 67–76. [CrossRef] [PubMed]

114. Mortensen, N.P.; Durham, P.; Hickey, A.J. The role of particle physico-chemical properties in pulmonary drug delivery for tuberculosis therapy. *J. Microencapsul.* **2014**, *31*, 785–795. [CrossRef] [PubMed]

115. Tatur, S.; Badia, A. Influence of Hydrophobic Alkylated Gold Nanoparticles on the Phase Behavior of Monolayers of DPPC and Clinical Lung Surfactant. *Langmuir* **2012**, *28*, 628–639. [CrossRef] [PubMed]

116. Farnoud, A.M.; Fiegel, J. Interaction of Dipalmitoyl Phosphatidylcholine Monolayers with a Particle-Laden Subphase. *J. Phys. Chem. B* **2013**, *117*, 12124–12134. [CrossRef]

117. Guzmán, E.; Ferrari, M.; Santini, E.; Liggieri, L.; Ravera, F. Effect of silica nanoparticles on the interfacial properties of a canonical lipid mixture. *Colloids Surf. B* **2015**, *136*, 971–980. [CrossRef]

118. Guzmán, E.; Santini, E.; Zabiegaj, D.; Ferrari, M.; Liggieri, L.; Ravera, F. Interaction of Carbon Black Particles and Dipalmitoylphosphatidylcholine at Water/Air Interface: Thermodynamics and Rheology. *J. Phys. Chem. C* **2015**, *119*, 26937–26947. [CrossRef]

119. Guzmán, E.; Santini, E.; Ferrari, M.; Liggieri, L.; Ravera, F. Effect of the Incorporation of Nanosized Titanium Dioxide on the Interfacial Properties of 1,2-Dipalmitoyl-sn-glycerol-3-phosphocholine Langmuir Monolayers. *Langmuir* **2017**, *33*, 10715–10725. [CrossRef]

120. Beck-Broichsitter, M.; Ruppert, C.; Schmehl, T.; Guenther, A.; Betz, T.; Bakowsky, U.; Seeger, W.; Kissel, T.; Gessler, T. Biophysical investigation of pulmonary surfactant surface properties upon contact with polymeric nanoparticles in vitro. *Nanomedicine* **2011**, *7*, 341–350. [CrossRef]

121. Santamaria, A.; Batchu, K.C.; Matsarskaia, O.; Prévost, S.F.; Russo, D.; Natali, F.; Seydel, T.; Hoffmann, I.; Laux, V.; Haertlein, M.; et al. Strikingly Different Roles of SARS-CoV-2 Fusion Peptides Uncovered by Neutron Scattering. *J. Am. Chem. Soc.* **2022**, *144*. [CrossRef]

122. Wang, R.; Guo, Y.; Liu, H.; Chen, Y.; Shang, Y.; Liu, H. The effect of chitin nanoparticles on surface behavior of DPPC/DPPG Langmuir monolayers. *J. Colloid Interface Sci.* **2018**, *519*, 186–193. [CrossRef] [PubMed]

123. Park, S.S.; Wexler, A.S. Size-dependent deposition of particles in the human lung at steady-state breathing. *J. Aerosol Sci.* **2008**, *39*, 266–276. [CrossRef]

124. Hofmann, W. Modelling inhaled particle deposition in the human lung—A review. *Aerosol Sci.* **2011**, *42*, 693–724. [CrossRef]

125. Lynch, I.; Elder, A. Disposition of nanoparticles as a function of their interactions with biomolecules. In *Nanomaterials: Risks and Benefits*; NATO Science for Peace and Security Series C: Environmental Security; Linkov, I., Steevens, J., Eds.; Springer: Dordrecht, Germany, 2009; pp. 31–41. [CrossRef]

126. Stachowicz-Kuśnierz, A.; Cwiklik, L.; Korchowiec, J.; Rogalska, E.; Korchowiec, B. The impact of lipid oxidation on the functioning of a lung surfactant model. *Phys. Chem. Chem. Phys.* **2018**, *20*, 24968–24978. [CrossRef] [PubMed]

127. Long, D.L.; Hite, R.D.; Grier, B.L.; Suckling, B.N.; Safta, A.M.; Morris, P.E.; Waite, B.M.; Seeds, M.C. Secretory Phospholipase A2-Mediated Depletion of Phosphatidylglycerol in Early Acute Respiratory Distress Syndrome. *Am. J. Med. Sci.* **2012**, *343*, 446–451. [CrossRef] [PubMed]

128. Guzmán, E.; Santini, E.; Ferrari, M.; Liggieri, L.; Ravera, F. Interaction of Particles with Langmuir Monolayers of 1,2-Dipalmitoyl-Sn-Glycero-3-Phosphocholine: A Matter of Chemistry? *Coatings* **2020**, *10*, 469. [CrossRef]

129. Muñoz-López, R.; Guzmán, E.; Velázquez, M.M.; Fernández-Peña, L.; Merchán, M.D.; Maestro, A.; Ortega, F.; Rubio, G.R. Influence of Carbon Nanosheets on the Behavior of 1,2-Dipalmitoyl-sn-glycerol-3-phosphocholine Langmuir Monolayers. *Processes* **2020**, *8*, 94. [CrossRef]

130. Gradon, L.; Podgorski, A. Hydrodynamical model of pulmonary clearance. *Chem. Eng. Sci.* **1989**, *44*, 741–749. [CrossRef]

131. Gradon, L.; Podgórski, A.; Sosnowski, T.R. Experimental and theoretical investigations of transport properties of DPPC monolayer. *J. Aerosol Med.* **1996**, *9*, 357–367. [CrossRef]

132. Wrobel, S.; Clements, J.A. Bubbles, babies and biology: The story of surfactant—Second breath: A medical mystery solved. *FASEB J.* **2004**, *18*, 1624e. [CrossRef]

133. Guzmán, E.; Santini, E.; Ferrari, M.; Liggieri, L.; Ravera, F. Interfacial Properties of Mixed DPPC–Hydrophobic Fumed Silica Nanoparticle Layers. *J. Phys. Chem. C* **2015**, *119*, 21024–21034. [CrossRef]

134. Orsi, D.; Rimoldi, T.; Guzmán, E.; Liggieri, L.; Ravera, F.; Ruta, B.; Cristofolini, L. Hydrophobic Silica Nanoparticles Induce Gel Phases in Phospholipid Monolayers. *Langmuir* **2016**, *32*, 4868–4876. [CrossRef] [PubMed]

135. Notter, R.H.; Wang, Z. Pulmonary surfactant: Physical chemistry, physiology, and replacement. *Rev. Chem. Eng.* **1997**, *13*, 1–118. [CrossRef]

136. Guzmán, E.; Liggieri, L.; Santini, E.; Ferrari, M.; Ravera, F. Influence of silica nanoparticles on phase behavior and structural properties of DPPC—Palmitic acid Langmuir monolayers. *Colloids Surf. A* **2012**, *413*, 280–287. [CrossRef]

137. Kondej, D.; Sosnowski, T.R. Changes in the Activity of the Pulmonary Surfactant after Contact with Bentonite Nanoclay Particles. *Chem. Eng. Trans.* **2012**, *26*, 531–536. [CrossRef]

138. Kramek-Romanowska, K.; Odziomek, M.; Sosnowski, T.R. Dynamic tensiometry studies on interactions of novel therapeutic inhalable powders with model pulmonary surfactant at the air-water interface. *Colloids Surf. A* **2015**, *480*, 149–158. [CrossRef]

139. Maestro, A.; Guzmán, E.; Ortega, F.; Rubio, R.G. Contact angle of micro- and nanoparticles at fluid interfaces. *Curr. Opin. Colloid Interface Sci.* **2014**, *19*, 355–367. [CrossRef]

140. Ariki, S.; Nishitani, C.; Kuroki, Y. Diverse functions of pulmonary collectins in host defense of the lung. *J. Biomed. Biotechnol.* **2012**, *2012*, 532071. [CrossRef]

141. Raesch, S.S.; Tenzer, S.; Storck, W.; Rurainski, A.; Selzer, D.; Ruge, C.A.; Perez-Gil, J.; Schaefer, U.F.; Lehr, C.-M. Proteomic and Lipidomic Analysis of Nanoparticle Corona upon Contact with Lung Surfactant Reveals Differences in Protein, but Not Lipid Composition. *ACS Nano* **2015**, *9*, 11872–11885. [CrossRef]

142. Hidalgo, A.; Cruz, A.; Pérez-Gil, J. Barrier or carrier? Pulmonary surfactant and drug delivery. *Eur. J. Pharm. Biopharm.* **2015**, *95*, 117–127. [CrossRef]

143. Hidalgo, A.; Cruz, A.; Pérez-Gil, J. Pulmonary surfactant and nanocarriers: Toxicity versus combined nanomedical applications. *Biochim. Biophys. Acta* **2017**, *1859*, 1740–1748. [CrossRef] [PubMed]

144. Schüer, J.J.; Arndt, A.; Wölk, C.; Pinnapireddy, S.R.; Bakowsky, U. Establishment of a Synthetic In Vitro Lung Surfactant Model for Particle Interaction Studies on a Langmuir Film Balance. *Langmuir* **2020**, *36*, 4808–4819. [CrossRef] [PubMed]

145. Barrow, R.E.; Hills, B.A. A critical assessment of the Wilhelmy method in studying lung surfactants. *J. Physiol.* **1979**, *295*, 217–227. [CrossRef] [PubMed]

146. Mendoza, A.J.; Guzmán, E.; Martínez-Pedrero, F.; Ritacco, H.; Rubio, R.G.; Ortega, F.; Starov, V.M.; Miller, R. Particle laden fluid interfaces: Dynamics and interfacial rheology. *Adv. Colloid Interface Sci.* **2014**, *206*, 303–319. [CrossRef]

147. Bykov, A.G.; Loglio, G.; Ravera, F.; Liggieri, L.; Miller, R.; Noskov, B.A. Dilational surface elasticity of spread monolayers of pulmonary lipids in a broad range of surface pressure. *Colloids Surf. A* **2018**, *541*, 137–144. [CrossRef]

148. Ulman, A. *An Introduction to Ultrathin Organic Films: From Langmuir-Blodgett to Self-Assembly*; Academic Press: Cambridge, MA, USA, 1991.

149. Rubinger, C.P.L.; Moreira, R.L.; Cury, L.A.; Fontes, G.N.; Neves, B.R.A.; Meneguzzi, A.; Ferreira, C.A. Langmuir–Blodgett and Langmuir–Schaefer films of poly(5-amino-1-naphthol) conjugated polymer. *Appl. Surf. Sci.* **2006**, *253*, 543–548. [CrossRef]

150. Hui, S.W.; Viswanathan, R.; Zasadzinski, J.A.; Israelachvili, J.N. The structure and stability of phospholipid bilayers by atomic force microscopy. *Biophys. J.* **1995**, *68*, 171–178. [CrossRef]

151. Alonso, C.; Alig, T.; Yoon, J.; Bringezu, F.; Warriner, H.; Zasadzinski, J.A. More Than a Monolayer: Relating Lung Surfactant Structure and Mechanics to Composition. *Biophys. J.* **2004**, *87*, 4188–4202. [CrossRef]

152. Enhorning, G. Pulmonary surfactant function studied with the pulsating bubble surfactometer (PBS) and the capillary surfactometer (CS). *Comp. Biochem. Physiol. Part A Mol. Integr. Physiol.* **2001**, *129*, 221–226. [CrossRef]

153. Zuo, Y.Y.; Li, D.; Acosta, E.; Cox, P.N.; Neumann, A.W. Effect of Surfactant on Interfacial Gas Transfer Studied by Axisymmetric Drop Shape Analysis—Captive Bubble (ADSA-CB). *Langmuir* **2005**, *21*, 5446–5452. [CrossRef]

154. Chen, Z.; Zhong, M.; Luo, Y.; Deng, L.; Hu, Z.; Song, Y. Determination of rheology and surface tension of airway surface liquid: A review of clinical relevance and measurement techniques. *Respir. Res.* **2019**, *20*, 274. [CrossRef] [PubMed]

155. Echaide, M.; Autilio, C.; López-Rodríguez, E.; Cruz, A.; Pérez-Gil, J. In Vitro Functional and Structural Characterization of a Synthetic Clinical Pulmonary Surfactant with Enhanced Resistance to Inhibition. *Sci. Rep.* **2020**, *10*, 1385. [CrossRef] [PubMed]

156. Zuo, Y.Y.; Ding, M.; Li, D.; Neumann, A.W. Further development of Axisymmetric Drop Shape Analysis-captive bubble for pulmonary surfactant related studies. *Biochim. Biophys. Acta* **2004**, *1675*, 12–20. [CrossRef] [PubMed]

157. Autilio, C.; Pérez-Gil, J. Understanding the principle biophysics concepts of pulmonary surfactant in health and disease. *Arch. Dis. Child. Fetal Neonatal Ed.* **2019**, *104*, F443–F451. [CrossRef] [PubMed]

158. Bakshi, M.S.; Zhao, L.; Smith, R.; Possmayer, F.; Petersen, N.O. Metal Nanoparticle Pollutants Interfere with Pulmonary Surfactant Function In Vitro. *Biophys. J.* **2008**, *94*, 855–868. [CrossRef] [PubMed]

159. Kondej, D.; Sosnowski, T. Physicochemical mechanisms of mineral nanoparticles effects on pulmonary gas/liquid interface studied in model systems. *Physicochem. Probl. Miner. Process.* **2014**, *50*, 57–69. [CrossRef]

160. Dobrowolska, K.; JabBczzDska, K.; Kondej, D.; Sosnowski, T.R. Interactions of insoluble micro- and nanoparticles with the air-liquid interface of the model pulmonary fluids. *Physicochem. Probl. Miner. Process.* **2017**, *54*, 151–162. [CrossRef]

161. Kondej, D.; Sosnowski, T.R. Interactions of Carbon Nanotubes and Carbon Nanohorns with a Model Membrane Layer and Lung Surfactant In Vitro. *J. Nanomater.* **2019**, *2019*, 9457683. [CrossRef]

162. Llamas, S.; Fernández-Peña, L.; Akanno, A.; Guzmán, E.; Ortega, V.; Ortega, F.; Csaky, A.G.; Campbell, R.A.; Rubio, R.G. Towards understanding the behavior of polyelectrolyte–surfactant mixtures at the water/vapor interface closer to technologically-relevant conditions. *Phys. Chem. Chem. Phys.* **2018**, *20*, 1395–1407. [CrossRef]

163. Notter, R.H.; Taubold, R.; Mavis, R.D. Hysteresis in Saturated Phospholipid Films and its Potential Relevance for Lung Surfactant Function In Vivo. *Exp. Lung Res.* **1982**, *3*, 109–127. [CrossRef]

164. Sosnowski, T.R.; Koliński, M.; Gradoń, L. Interactions of benzo[a]pyrene and diesel exhaust particulate matter with the lung surfactant system. *Ann. Occup. Hyg.* **2011**, *55*, 329–338. [CrossRef] [PubMed]

165. Guzmán, E.; Liggieri, L.; Santini, E.; Ferrari, M.; Ravera, F. Influence of silica nanoparticles on dilational rheology of DPPC–palmitic acid Langmuir monolayers. *Soft Matter* **2012**, *8*, 3938–3948. [CrossRef]

166. Kondej, D.; Sosnowski, T.R. Interfacial rheology for the assessment of potential health effects of inhaled carbon nanomaterials at variable breathing conditions. *Sci. Rep.* **2020**, *10*, 14044. [CrossRef]

167. Yang, J.; Yu, K.; Tsuji, T.; Jha, R.; Zuo, Y.Y. Determining the surface dilational rheology of surfactant and protein films with a droplet waveform generator. *J. Colloid Interface Sci.* **2019**, *537*, 547–553. [CrossRef]

168. Sørli, J.B.; Låg, M.; Ekeren, L.; Perez-Gil, J.; Haug, L.S.; Da Silva, E.; Matrod, M.N.; Gützkow, K.B.; Lindeman, B. Per- and polyfluoroalkyl substances (PFASs) modify lung surfactant function and pro-inflammatory responses in human bronchial epithelial cells. *Toxicol. In Vitro* **2020**, *62*, 104656. [CrossRef]

169. Liekkinen, J.; de Santos Moreno, B.; Paananen, R.O.; Vattulainen, I.; Monticelli, L.; Bernardino de la Serna, J.; Javanainen, M. Understanding the Functional Properties of Lipid Heterogeneity in Pulmonary Surfactant Monolayers at the Atomistic Level. *Front. Cell Dev. Biol.* **2020**, *8*, 581016. [CrossRef]

170. Arroyo, R.; Martín-González, A.; Echaide, M.; Jain, A.; Brondyk, W.H.; Rosenbaum, J.; Moreno-Herrero, F.; Pérez-Gil, J. Supramolecular Assembly of Human Pulmonary Surfactant Protein SP-D. *J. Mol. Biol.* **2018**, *430*, 1495–1509. [CrossRef]

171. Finot, E.; Markey, L.; Hane, F.; Amrein, M.; Leonenko, Z. Combined atomic force microscopy and spectroscopic ellipsometry applied to the analysis of lipid-protein thin films. *Colloids Surf. B* **2013**, *104*, 289–293. [CrossRef]

172. Przybyla, R.J.; Wright, J.; Parthiban, R.; Nazemidashtarjandi, S.; Kaya, S.; Farnoud, A.M. Electronic cigarette vapor alters the lateral structure but not tensiometric properties of calf lung surfactant. *Respir. Res.* **2017**, *18*, 193. [CrossRef]

173. Guzmán, E.; Liggieri, L.; Santini, E.; Ferrari, M.; Ravera, F. DPPC–DOPC Langmuir monolayers modified by hydrophilic silica nanoparticles: Phase behaviour, structure and rheology. *Colloids Surf. A* **2012**, *413*, 174–183. [CrossRef]

174. Arora, S.; Kappl, M.; Haghi, M.; Young, P.M.; Traini, D.; Jain, S. An investigation of surface properties, local elastic modulus and interaction with simulated pulmonary surfactant of surface modified inhalable voriconazole dry powders using atomic force microscopy. *RSC Adv.* **2016**, *6*, 25789–25798. [CrossRef]

175. Carrascosa-Tejedor, J.; Santamaria, A.; Pereira, D.; Maestro, A. Structure of DPPC Monolayers at the Air/Buffer Interface: A Neutron Reflectometry and Ellipsometry Study. *Coatings* **2020**, *10*, 507. [CrossRef]

176. Lee, K.Y.C.; Majewski, J.; Kuhl, T.L.; Howes, P.B.; Kjaer, K.; Lipp, M.M.; Waring, A.J.; Zasadzinski, J.A.; Smith, G.S. Synchrotron X-Ray Study of Lung Surfactant-Specific Protein SP-B in Lipid Monolayers. *Biophys. J.* **2001**, *81*, 572–585. [CrossRef]

177. Liu, X.; Counil, C.; Shi, D.; Mendoza-Ortega, E.E.; Vela-Gonzalez, A.V.; Maestro, A.; Campbell, R.A.; Krafft, M.P. First quantitative assessment of the adsorption of a fluorocarbon gas on phospholipid monolayers at the air/water interface. *J. Colloid Interface Sci.* **2021**, *593*, 1–10. [CrossRef]

178. Orsi, D.; Guzmán, E.; Liggieri, L.; Ravera, F.; Ruta, B.; Chushkin, Y.; Rimoldi, T.; Cristofolini, L. 2D dynamical arrest transition in a mixed nanoparticle-phospholipid layer studied in real and momentum spaces. *Sci. Rep.* **2015**, *5*, 17930. [CrossRef]

179. Guzmán, E.; Orsi, D.; Cristofolini, L.; Liggieri, L.; Ravera, F. Two-Dimensional DPPC Based Emulsion-like Structures Stabilized by Silica Nanoparticles. *Langmuir* **2014**, *30*, 11504–11512. [CrossRef]

180. Mandal, P.; Bhatta, F.; Kooijman, E.E.; Allender, D.W.; Mann, E.K. Combined Brewster Angle and Fluorescence Microscopy of DMPC/D-Cholesterol Mixed Langmuir Monolayers. *Biophys. J.* **2012**, *102*, 96a–97a. [CrossRef]

181. Mandal, P.; Noutsi, P.; Chaieb, S. Cholesterol Depletion from a Ceramide/Cholesterol Mixed Monolayer: A Brewster Angle Microscope Study. *Sci. Rep.* **2016**, *6*, 26907. [CrossRef]

182. Mendelsohn, R.; Mao, G.; Flach, C.R. Infrared reflection–absorption spectroscopy: Principles and applications to lipid–protein interaction in Langmuir films. *Biochim. Biophys. Acta* **2010**, *1798*, 788–800. [CrossRef]

183. Can, S.Z.; Chang, C.F.; Walker, R.A. Spontaneous formation of DPPC monolayers at aqueous/vapor interfaces and the impact of charged surfactants. *Biochim. Biophys. Acta* **2008**, *1778*, 2368–2377. [CrossRef]

184. Phillips, M.C.; Chapman, D. Monolayer characteristics of saturated 1,2-diacyl phosphatidylcholines (lecithins) and phosphatidylethanolamines at the air-water interface. *Biochim. Biophys. Acta* **1968**, *163*, 301–313. [CrossRef]

185. Klopfer, K.J.; Vanderlick, T.K. Isotherms of Dipalmitoylphosphatidylcholine (DPPC) Monolayers: Features Revealed and Features Obscured. *J. Colloid Interface Sci.* **1996**, *182*, 220–229. [CrossRef]

186. Arriaga, L.R.; López-Montero, I.; Ignés-Mullol, J.; Monroy, F. Domain-Growth Kinetic Origin of Nonhorizontal Phase Coexistence Plateaux in Langmuir Monolayers: Compression Rigidity of a Raft-Like Lipid Distribution. *J. Phys. Chem. B* **2010**, *114*, 4509–4520. [CrossRef] [PubMed]

187. Hifeda, Y.F.; Rayfield, G.W. Evidence for first-order phase transitions in lipid and fatty acid monolayers. *Langmuir* **1992**, *8*, 197–200. [CrossRef]

188. Guzmán, E.; Liggieri, L.; Santini, E.; Ferrari, M.; Ravera, F. Mixed DPPC–cholesterol Langmuir monolayers in presence of hydrophilic silica nanoparticles. *Colloids Surf. B* **2013**, *105*, 284–293. [CrossRef]
189. Behyan, S.; Borozenko, O.; Khan, A.; Faral, M.; Badia, A.; DeWolf, C. Nanoparticle-induced structural changes in lung surfactant membranes: An X-ray scattering study. *Environ. Sci. Nano* **2018**, *5*, 1218–1230. [CrossRef]
190. Stachowicz-Kuśnierz, A.; Trojan, S.; Cwiklik, L.; Korchowiec, B.; Korchowiec, J. Modeling Lung Surfactant Interactions with Benzo[a]pyrene. *Chem. Eur. J.* **2017**, *23*, 5307–5316. [CrossRef]
191. Rose, D.; Rendell, J.; Lee, D.; Nag, K.; Booth, V. Molecular dynamics simulations of lung surfactant lipid monolayers. *Biophys. Chem.* **2008**, *138*, 67–77. [CrossRef]
192. Klenz, U.; Saleem, M.; Meyer, M.C.; Galla, H.-J. Influence of Lipid Saturation Grade and Headgroup Charge: A Refined Lung Surfactant Adsorption Model. *Biophys. J.* **2008**, *95*, 609–799. [CrossRef]
193. Zhao, Q.; Li, Y.; Chai, X.; Zhang, L.; Xu, L.; Huang, J.; Ning, P.; Tian, S. Interaction of nano carbon particles and anthracene with pulmonary surfactant: The potential hazards of inhaled nanoparticles. *Chemosphere* **2019**, *215*, 746–752. [CrossRef]
194. Suri, L.N.M.; McCaig, L.; Picardi, M.V.; Ospina, O.L.; Veldhuizen, R.A.W.; Staples, J.F.; Possmayer, F.; Yao, L.-J.; Perez-Gil, J.; Orgeig, S. Adaptation to low body temperature influences pulmonary surfactant composition thereby increasing fluidity while maintaining appropriately ordered membrane structure and surface activity. *Biochim. Biophys. Acta* **2012**, *1818*, 1581–1589. [CrossRef] [PubMed]
195. Walther, F.J.; Gordon, L.M.; Waring, A.J. Advances in synthetic lung surfactant protein technology. *Exp. Rev. Med. Respir.* **2019**, *13*, 499–501. [CrossRef] [PubMed]
196. Robertson, B. Lung surfactant for replacement therapy. *Clin. Physiol.* **1983**, *3*, 97–110. [CrossRef] [PubMed]
197. Egan, E.A.; Notter, R.H.; Kwong, M.S.; Shapiro, D.L. Natural and artificial lung surfactant replacement therapy in premature lambs. *J. Appl. Physiol. Respir. Environ. Exerc. Physiol.* **1983**, *55*, 875–883. [CrossRef]
198. Mirastschijski, U.; Dembinski, R.; Maedler, K. Lung Surfactant for Pulmonary Barrier Restoration in Patients With COVID-19 Pneumonia. *Front. Med.* **2020**, *7*, 254. [CrossRef]
199. Farnoud, A.M.; Fiegel, J. Calf Lung Surfactant Recovers Surface Functionality After Exposure to Aerosols Containing Polymeric Particles. *J. Aerosol Med. Pulm. Drug Deliv.* **2015**, *29*, 10–23. [CrossRef]
200. Guzmán, E.; Liggieri, L.; Santini, E.; Ferrari, M.; Ravera, F. Effect of Hydrophilic and Hydrophobic Nanoparticles on the Surface Pressure Response of DPPC Monolayers. *J. Phys. Chem. C* **2011**, *115*, 21715–21722. [CrossRef]
201. Baoukina, S.; Rozmanov, D.; Mendez-Villuendas, E.; Tieleman, D.P. The Mechanism of Collapse of Heterogeneous Lipid Monolayers. *Biophys. J.* **2014**, *107*, 1136–1145. [CrossRef]
202. Phan, M.D.; Lee, J.; Shin, K. Collapsed States of Langmuir Monolayers. *J. Oleo Sci.* **2016**, *65*, 385–397. [CrossRef]
203. Kuo, C.-C.; Kodama, A.T.; Boatwright, T.; Dennin, M. Particle Size Effects on Collapse in Monolayers. *Langmuir* **2012**, *28*, 13976–13983. [CrossRef]
204. Tanaka, Y.; Takei, T.; Aiba, T.; Masuda, K.; Kiuchi, A.; Fujiwara, T. Development of synthetic lung surfactants. *J. Lipid Res.* **1986**, *27*, 475–485. [CrossRef]
205. Casals, C.; Cañadas, O. Role of lipid ordered/disordered phase coexistence in pulmonary surfactant function. *Biochim. Biophys. Acta* **2012**, *1818*, 2550–2562. [CrossRef] [PubMed]
206. Miguel Diez, M.; Buckley, A.; Tetley, T.D.; Smith, R. The method of depositing CeO$_2$ nanoparticles onto a DPPC monolayer affects surface tension behaviour. *NanoImpact* **2019**, *16*, 100186. [CrossRef]
207. Alhakamy, N.A.; Elandaloussi, I.; Ghazvini, S.; Berkland, C.J.; Dhar, P. Effect of Lipid Headgroup Charge and pH on the Stability and Membrane Insertion Potential of Calcium Condensed Gene Complexes. *Langmuir* **2015**, *31*, 4232–4245. [CrossRef]
208. Mousseau, F.; Berret, J.F. The role of surface charge in the interaction of nanoparticles with model pulmonary surfactants. *Soft Matter* **2018**, *14*, 5764–5774. [CrossRef]
209. Yang, Y.; Wu, Y.; Ren, Q.; Zhang, L.G.; Liu, S.; Zuo, Y.Y. Biophysical Assessment of Pulmonary Surfactant Predicts the Lung Toxicity of Nanomaterials. *Small Methods* **2018**, *2*, 1700367. [CrossRef]
210. Sosnowski, T.R.; Jabłczyńska, K.; Odziomek, M.; Schlage, W.K.; Kuczaj, A.K. Physicochemical studies of direct interactions between lung surfactant and components of electronic cigarettes liquid mixtures. *Inhal. Toxicol.* **2018**, *30*, 159–168. [CrossRef]
211. Taneva, S.G.; Stewart, J.; Taylor, L.; Keough, K.M.W. Method of purification affects some interfacial properties of pulmonary surfactant proteins B and C and their mixtures with dipalmitoylphosphatidylcholine. *Biochim. Biophys. Acta* **1998**, *1370*, 138–150. [CrossRef]
212. Beck-Broichsitter, M.; Ruppert, C.; Schmehl, T.; Günther, A.; Seeger, W. Biophysical inhibition of synthetic vs. naturally-derived pulmonary surfactant preparations by polymeric nanoparticles. *Biochim. Biophys. Acta* **2014**, *1838*, 474–481. [CrossRef]
213. Sachan, A.K.; Harishchandra, R.K.; Bantz, C.; Maskos, M.; Reichelt, R.; Galla, H.-J. High-Resolution Investigation of Nanoparticle Interaction with a Model Pulmonary Surfactant Monolayer. *ACS Nano* **2012**, *6*, 1677–1687. [CrossRef]
214. Zhang, K.; Liu, L.; Bai, T.; Guo, Z. Interaction Between Hydrophobic Au Nanoparticles and Pulmonary Surfactant (DPPC) Monolayers. *J. Biomed. Nanotechnol.* **2018**, *14*, 526–535. [CrossRef] [PubMed]
215. Anderson, D.S.; Patchin, E.S.; Silva, R.M.; Uyeminami, D.L.; Sharmah, A.; Guo, T.; Das, G.K.; Brown, J.M.; Shannahan, J.; Gordon, T.; et al. Influence of Particle Size on Persistence and Clearance of Aerosolized Silver Nanoparticles in the Rat Lung. *Toxicol. Sci.* **2015**, *144*, 366–381. [CrossRef] [PubMed]

216. Chen, Y.; Yang, Y.; Xu, B.; Wang, S.; Li, B.; Ma, J.; Gao, J.; Zuo, Y.Y.; Liu, S. Mesoporous carbon nanomaterials induced pulmonary surfactant inhibition, cytotoxicity, inflammation and lung fibrosis. *J. Environ. Sci.* **2017**, *62*, 100–114. [CrossRef]
217. Curtis, E.M.; Bahrami, A.H.; Weikl, T.R.; Hall, C.K. Modeling nanoparticle wrapping or translocation in bilayer membranes. *Nanoscale* **2015**, *7*, 14505–14514. [CrossRef]
218. Ku, T.; Gill, S.; Löbenberg, R.; Azarmi, S.; Roa, W.; Prenner, E.J. Size Dependent Interactions of Nanoparticles with Lung Surfactant Model Systems and the Significant Impact on Surface Potential. *J. Nanosc. Nanotechnol.* **2008**, *8*, 2971–2978. [CrossRef]
219. Kodama, A.T.; Kuo, C.-C.; Boatwright, T.; Dennin, M. Investigating the Effect of Particle Size on Pulmonary Surfactant Phase Behavior. *Biophys. J.* **2014**, *107*, 1573–1581. [CrossRef] [PubMed]
220. Lundqvist, M.; Stigler, J.; Elia, G.; Lynch, I.; Cedervall, T.; Dawson, K.A. Nanoparticle size and surface properties determine the protein corona with possible implications for biological impacts. *Proc. Natl. Acad. Sci. USA* **2008**, *105*, 14265–14270. [CrossRef]
221. Schleh, C.; Hohlfeld, J.M. Interaction of nanoparticles with the pulmonary surfactant system. *Inhal. Toxicol.* **2009**, *21*, 97–103. [CrossRef]
222. Kim, J.; Chankeshwara, S.V.; Thielbeer, F.; Jeong, J.; Donaldson, K.; Bradley, M.; Cho, W.-S. Surface charge determines the lung inflammogenicity: A study with polystyrene nanoparticles. *Nanotoxicology* **2016**, *10*, 94–101. [CrossRef]
223. Hao, C.; Li, J.; Mu, W.; Zhu, L.; Yang, J.; Liu, H.; Li, B.; Chen, S.; Sun, R. Adsorption behavior of magnetite nanoparticles into the DPPC model membranes. *Appl. Surf. Sci.* **2016**, *362*, 121–125. [CrossRef]
224. Adair, J.H.; Suvaci, E.; Sindel, J. Surface and Colloid Chemistry. In *Encyclopedia of Materials: Science and Technology*; Buschow, K.H.J., Cahn, R.W., Flemings, M.C., Ilschner, B., Kramer, E.J., Mahajan, S., Veyssière, P., Eds.; Elsevier: Oxford, UK, 2001; pp. 1–10.
225. Chen, P.; Zhang, Z.; Gu, N.; Ji, M. Effect of the surface charge density of nanoparticles on their translocation across pulmonary surfactant monolayer: A molecular dynamics simulation. *Mol. Simul.* **2018**, *44*, 85–93. [CrossRef]
226. Beck-Broichsitter, M. Biophysical Activity of Impaired Lung Surfactant upon Exposure to Polymer Nanoparticles. *Langmuir* **2016**, *32*, 10422–10429. [CrossRef] [PubMed]
227. Valle, R.P.; Huang, C.L.; Loo, J.S.C.; Zuo, Y.Y. Increasing Hydrophobicity of Nanoparticles Intensifies Lung Surfactant Film Inhibition and Particle Retention. *ACS Sustain. Chem. Eng.* **2014**, *2*, 1574–1580. [CrossRef]
228. Konduru, N.V.; Damiani, F.; Stoilova-McPhie, S.; Tresback, J.S.; Pyrgiotakis, G.; Donaghey, T.C.; Demokritou, P.; Brain, J.D.; Molina, R.M. Nanoparticle Wettability Influences Nanoparticle–Phospholipid Interactions. *Langmuir* **2018**, *34*, 6454–6461. [CrossRef]
229. Kondej, D.; Sosnowski, T.R. Effect of clay nanoparticles on model lung surfactant: A potential marker of hazard from nanoaerosol inhalation. *Environ. Sci. Poll. Res.* **2016**, *23*, 4660–4669. [CrossRef]
230. Xu, Y.; Deng, L.; Ren, H.; Zhang, X.; Huang, F.; Yue, T. Transport of nanoparticles across pulmonary surfactant monolayer: A molecular dynamics study. *Phys. Chem. Chem. Phys.* **2017**, *19*, 17568–17576. [CrossRef]
231. Sacanna, S.; Pine, D.J. Shape-anisotropic colloids: Building blocks for complex assemblies. *Curr. Opin. Colloid Interface Sci.* **2011**, *16*, 96–105. [CrossRef]
232. Yang, K.; Ma, Y.-Q. Computer simulation of the translocation of nanoparticles with different shapes across a lipid bilayer. *Nat. Nanotechnol.* **2010**, *5*, 579–583. [CrossRef]
233. Lin, X.; Zuo, Y.Y.; Gu, N. Shape affects the interactions of nanoparticles with pulmonary surfactant. *Sci. China Mater.* **2015**, *58*, 28–37. [CrossRef]
234. Aramrak, S.; Flury, M.; Harsh, J.B.; Zollars, R.L.; Davis, H.P. Does Colloid Shape Affect Detachment of Colloids by a Moving Air-Water Interface? *Langmuir* **2013**, *29*, 5770–5780. [CrossRef]
235. Tarn, D.; Ashley, C.E.; Xue, M.; Carnes, E.C.; Zink, J.I.; Brinker, C.J. Mesoporous Silica Nanoparticle Nanocarriers: Biofunctionality and Biocompatibility. *Acc. Chem. Res.* **2013**, *46*, 792–801. [CrossRef] [PubMed]
236. Beck-Broichsitter, M.; Ruppert, C.; Schmehl, T.; Günther, A.; Seeger, W. Biophysical inhibition of pulmonary surfactant function by polymeric nanoparticles: Role of surfactant protein B and C. *Acta Biomater.* **2014**, *10*, 4678–4684. [CrossRef] [PubMed]
237. Hossain, S.I.; Luo, Z.; Deplazes, E.; Saha, S.C. Shape matters—The interaction of gold nanoparticles with model lung surfactant monolayers. *J. R. Soc. Interface* **2021**, *18*, 20210402. [CrossRef] [PubMed]
238. Stenger, P.C.; Alonso, C.; Zasadzinski, J.A.; Waring, A.J.; Jung, C.-L.; Pinkerton, K.E. Environmental tobacco smoke effects on lung surfactant film organization. *Biochim. Biophys. Acta* **2009**, *1788*, 358–370. [CrossRef]
239. Da Silva, E.; Autilio, C.; Hougaard, K.S.; Baun, A.; Cruz, A.; Perez-Gil, J.; Sørli, J.B. Molecular and biophysical basis for the disruption of lung surfactant function by chemicals. *Biochim. Biophys. Acta* **2021**, *1863*, 183499. [CrossRef]
240. Kanno, S.; Hirano, S.; Kato, H.; Fukuta, M.; Mukai, T.; Aoki, Y. Benzalkonium chloride and cetylpyridinium chloride induce apoptosis in human lung epithelial cells and alter surface activity of pulmonary surfactant monolayers. *Chem. Biol. Interact.* **2020**, *317*, 108962. [CrossRef]
241. Nisoh, N.; Karttunen, M.; Monticelli, L.; Wong-Ekkabut, J. Lipid monolayer disruption caused by aggregated carbon nanoparticles. *RSC Adv.* **2015**, *5*, 11676–11685. [CrossRef]
242. Choe, S.; Chang, R.; Jeon, J.; Violi, A. Molecular Dynamics Simulation Study of a Pulmonary Surfactant Film Interacting with a Carbonaceous Nanoparticle. *Biophys. J.* **2008**, *95*, 4102–4114. [CrossRef]
243. Li, S.; Stein, A.J.; Kruger, A.; Leblanc, R.M. Head Groups of Lipids Govern the Interaction and Orientation between Graphene Oxide and Lipids. *J. Phys. Chem. C* **2013**, *117*, 16150–16158. [CrossRef]

244. Kanishtha, T.; Banerjee, R.; Venkataraman, C. Effect of particle emissions from biofuel combustion on surface activity of model and therapeutic pulmonary surfactants. *Environ. Toxicol. Pharmacol.* **2006**, *22*, 325–333. [CrossRef]
245. Li, L.; Xu, Y.; Li, S.; Zhang, X.; Feng, H.; Dai, Y.; Zhao, J.; Yue, T. Molecular modeling of nanoplastic transformations in alveolar fluid and impacts on the lung surfactant film. *J. Hazard. Mater.* **2022**, *427*, 127872. [CrossRef] [PubMed]

MDPI
St. Alban-Anlage 66
4052 Basel
Switzerland
Tel. +41 61 683 77 34
Fax +41 61 302 89 18
www.mdpi.com

*Coatings* Editorial Office
E-mail: coatings@mdpi.com
www.mdpi.com/journal/coatings